CONSTRUCTION MANAGEMENT

ESSENTIALS OF CONSTRUCTION PROJECT MANAGEMENT

Thomas E. Uher
and
Martin Loosemore

UNSW
PRESS

A UNSW Press book

Published by
University of New South Wales Press Ltd
University of New South Wales
Sydney NSW 2052
AUSTRALIA
www.unswpress.com.au

National Library of Australia
Cataloguing-in-Publication entry

Uher, Thomas E. (Thomas Edward).
    Essentials of construction project management.

    Includes index.
    ISBN 0 86840 733 X.

    1. Construction industry – Management.
    2. Project management.
    I. Loosemore, Martin, 1962– .
    II. Title. (Series : Construction management).

690.068

Printer Ligare, Australia

CONTENTS

PREFACE

This book is about project management with a focus on the construction industry. It defines and describes important issues in mainstream project management and demonstrates their relevance to the management of construction projects. We have written this book for practising project managers and as a textbook for students studying project management courses at both undergraduate and postgraduate levels. It presents the concept of project management in four separate parts.

In Part 1, we discuss the essentials of project management including organisational theory, the project management system and its functions, the context within which project management operates in the construction industry, and the roles performed by the members of a project team. We also discuss organisational structure and culture in project management, and suggest how best to manage cultural diversity in the construction industry.

Part 2 examines how to manage individual stages of the project lifecycle from concept to completion. The description of each stage provides a detailed account of important tasks and their management and also discusses the roles performed by individual project team members.

Part 3 focuses on people. In particular, it addresses the important topics of motivation, leadership, communication, conflict and negotiation.

Part 4 outlines a process of risk management, and the management of health and safety in the construction industry. Risk management is examined from both qualitative and quantitative perspectives, and its benefits are demonstrated through practical examples.

In writing the book, we have combined our individual knowledge and experiences in the field of project management, which is reflected in a broad coverage of many project management topics. Chapters 1, 2, 5–9, 11 and 15 were largely written by Thomas Uher while Martin Loosemore is the author of Chapters 3, 4, 10, 12–14 and 16. Some of the text in Chapter 15, which describes the concept of risk management and some of its applications, has been reproduced from Uher's book *Programming and Scheduling Techniques*.

We believe that this book provides a different perspective from the many other books written about project management and hope that you find it interesting and useful.

We wish to express our sincere gratitude to Brian Farmer for his invaluable contribution to this book.

TE Uher
M Loosemore

AUTHORS' NOTE

In the text, the authors often refer to project manager, client, contractor, or members of an organisation as 'he'. This is done purely for convenience, to sidestep the pronoun problem in English, and no sexism is intended.

ABBREVIATIONS

ABS	Australian Bureau of Statistics
ADR	Alternative dispute resolution
AGC	Associated General Contractors of America
AIQS	Australian Institute of Quantity Surveyors
AS/NZS	Australian/New Zealand Risk Management Standard
BA	Building approval
BOOT	Build, own, operate, transfer
CATRAP	Cost and time risk analysis program
CDM	Construction (Design and Management)
CIDA	Construction Industry Development Agency
CIDB	Construction Industry Development Board
CII	Construction Industry Institute, Texas
CIM	Control interval and memory
CIOB	Chartered Institute of Building
DA	Development approval
DIMA	Department of Immigration and Multicultural Affairs, Canberra
DPWS	Department of Public Works and Services (now Department of Commerce)
EIS	Environmental impact statement
HAZOP	Hazard and operability
HRM	Human resources management
HSE	Health and Safety Executive
IOD	Institute of Directors
IRR	Internal rate of return
MBTI	Myers-Briggs type indicator
MCT	Management confidence technique
MIS	Management information system
MLSA	Ministry of Labour and Social Affairs, Riyadh
NEDO	Network Economic Development
NPV	Net present value
OHS	Occupational health and safety
OHS&R	Occupational health, safety and rehabilitation
PAN	Probabilistic Schedule Appraisal Program
PC	Prime cost
PERT	Program evaluation and review technique
PFI	Private finance initiative
PMBOK	Project Management Body of Knowledge
PPP	Public private partnership
PS	Provisional sum
PV	Present value
RCBCI	Royal Commission into the Australian Building and Construction Industry
RDA	*Racial Discrimination Act*
RICS	Royal Institute of Chartered Surveyors
SA	Standards Australia
SCERT	Synergistic contingency evaluation and review technique
SD	Standard deviation
TQM	Total quality management
UNSW	University of New South Wales
USACE	US Army Corps of Engineers
WBS	Work breakdown structure

PART 1

THE PROJECT MANAGEMENT FRAMEWORK

CHAPTER 1

INTRODUCTION TO PROJECT MANAGEMENT

INTRODUCTION

The discipline of project management provides the necessary processes, techniques and tools for accomplishing successful project outcomes. Construction projects, particularly those of a complex nature, have benefited from the concept of project management through improved cost and time performance. Project management applies to many different kinds of projects, but this book will focus on large construction projects.

Construction projects symbolise the social, cultural and technological developments of succeeding civilisations. Many ancient structures such as the pyramids of ancient Egypt and Peru, the Great Wall of China, the temples of the Hindus, Greeks and Romans, the Gothic cathedrals of Europe and the like which have survived for centuries serve as examples of human ingenuity and endeavour. Despite their huge size and often intricate designs, they were successfully completed without the need for project management skills. This is because they were built under substantially different conditions from those that prevail today. For example, completion times were often open-ended and frequently extended over years or even generations. The labour force was plentiful and projects were rarely subjected to strict budgetary constraints. Furthermore, the construction process was largely unhindered by regulatory controls or environmental, social and political pressures. And such structures were commonly designed and built by the same organisation, a practice that continued until the 18th century.

The Industrial Revolution and its technological developments greatly influenced the design of construction projects. New materials such as steel revolutionised design and allowed the development of new structures such as suspended bridges, observation towers and high-rise buildings. The increased complexity of designs led to the emergence of specialist design professionals, and by about the mid-18th century design activities became contractually separate from construction activities. This changed the manner in which construction projects were managed. In addition to design duties, the design professional was often involved in overseeing the construction process, administering a contract between the client and the builder, and meeting the client's expectation of project cost, time and quality. The project organisation structure became more refined to reflect the separation of design and construction activities. This approach to project delivery is known as the 'traditional method', which continues to be the most popular option of project delivery even today (see Uher & Davenport 2002).

With the emergence of the traditional method of delivery came the need to exert greater control over the cost of a project and the time taken to complete it. However, in the absence of the tools and techniques of what we have come to call project management, projects were ineffectively planned and controlled. Decision-making at project level was based on intuition and may be characterised as a hit-and-miss approach.

The evolution of project management was triggered by the development of modern management theory in the nineteenth and twentieth centuries. Scientific management contributed by developing better work methods through time and motion studies. It also gave birth to the Gantt chart, the first systematic planning technique. The main contribution of classical management theory was in the area of organisation bureaucracy. Both scientific and classical management theories saw organisations as machines and managers as engineers. This simplistic machine-model view was rejected by the adherents of behavioural management theories. These supported a view that interpersonal relationships and social systems in the workplace were largely responsible for influencing an organisation's success. The systems approach provided the basis for seeing a project organisation as a set of interrelated and interdependent parts that were to be closely co-ordinated and integrated within the overall system. The systems approach assisted in the development of the concept of work breakdown structure (WBS), through which complex projects are logically broken up into subsystem components that are then managed in a co-ordinated and integrated way within the overall system.

The first major examples of projects that have displayed effective project management and achieved the desired objectives emerged in World War II. The invasion of Normandy (D-Day) in 1944 and the development of an atom bomb (the Manhattan Project) (Morris 1994) are two notable examples of project management in action.

The post-World War II period is synonymous with highly complex and technologically sophisticated military, space and economically strategic projects, among which the Polaris nuclear submarine system, the Apollo moon landing, the Alaska pipeline and the Concord supersonic aircraft project are well-known examples. Their successful completion signified not only great technological accomplishments but also the birth of a formal management process that has become known as project management.

This chapter will define project management and identify the key characteristics of projects. It will briefly review organisation theory in project management and examine project management functions. It will also summarise the main knowledge areas in Project Management Body of Knowledge (PMBOK) (see pages 20–23).

WHAT IS PROJECT MANAGEMENT?

A project management approach to the management of construction projects represents a fundamental shift from the traditional or functional management philosophy, which is concerned with managing ongoing operations such as manufacturing products or providing services. These operations are expected to continue over a very long time with minimal change. They are staffed by a fairly stable workforce in a fixed physical environment, which can be accurately monitored and controlled. Functional organisations comprise a hierarchical system of departments, each with a different function. The roles of managers and staff are well defined, and communication links between management and the workforce is clearly delineated. Graphically, the functional management structure is often depicted as a pyramid with each management position pigeon-holed in broader levels down the pyramid. Communication patterns, responsibilities and authority are rigid and generally transparent, making it fairly easy for everyone to know where they stand in the organisation.

Introduction to Project Management

In contrast to the functional approach, the focus of project management is on delivering projects that are finite in time and have a fairly short life span, using a team drawn from various functional departments of participating organisations. A project team under the leadership of a project manager then manages the project from inception to completion.

Team membership is not stable. People with specialised knowledge and expertise join the team at different stages of the project lifecycle and leave when their work has ended. When the project is finished, the residual team is disbanded, with individual members moving to other projects or returning to functional departments of their organisations. What is unique to construction project teams is that their membership varies substantially from project to project. This makes it difficult for clients to transfer strong team synergy from the current project to other projects. A process of strategic alliance alleviates this problem to some extent by keeping two or more alliance members together as team members from project to project (Uher & Davenport 2002).

The concept of project management has been applied in the construction industry since the 1960s, but its use has intensified only from the mid-1980s. Its rising popularity reflects improved performance in construction projects. Numerous definitions exist: most people think of project management as managing a project in order to meet its objectives in a timely manner, within a cost budget, to the required quality standard, and meeting the requirements of functionality and utility. This definition can be broadened by adding elements such as the project lifecycle — its course from concept to completion — as well as the fundamental management functions of planning, organising, control and leading. Such a broad definition of project management is illustrated graphically in Figure 1.1.

Although the success of project management is measured in terms of time and cost performance, the achievement of such an objective is totally dependent on the management of resources, particularly people, technologies and systems. Project management thus embodies the management functions, people, systems, techniques and technologies necessary to carry the project to successful completion within the constraints of time, budget and performance (Kezsbom et al. 1989).

A comprehensive definition of project management would therefore be: 'Project management is concerned with management of people who form a short-term project team for the purpose of achieving project objectives in terms of cost, time, quality, function and utility

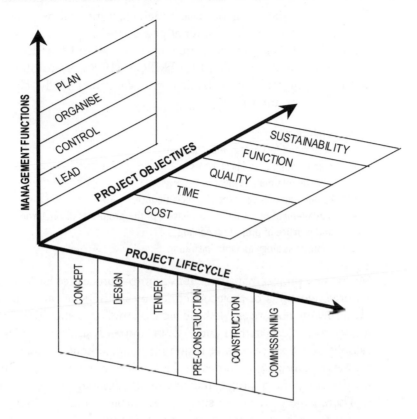

across all the stages of a project life cycle through the application of the fundamental management functions of planning, organising, control and leading'.

A classic definition of project management can be found in PMBOK (1996: 6): 'Project management is the application of knowledge, skills, tools and techniques to project activities in order to meet or exceed stakeholder needs and expectations from a project'. PMBOK related project management to nine knowledge areas:

- project integration management
- project scope management
- project time management
- project cost management
- project quality management
- project human resource management
- project communication management

- project risk management
- project procurement management.

These knowledge areas of project management will be discussed in more detail later in this chapter at pages 20–23.

Morris (1994) argued that the current view and practices of project management, formalised by PMBOK (1996), were often inadequate to the task of managing projects successfully since they ignored important aspects of:

- strategic planning
- design and technology management
- management of political forces
- cost–benefit management
- raising and management of project finance
- management of the timing or phasing of the project as distinct from the theory and practice of project scheduling
- contract strategy and administration.

Morris saw project management as a simple process of integrating prescriptive elements of projects defined by PMBOK in order to accomplish specific objectives. He believed that the focus should be more on managing projects in their entirety, starting with strategic planning and establishing objectives, and including management of finance, environment, technology and people. He argued that a more precise name for the discipline of project management is 'management of projects'.

There is no evidence to suggest that project management practice in Australia follows diligently the narrowly defined spectrum of activities laid out in PMBOK. While some project management practices and some academic institutions may prefer to work to the PMBOK framework, the authors' experience in Australia and the United Kingdom suggests that most project management firms and academic institutions have adopted a much broader view of project management, and the authors endorse Morris' distinction in referring to it as 'management of projects'.

Project management as distinct from functional management displays its unique characteristics. These are commonly summarised as:

- task orientation (unique non-repetitive tasks, the end product definable at the beginning)
- system orientation
- interaction with other management systems
- service focus
- people focus

- short-term organisation (finite for both the project and the project management organisation)
- objectives orientation (cost, time, quality, function, etc.)
- dynamic character
- use of interdisciplinary teams
- horizontal management structure.

Table 1.1 attempts to highlight the main distinctions between project management and functional management.

TABLE 1.1 COMPARISON OF PROJECT AND FUNCTIONAL MANAGEMENT PRINCIPLES

Phenomenon	Project management	Functional management
Line staff organisational dichotomy	Vestiges of the hierarchical model remain but line functions are placed in a support position. A web of authority and responsibility relationship exists.	Line functions have direct responsibility for accomplishing the objectives; line commands and staff advises.
Superior–subordinate relationship	Peer-to-peer, manager-to-technical expert, associate-to-associate, etc., relationships are used to conduct much of the salient business.	This is the most important relationship; if kept healthy, success will follow. All-important business is conducted through a pyramid structure of superior and subordinates.
Organisational objectives	Management of a project becomes a joint venture of many relatively independent organisations. Thus the objective becomes multilateral.	Organisational objectives are sought by the parent unit (an assembly of sub-organisations) working within its environment. The objective is unilateral.
Unity of direction	The project manager manages across functional and organisational lines to accomplish a common inter-organisational objective.	The general manager acts as the one head for a group of activities having the same plan.
Parity of authority and responsibility	Considerable opportunity exists for the project manager's responsibility to exceed authority. Support people are often responsible to other managers (functional) for pay, performance, reports, promotions, etc.	Consistent with functional management, the integrity of the superior–subordinate relationship is maintained through functional authority and advisory staff services.
Time duration	The project and the project organisation are finite in duration.	Tends to perpetuate itself to provide continuing facilitative support.
Change	Creating change is central to project management.	Change evolves slowly.
Resources	Highly variable and constantly changing.	Largely static.

SOURCE adapted from Levido 1990

MANAGEMENT AND ORGANISATION

The key characteristics of construction projects are complexity and a high level of human interaction. No matter how large or small, construction projects require careful planning, appropriate allocation of resources, and a tight control of the production process across sequential stages of a project lifecycle. At every stage of a project lifecycle, people with various skills form co-operative teams to accomplish specific goals and to meet the client's expectation of satisfactory performance in time, cost and quality. Human interaction across the entire project lifecycle provides the necessary dynamic that drives the production process.

The production process employed in the construction industry is not at first glance unlike a typical manufacturing production process. In both cases a production line is used for assembly of components into a final product. But a typical factory production line produces many identical products over a long time, while a construction production line is geared towards accomplishing one finite project. Another significant difference exists in the type of organisation employed. Most manufacturing enterprises rely on the general or functional organisation structure, while an organisation established for the delivery of a construction project may take one of many different forms, all of which embrace the principles of project management.

Before proceeding further it may be useful to explain what the terms 'management' and 'organisation' mean. A useful explanation of these two terms can be found in Cleland and King (1975), who found management and organisation to be intrinsically interlinked. Their definition of 'management' identifies the criteria of organised activity, objectives, relationships among resources, working through others, and decisions, while 'organisation' is associated with organised activity, objectives, some pattern of authority, and responsibility between the participants with some non-human elements involved.

Walker (1996: 5) extended the above definitions of management and organisation to a construction project, where 'an organisation can be said to be the pattern of interrelationships, authority and responsibility that is established between the contributors to achieve the construction client's objective. Management is the dynamic input that makes the organisation work'. Project team members work together within an established project organisation to generate information and make appropriate decisions to accomplish project goals. Walker further stresses the importance of establishing an effective organisation structure necessary for the production of quality information, and on which decisions will be made, and the decision-making process itself.

DEFINITION AND CHARACTERISTICS OF A PROJECT

The term 'project' implies a particular and unique activity that has a definite start and a definite finish. It suggests that this activity is finite in time. Projects may be physical in nature, such as the construction of a building, the building of a space exploration program or the development of computer software. They may also be non-physical, for example the development of a new information system.

Apart from their definite start and finish, projects are characterised in many different terms (Burke 1999):

- Projects pass through a series of sequential stages between the start and the end commonly known as a project lifecycle.
- Projects have their own budget.
- Projects activities are generally unique and non-repetitive.
- Projects extract limited resources from functional departments of project team members' firms.
- The project manager assumes sole responsibility over the project (a 'single-point responsibility').
- Projects are executed by specialised project teams.
- Projects are goal-oriented.

Construction projects have their own project organisation structure and are managed by a project team. Individual members of a team are employees of stakeholder organisations involved in the project. These organisations, including the client, contractor, design consultants and subcontractors, have a mainly general or functional organisation structure. Through their representatives on a team they manage a project that is external to their own organisations. It is a well-known fact that the ability of team members to work together as a cohesive group (that is, synergy) is a condition for successful project outcomes. Since individual team members come from different stakeholder organisations, achieving team synergy is often difficult.

Organisations may also be involved in generating and managing their own internal projects. For example, a contractor may initiate a project of developing a new IT system for its head office operations. Most members of a team would be drawn from the contractor's own organisation. They would work alongside an expert IT consultant. In comparison with a typical construction project, developing a team synergy within such an internal project should be a much simpler task.

ORGANISATION THEORY IN PROJECT MANAGEMENT

Organisation theory has taken many paths in its development from classical theory pioneered by Fayol and Weber, through behaviour and social system theory (Mayo 1933; Mintzberg 1973) to the systems theory (Bertalanffy 1972; Katz & Kahn 1978) and contingency theory (Lawrence & Lorsch 1967) in common use at present.

While project management may be regarded as a subset of general management, its development and in particular the development of project organisation structure have largely lacked a framework of organisation theory relevant to construction projects. Systems theory has been found by many to be an ideal vehicle for developing such a framework (see Handler 1970; Napier 1970; Morris 1972, 1994; Ireland 1985; Kezsbom et al. 1989; Woodward 1997).

Ackoff (1969: 34) defined a system as 'any entity, conceptual or physical that consists of interdependent parts. Each of a system's elements is connected to every other element, directly or indirectly, and no sub-set of elements is unrelated to any other sub-set'.

The fundamental issue of systems theory is relationships, not only between the parts of the system but also within the parts themselves. Construction projects are characterised by such relationships, which reflect a multitude of contracts and subcontracts, and systems theory may be a useful way to study them. A systematic approach to project management requires the breakdown of the total system represented by the project into logical subsystem components that may comprise resources (including people), information, organisation and other components. The next phase is to integrate and co-ordinate activities within and across subsystems to achieve smooth and efficient performance of the overall system.

The classical theory of organisation design, of which Fayol and Weber are the best-known proponents, supported a bureaucratic and highly rigid form of organisation characterised by a pyramidal structure; unity of command, line and staff; and span of control. This theory largely ignored the impact of people and the outside environment on an organisation.

The shortcomings of traditional management theory were uncovered by the Hawthorne Studies of 1924–32 at the Western Electric Company plant in the United States (Mayo 1933; Roethlisberger & Dickson 1939). The intention of the study was to examine the effect of various illumination levels on worker productivity. It was assumed that illumination intensity would be directly related to group productivity. Surprisingly, the level of productivity achieved was found to be unrelated to the intensity of illumination. Further experiments by Mayo and his associates concluded that behaviour and sentiments are closely related, that group influences significantly affect individual behaviour, that group standards establish individual worker output, and that money is less of a factor in determining output than are group standards, group sentiments and security.

The Hawthorne studies played a significant role in developing behavioural and social system theory, which advocates founding the study of management on interpersonal relations.

Contingency theory states that there is no one best way to organise but rather that organisation is a function of the nature of the task to be carried out and its environment. Since it is based on the notion that an organisation is constantly interacting with its environment, which requires understanding of the relationships between subsystems as well as the relationship between organisation and environment, it is also relevant to construction projects.

Unlike the other theories discussed earlier, contingency theory is not based on simple principles; rather it attempts to identify situational variables that affect managerial actions and organisational performance. Contingency theory may be seen as a major contributor to the integration of much of management theory (Fry & Smith 1987).

PROJECT MANAGEMENT SYSTEM

Walker (1996) defined the construction process in terms of the operating system and the management system. The operating system is concerned with professional and technical tasks such as developing design and full documentation, preparing contract documentation or building a project, while the management system integrates and controls the work across different stages of the project lifecycle. Project management activities are those directly linked with the activities of the management system. This implies that the task of project management or the role of a project manager is primarily to co-ordinate, integrate and control a range of activities within the operating system rather than performing any of these activities directly. In this book, project management and tasks performed by a project manager should be seen in this context.

Walker further identified the key activities of project management as approval and recommendation, boundary control, monitoring and maintenance, and general and direct oversight. These will now be briefly discussed.

Approval and recommendation

The project manager has the power to approve the output of project team members who work within the operating system, but the amount of this power is limited by the authority vested in the project manager by the client. Since the manager makes recommendations to the client on various courses of action, his ability to work closely with the client and influence the client's decision-making process will strengthen his authority. The project manager's extent of authority needs to be clearly defined and communicated to the team members to prevent misunderstandings and potential conflict. Only in this way will the project manager be able to extract the necessary level of performance from the team.

Boundary control, monitoring and maintenance

Since a typical construction project may be seen as a system consisting of numerous subsystems operating throughout different stages of the project lifecycle, one of the key activities of project management is boundary control within and across such various subsystems. The reason is to ensure the functional compatibility of activities performed by different project team members, to minimise duplication and overlaps, and to ensure integration and co-ordination across the entire spectrum of activities. Boundary control is closely coupled with monitoring and maintenance. Monitoring ensures that specific tasks performed in different operating subsystems are integrated and meet the required specification requirements. Maintenance ensures that processes employed in such operating subsystems work efficiently and are adequately resourced.

General and direct oversight

These are two classes of supervision or oversight relevant to project management: general and direct. General oversight refers to the establishment and implementation of policy guidance for the project and the team members. Policy guidance is established within the client organisation at the conceptual stage; as the project progresses beyond this stage it is passed to the project manager to implement. Direct oversight is concerned with a focused supervision of specific skills engaged in different operating subsystems.

FUNCTIONS OF PROJECT MANAGEMENT

Within the management system defined by Walker, it is possible to define specific project management functions for which the project manager is responsible. Most textbooks on project management provide a comprehensive list of these functions, of which a condensed summary will now be presented. It has been extracted from Kerzner (1989), Walker (1996), Woodward (1997) and Hamilton (1997).

Establishment of the client's objectives and priorities

This activity takes place during the conceptual stage. The role of the project manager is to advise the client on the viability of the proposed development strategy within a context of economic, environmental and social influences, and on the attractiveness of potential alternatives. This process leads to the client's establishment of appropriate objectives and priorities. This is the first step towards developing a project brief.

Design of a project organisation structure

Who is responsible for the design of a project organisation structure? In most cases it is the client with or without the project manager's input. Clients often prefer to rely on a structure that has produced successful outcomes in the past. However, since the client's objectives may change substantially from project to project, the client's preferred and trusted structure may be too inflexible to cope with changing requirements and priorities. The role of the project manager is then to assist the client in defining the most appropriate organisation structure for a project.

In designing a project organisation structure, the paramount factor is the client's objectives. If time were the critical objective, a project structure would commonly embrace some form of fast-tracking (overlapping of design and construction stages). However, with sufficient time available to complete the design and compile full documentation, the traditional method of project delivery or procurement would be difficult to ignore (Uher & Davenport 2002).

Apart from the client's objectives, a wide range of factors needs to be considered in designing a project organisation structure. Some of these are listed below:

- availability of resources both human and material
- specific contract provisions such as staged delivery
- potentially difficult technical and technological issues
- communication requirements
- economic issues. For example, what will be the impact of interest rates, currency exchange levels and inflation on the project? Should the design of the organisation structure reflect such potential risks?
- political issues. Some projects are politically induced. If unsuccessful, the risk of a major political fallout must be expected. Should the design of the organisation structure reflect such political risks?
- social issues. Some other projects such as airports, nuclear power plants, dams and freeways are commonly strongly opposed not only by the people directly affected but also by the general population. Such projects have become socially unacceptable despite providing essential services to the community. Should this risk be reflected in the design of the project organisation structure?

Integration of the client into the project

Some clients want to be directly involved on a project team, while others prefer to be represented by a consultant such as the architect or the project manager. Common sense and past experiences with partnering projects suggest that direct involvement of the client on a project team is beneficial in facilitating open communication and speedily resolving issues or problems. The project manager's task is to persuade the client to integrate actively into the project team.

Advice on the selection and appointment of the project team members

Assuming that the client has already appointed the project manager on terms and conditions mutually acceptable, he will then proceed to select other project team members. This is commonly done by the client in consultation with the project manager. In less frequent cases, the client may delegate the responsibility for selecting team members solely to the project manager. Alternatively, the experienced client may feel confident to choose the team members without the project manager's input.

However the team members may have been selected, a crucial issue that the project manager needs to pursue with the client is a careful definition of the terms of their appointment, particularly with regard to authority, responsibility and communication. All team members should be informed of the role of the project manager and the project manager's authority.

Translation of the client's objectives into a brief

The development of a strategic plan is one of the critical tasks performed by the project manager in conjunction with the other team members. In developing a strategic plan, the project manager assesses risk, alternative project strategies, alternative contract strategies, and sources and availability of finance. The project manager then carries out a feasibility study and selects the preferred strategy. The main outcomes of strategic planning are expressed in the form of a brief. A brief is a factual statement that fully defines the project in terms of its purpose, function, cost, time and quality. It must be complete in all respects and must be fully understood by all the project team members, particularly by the design consultant(s), whose task is to design and document the project from the brief.

Preparation of the program for the project

The project manager is responsible for planning and programming of the project throughout its lifecycle. Programs are essential for organising the work and monitoring progress. While extensively used in the construction stage, the detailed programming of the conceptual and design stages is less frequent. Yet these are the stages that offer the greatest opportunity for cost and time savings. The task of the project manager is to create a comprehensive and integrated program from the project's inception until its completion. This may be done directly by the project manager, who may have the necessary technical skills, or by a specialist consultant. For such a program to be actionable, it must be 'owned' by all the team members.

Activation of the framework of relationships established for the team members

In designing the project's organisation structure, a range of relationships involving the team members has been established. It is now the responsibility of the project manager to activate such relationships and monitor their performance.

Establishment of appropriate information and communication

The project organisation structure defines in broad terms the flow of information and the communication process. The project manager is required to refine this process to ensure it works efficiently. Where possible, he should put in place a suitable information system for ease and speed of accessing information by the team members.

Introduction to Project Management

Convening and chairing meetings

Open lines of communication are essential for identifying problems and for their speedy resolution. Meetings play an important role in achieving such outcomes. They provide a forum for open debate, brainstorming, and the identification and resolution of issues or problems. Their conduct and administration is the responsibility of the project manager.

Monitoring and controlling the work at different stages of the project lifecycle

The project manager's role is to develop a project strategy (a plan of action) followed by its close monitoring and control to ensure that the plan is adhered to. Walker (1996) has summarised the tasks over which the project manager is required to exercise control:

* land acquisition
* applications for planning consent
* outline of design strategies
* budget and investment strategies
* advice on finance, taxation and grants
* detailed design
* design cost control
* disposal of the completed facility strategies
* contractual arrangements proposals
* appointment of most project team members including the contractor and sub-contractors
* construction
* cost control during construction
* disposal of the facility.

Contribution to primary and key decisions and to making operational decisions

The project manager presents recommendations to the client on alternative proposals and works closely with the client in arriving at key decisions. At the same time he monitors and controls the work of the team and makes operational decisions as required.

Developing and controlling a strategy for disposal or management of the completed project

The decision to sell or retain the completed project is made by the client at the conceptual stage of the project lifecycle. The project manager's task is to implement such a decision. If the decision is to sell off the project, the project manager will advise the client on the best timing and the most effective means of the sale.

The decision to retain the completed project could, in the case of a building, involve the client occupying it or leasing it out. The project manager's role does not commonly extend to the management of the completed project. This is the task usually performed by a facility manager.

Controlling commissioning of the project

The client takes legal possession of the project, such as a building, when the date of 'practical completion' has been reached. Although practically complete, the project may not yet be ready for its intended use. For example, internal services such as air-conditioning, lifts, security and telecommunication may need to be activated and fine-tuned. Some fixtures and fittings may be yet to be installed and any defects in the structure or its services may need to be rectified or fixed. It is the project manager's role to ensure that the project is commissioned and that it meets its intended purpose.

Developing a plan for maintaining the completed project

The completed project and its services will require regular maintenance in the post-commissioning period (after the contract between the client and the contractor had been concluded). The project manager will develop a maintenance strategy, but its implementation is usually the responsibility of the facility manager.

The final project evaluation

When the project has been completed, the project manager is required to compile a final report to the client which details assessment of the performance, measured against the objectives, and comments on both positive and negative outcomes.

Summary The project management functions listed above are not exhaustive and the reader is encouraged to expand and refine them from his or her experience.

PROJECT MANAGEMENT ORGANISATIONAL ALTERNATIVES

Some construction client organisations have in-house project management expertise that they may employ in managing some or all of the stages of the project lifecycle. Others may prefer to source project management expertise by bringing in specialist consultants. The decision on whether or not to use internal or external project management resources largely depends on the size, type and complexity of the project.

Organisations most likely to develop in-house management expertise are those with an ongoing program of projects. Their main objective in doing so is to have total control over the project. An in-house project manager would have the responsibility for co-ordinating the work of internal departments, external consultants, contractors and subcontractors. A potential problem may be the lack of authority or the lack of clarity of authority vested in the project manager. This may prevent the project manager from gaining the full co-operation of various functional managers. In the absence of line control, the project manager may experience difficulties in stimulating and motivating people within the same organisation.

If the project is large enough and there are sufficient resources available, the project manager may lead a self-contained organisation within the parent company with all the human and physical resources necessary to complete the project. This micro-organisation is established for the duration of the project. Upon its completion, it is transferred to another project or disbanded.

PROJECT MANAGEMENT BODY OF KNOWLEDGE

The US Project Management Institute prepared a comprehensive report titled the *Project Management Body of Knowledge* (PMBOK 1996) which describes the sum of knowledge within the profession of project management. The aim of the report was to provide a basic reference for anyone interested in project management. PMBOK is organised into nine knowledge areas that are claimed to correspond widely to those applied in practice (Figure 1.2). These are now listed.

- Project integration management includes the processes required to ensure that the various elements of the project are co-ordinated properly. It integrates the three main processes of

  - project development planning
  - project plan execution
  - and overall control.

- Project scope management is concerned with defining the project, its purpose and objectives, and identifying all the relevant work to be accomplished. The major processes of project scope management are

- initiation
- scope planning
- scope definition
- scope verification
- scope change control.

Figure 1.2 Project Management Body of Knowledge

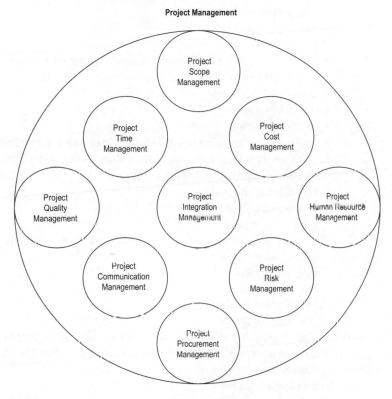

Project Management

Project Scope Management

Project Time Management

Project Cost Management

Project Quality Management

Project Integration Management

Project Human Resource Management

Project Communication Management

Project Risk Management

Project Procurement Management

Body of Knowledge

SOURCE PMBOK 1996.

- *Project time management* is concerned with processes required to achieve timely completion of the project:
  - activity definition
  - activity sequencing
  - activity duration estimating
  - schedule development
  - schedule control.

- *Project cost management* includes the processes required to ensure that the project is completed within the approved budget:
  - resource planning
  - cost estimating
  - cost budgeting
  - cost control.
- *Project quality management* is concerned with the processes required to ensure that the project will be completed within the required quality standards and that the needs for which it is being undertaken are satisfied:
  - quality planning
  - quality assurance
  - quality control.
- *Project human resource management* includes the processes required to make the most effective use of the people involved with the project:
  - organisational planning
  - staff acquisition
  - team development.
- *Project communications management* is concerned with ensuring timely and appropriate generation, collection, dissemination, storage and ultimate disposition of project information. The key processes are
  - communications planning
  - information distribution, performance reporting
  - administrative closure.
- *Project risk management* is concerned with the processes of identifying, analysing and responding to project risk:
  - risk identification
  - risk quantification
  - risk response development
  - risk response control.
- *Project procurement management* includes the processes required to acquire goods and services from outside the performing organisation. The key processes are:
  - procurement planning
  - solicitation planning
  - solicitation
  - source selection
  - contract administration
  - contract close-out.

Summary While useful in defining areas of knowledge in project management across a wide spectrum of industries, PMBOK is too broadly focused and ignores a number of crucial issues relevant to the construction industry, such as project finance, project delivery methods and contract strategy. Furthermore, not many construction projects today would be immune from exposure to economic, political, social and environmental influences. These are often the most critical issues that make or break construction projects, and they need to be seriously assessed during the conceptual stage of the life-cycle.

Morris' (1994) criticism of PMBOK as being incomplete (see above, page 8) was reinforced by Woodward (1997), who asserted that a comprehensive list of knowledge areas in project management should also include measurement and evaluation of project success or failure, and the use of facilities or capital assets.

CONCLUSION

This chapter defined the concept of project management and the specific characteristics of projects. It reviewed organisation theory in project management and described the project management system and its functions. It also summarised the main knowledge areas defined in the Project Management Body of Knowledge.

EXERCISES

1 What are major similarities and differences between project management and functional management?

2 Identify the main differences in managing the ongoing production of prefabricated houses (kit houses) and the construction of a one-off house.

3 Give examples of project management being employed by functional organisations.

4 What steps should the client's directly employed project manager take to ensure support from various functional managers in the client's organisation?

5 How relevant are systems theory and contingency theory to project management? What functions of project management do you consider to be the most critical? Is it possible to prioritise a list of such functions?

CHAPTER 2

THE CONTEXT OF PROJECT MANAGEMENT

INTRODUCTION

Management of construction projects is a challenging and often difficult task that extends well beyond the boundaries of the project itself into areas of economic, political, social and environmental influences. Focusing on managing the day-to-day activities of the project is not enough to achieve successful project outcomes, particularly in the presence of adverse external factors. For example, a proposed construction development may be well thought out and may show a healthy financial return, but its realisation and financial success will almost certainly depend on the presence of external risk. It is now almost impossible to build construction projects, particularly infrastructure projects located in or near urban areas, without somebody opposing them. The opposition may come from individuals, groups of residents, small but powerful action groups, a local newspaper, TV or radio stations, unions or even houses of parliament. Once the opposition to a project attracts the attention of the media, it may never get off the ground. If it does, its scope may be constrained and its financial viability eroded. In this situation the client gets the full value from employing a professional project manager with the necessary skills to foresee such external risks, for which he then develops mitigation strategies that reduce or deflect the risk.

Public infrastructure projects are particularly vulnerable to external risks. A proposal to build a second international airport in Sydney is a typical example of how social pressures can frustrate the development of an important infrastructure project. The airport has been planned since the late 1960s on a number of different sites around Sydney but its development has continually been postponed due to the strong opposition from local communities. It will probably never be built as neither the Commonwealth nor the State governments have the political will to proceed with the project in the face of ongoing public opposition.

This chapter examines the context within which project management operates. External factors in the form of physical, social, political, legal, economic and environmental influences and their likely impact on a project are examined first. Since the concept of project management is applied to each stage of a project's lifecycle, a brief review of stages of a project lifecycle will be discussed next. Lastly, the discussion will focus on the composition of a project team and the roles of individual project team members.

EXTERNAL INFLUENCES ON A PROJECT

No construction project can be developed successfully without effective management of a range of external influences or risks. The project manager, by applying the concept of risk management, will identify, assess and treat such risks in the most appropriate manner. For more information on risk management, see Chapter 15. Following is a brief review of common external project influences.

Physical influences

The physical environment within which a construction project is sited may impact considerably on its development. The geographical location of a project, ground conditions and weather patterns are the most common examples of physical influences. They tend to be highly unpredictable and no amount of management action can prevent their occurrence altogether.

When given a choice, the client will select the most geographically suitable location for a project to avoid extremes in weather patterns and to take advantage of available infrastructure and resources. But projects such as the Alaska pipeline or the International Space Station are geographically fixed in location and the client's only choice is either to commit to their construction or to abandon them.

Extensive soil testing can substantially reduce the risk of finding unfavourable ground conditions on a proposed construction site. However, some residual risk will remain since the true condition of the soil can only be established after the site has been excavated.

Construction projects are always exposed to physical influences. In comparison, other types of projects, particularly those generated internally within an organisation such as software development, are unlikely to be greatly affected by physical influences.

Social influences

Societies in the developed world, and increasingly also in the developing world, are very much concerned with the quality of the built environment. These concerns are often related to the type of new developments, their location and their impact on the surrounding environment, rehabilitation of existing facilities, and preservation of structures that have historical or cultural value. Securing the necessary approvals from relevant authorities does not guarantee that a project will go unchallenged by individuals, community groups or even governments. For example, approval by the local council to allow a developer to build a high-rise apartment building within a medium to high-density residential area is likely to be vigorously challenged by local residents. Although the new development may satisfy the planning laws and regulations, the local residents may oppose it on the grounds that it would destroy the local heritage values, increase traffic flow through residential streets, increase noise, and destroy the character of the neighbourhood. Strong community opposition to such a project is a significant risk with potentially serious consequences on the profitability of the proposed development. In some cases, the developer may even consider abandoning the project.

There are many examples of developments that have been either abandoned or dramatically altered in scope under strong opposition from local communities and various action groups. Perhaps the best-known example of community power and its ability to stop new developments is the famous 'green bans' movement of the 1970s imposed by building unions on the redevelopment of Woolloomooloo and the Rocks in support of local residents who vigorously opposed the projects. These two historically significant residential suburbs of Sydney were saved from high-rise redevelopment by the imposition of bans that stopped the work indefinitely. After a prolonged stand-off, the developers finally agreed to abandon the original high-rise scheme in favour of a low to medium-rise residential and commercial scheme that retained the historical character of the area. The Rocks area of Sydney has since become one of the most popular precincts of Sydney for its unique beauty, preserved history, and its mixture of specialty shops, galleries and restaurants.

Political influences

Governments play a significant role in the construction industry as:

- construction clients
- financiers
- regulators of the national economy
- regulators of the construction environment.

Governments are thus able to influence activities of the construction industry. As a construction client, a financier and a regulator of the national economy, governments can significantly increase or decrease the demand for construction services through budgetary measures and monetary policies. In their capacity as regulators of the construction environment, governments influence the development and building approval processes, and enforce compliance with acts and regulations. Governments may also invoke their powers to initiate or stop projects that are opposed on political, social or environmental grounds. For example, the NSW State Government passed special legislation under which the Sydney Harbour Tunnel project was eventually built. On the other hand, in the early 1990s the Commonwealth Government used its constitutional power to stop temporarily the construction of the Hindmarsh Bridge in South Australia because of objections from the local community. These are just two examples of political interference of governments in the construction industry.

Legal influences

The construction industry operates within a maze of planning and environmental regulations, codes of practice, labour laws, safety regulations, licensing, insurances and taxation laws. These laws, codes and regulations are generally well defined, making it possible to predict with reasonable accuracy their impact on construction projects. But periodic changes to industrial, safety, taxation and environmental laws are not uncommon and problems may arise when the law changes during the life of a project.

Economic influences

A challenging task for any project manager is to ensure that a project is financially viable within a fluctuating economic environment. Since periodic economic cycles significantly affect the activities of the construction industry, accurate forecasting of economic trends, both local and global, is very important. Fluctuating economic cycles affect the demand for construction work, money supply, availability of resources, and the value of construction facilities. They are also likely to affect the rate of inflation, and interest and exchange rates. To safeguard against the impact of economic influences, a project manager should seek expert advice from an economist on the likely trends in the national and the global economy.

Environmental influences

Reducing the reliance on non-renewable resources, preservation of the natural environment and curtailing emission of greenhouse gases have become the key priorities of sustainable construction (Spence & Mulligan 1995; Hill & Bowen 1997; Toakley & Aroni 1997). A new challenge for the construction industry and all of its participants will be to embrace the key principles of sustainable construction and incorporate them in building sustainable projects. While at present environmental regulations are scarce, with those in existence being mainly concerned with local issues such as traffic flow, air and water quality, noise pollution and waste, it is only a matter of time before laws and regulations will be imposed on the construction industry in an effort to achieve sustainable construction. The ban of CFC gasses and the ratification of the Kyoto Protocol are the first examples of many future international environmental treaties that impact directly on the industry. Meeting the greenhouse emission limits set down by the Kyoto Protocol for each participating country is likely to change the mode of operation of the entire economy, including the construction industry. This may involve better management of renewable and non-renewable natural resources such as water and land, and a more efficient management of waste.

PROJECT LIFECYCLE

Construction projects receive the greatest amount of public attention at the construction stage when their physical shape starts to emerge. Local communities are particularly concerned about the likely impact that projects under construction may have on the environment, while others in the community often comment on design features, both positively and negatively. The media, apart from amplifying community concerns, are also attracted to projects that experience technical or industrial relations problems during construction.

Delivering a successful construction project involves a lot more than achieving its trouble-free construction. Before a project is constructed, months or even years of work are devoted to defining its scope, arranging the funding, developing the design and preparing the necessary contract documentation. It is the quality of work associated with these pre-construction stages that often determines how successful or otherwise project outcomes are. Bromilow (1970, 1971) and Levido and colleagues (1981) showed that the greatest opportunity for significant cost efficiencies is at the conceptual and design stages of a project lifecycle. Conversely, they found that faults in the design and documentation which only become apparent at the construction stage could substantially increase the cost and time of projects.

A project's life extends well beyond its construction. A completed project must first be commissioned to ensure that it meets all of its designed performance criteria and that it is free of defects. A client will then either sell the project or retain it and manage it as a valuable asset.

The foregoing discussion alludes to the fact that construction projects progress through a series of stages from inception to completion. This is what is known as the project lifecycle. The stages may be defined broadly as:

- concept
- design
- tendering
- pre-construction
- construction
- commissioning
- asset management.

A successful completion of construction projects thus depends on the effective management of individual project stages, and their co-ordination and integration over the entire project lifecycle. The membership of a project team varies from stage to stage, reflecting a unique set of activities associated with each stage. The stages are largely sequential, that is, a preceding stage must be successfully completed before the next can start. However, when time is the most important constraint, it is possible to 'fast-track' or overlap the design and construction stages (Uher & Davenport 2002).

The management of project lifecycle stages will be examined in detail in Chapters 5, 6, 7, 8 and 9. The following sections provide only a brief overview of such individual stages and the main activities associated with them.

The concept stage

The concept stage is the first and probably the most important stage in the entire project lifecycle. Here the client articulates his needs for a project, defines specific objectives and sets the maximum amount of available funds. The project manager in co-operation with the client converts this information into a scope statement from which he formulates possible development alternatives. These are assessed through a feasibility study which identifies a preferred development alternative.

A strategic plan is then prepared for a preferred development alternative. It provides a blueprint for a successful project development. Apart from defining the scope of a project, its organisation structure, staffing and budgets, it also defines in detail its design requirements. These are then compiled into a brief from which a design consultant designs a project.

The design stage

A design team appointed by the client translates the project design requirements specified in the brief into an overall design scheme. Such a design scheme or schematic design is characterised in terms of site aspect and orientation, the size, shape and height of the proposed project, its appearance and its suitability for its purpose or function.

If accepted by the client, a design team then formalises the design scheme into a coherent design concept that includes more detailed sketches of the architectural, structural and services components of the proposed project. This stage of a design development is referred to as the preliminary design.

A final detailed design and the production of design documentation begins after the local planning authority has granted an approval to proceed with the development of the project.

The tender stage

The tender stage serves the purpose of selecting the best contractor to build a project. The most commonly employed selection method is competitive tendering, which is based on the available design documentation. Tendering is a form of competition among bidding contractors, with a contract commonly being awarded to the most competitive, though not necessarily the lowest, tenderer.

While tendering is the most popular method of selecting a contractor, the client may prefer to enter into a contract with a preferred contractor through negotiation.

The pre-construction stage

The award of a contract to the contractor does not constitute the start of the construction stage. That will occur some time in the future when the date for site possession, agreed to between client and contractor, has been reached. The period between the awarding of a contract to the contractor and the date for site possession stipulated in the contract is known as the pre-construction stage. During this stage the contractor will, among other things, prepare a detailed program of construction activities, mobilise the necessary resources, establish time and cost budgets, and develop a project control system to ensure completion within the time and cost budgets and to the required standards.

The construction stage

The construction stage starts from the date of site possession or from a specific date agreed to between client and contractor and ends when the date for practical completion has been reached. The contractor implements the programs and strategies devised in the pre-construction stage, and commits resources in order to build the project in accordance with the contract documentation and to accomplish the agreed time schedule and cost budget.

The commissioning stage

When a project reaches the practical completion stage, the client takes possession of the completed project. However, the contractor is required under the terms of the contract to fully commission the services of the project such as air-conditioning, lifts, fire protection and security system, and rectify any defects to the satisfaction of the client.

If the client intends to derive income from the project by leasing it out, the project manager would assist the client in arranging tenancy agreements and fitouts. Alternatively, the client may proceed to sell the project.

The asset management stage

The completed project is a valuable asset which, apart from producing income, is also likely to appreciate in value over the years provided it is properly maintained. It is therefore necessary for the client's project manager to develop a long-term asset management and maintenance plan to enhance the project's value and ensure its economic viability in the future.

THE PROJECT TEAM

In the context of project management, people play a vital role in achieving successful project outcomes. They form teams through which each stage of a project lifecycle is managed under the leadership of a project manager.

People form groups for a variety of purposes. They can be formal with specific goals or informal and of a social nature. A group may be defined as 'two or more interacting individuals who work together to achieve a specific goal'. A team, however, is a group of a higher order. It may be defined as 'a collection of committed people with specific skills, abilities and interdependent roles who work together in an environment of trust, openness and co-operation towards achieving common goals'. Effective teams develop solutions to problems that are consistently superior to those developed by individual team members. Such teams are said to have 'synergy'. A detailed examination of teams, teamwork and team-building is presented in Chapter 10. Only a brief overview of the importance of teams and teamwork in construction will now be given.

The unique feature of construction teams is that their composition varies from stage to stage of a project lifecycle. Their membership may also vary according to the size, type and complexity of a project. Furthermore, different types of delivery methods usually require a different composition of a project team.

Only the client and the project manager remain involved on a construction team from inception to completion. Other important team members such as designers, specialist consultants, a contractor and subcontractors join a team when their expertise is required. On completion of their specific tasks, they leave. For example, a contractor joins a project team in the pre-construction stage after being awarded a contract and leaves at the end of the commissioning stage, while a quantity surveyor commonly joins a team at a later part of the design stage and departs at the end of the construction stage.

The membership of a project team is not always restricted to organisations contracted to develop and build the project. For example, partnering projects that require team members to work together in an environment of trust and openness would certainly benefit from having a partnering facilitator on a team. Similarly, if a project is exposed to significant social or environmental influences, their intensity may be reduced by inviting a local community representative or an expert environmentalist to join a team. The ability to dis-

cuss contentious issues from different viewpoints in a[r]
operative team environment may then lead to a win
for the project, the local community and the environment.

Teamwork must first be developed and then sustained. It cannot be assumed to exist. It cannot even be imposed. It needs to be understood that most people who form a team are representatives of organisations contracted to a project. Their principal allegiance is to their respective organisations and their main aim is to make a profit for their organisations. However, the client expects a project team to focus on meeting the specific objectives of the project. Clearly, one of the most important tasks of a project manager is to shift the team's attention away from individual to project goals. This can only be achieved when real teamwork is developed.

Covey (1989) established a simple sequence of events required for achieving effective teamwork and team synergy. They are:

RESPECT ➤ TRUST ➤ OPENNESS ➤ SYNERGY = TEAMWORK

Respect for each other will lead to the development of trust, and conversely, disrespect will stifle trust. A rising level of trust promotes open communication, which in turn leads to genuine teamwork that gets results. While logical and simple in theory, developing teamwork in the construction industry is often difficult. Entrenched professional rivalry, a master–servant mentality, and a chronic level of disrespect and mistrust among different participants are the main barriers to teamwork. There are ways of overcoming such barriers, one of which is a process of partnering (CII 1989; USACE 1990; Fehlig 1995; Lendrum 1995). Experiences with partnering in the construction industry throughout the 1990s suggest that it is possible to develop and maintain teamwork for the entire project period. Partnering has shown that cohesive and harmonious teams achieve better project outcomes (CII 1996; NSW DPWS 1996; Uher 1999).

Kezsbom and colleagues (1989: 272) identified the essential elements that lead to successful team performance as:

- a mission or a reason for working together
- a sense of ownership, commitment and interdependence of each team member
- commitment to the benefits of group problem-solving and group decision-making
- accountability as a functioning unit.

A comprehensive summary of characteristics of effective teams is given by Robbins and colleagues (2003). These include:

- setting clear and achievable goals
- requiring team members to have appropriate skills
- developing mutual trust
- unifying commitment and loyalty of team members to a project
- achieving open communication throughout a team
- developing negotiation skills to confront and reconcile differences
- providing effective leadership for achieving synergy and teamwork
- supporting activities of a team by training and adequate resourcing.

With the characteristics of effective teams being well defined and clearly understood, the next important step is to achieve them. This is commonly done by a process known as team-building. Team-building may involve one or more carefully structured and facilitated brainstorming workshops involving team members which are conducted in the early stages of a project. Through direct participation in a range of specific project-related activities, team members not only develop solutions for important issues but also learn about each other, learn to appreciate the value of different viewpoints, develop a sense of interdependence, and develop an understanding of the value of team decisions. Specific tasks commonly undertaken in such workshops are:

- defining project goals
- agreeing on a mission statement
- identifying roadblocks to achieving project goals and developing solutions for overcoming such roadblocks
- defining roles and responsibilities of each team member
- developing a staffing management plan that describes when and how team members will be brought onto and taken off the team
- developing an organisation chart showing project reporting relationships and communication links
- developing a decision-making process
- developing a process for resolving issues or problems
- developing a performance evaluation process.

The effort that has gone into developing teamwork needs to be matched with a similar amount of effort to keep the momentum going. Subsequent refresher team-building workshops serve the purpose of ensuring maintenance of teamwork. Refresher workshops also serve as a vehicle for inducting new entrants to a project team.

A typical construction project team comprises a large number of participants:

- a client
- a project manager
- a financier
- a legal consultant
- a design leader (architect or structural engineer)
- other design consultants
- a main contractor
- subcontractors
- a cost consultant
- other consultants (depending on the project needs)
- an end user of the completed project (where appropriate).

This list is not exhaustive and additional participants may join the team depending on the size and type of project and the method of its delivery. The most commonly involved team participants will now be briefly examined.

A client

Construction industry clients are the single most important group on a project team. They initiate projects and thus stimulate demand for construction work. Some clients assume the role of a project leader, while others prefer to delegate the leadership role to another party, such as a project manager or an architect. The degree of their involvement on a team is largely influenced by their usual mode of operation, the extent of experience with a particular type of project, and the amount of risk associated with a project.

Construction clients are both public (government) and private, with private clients being either corporate organisations or individuals. In Australia, public clients generate over 60 per cent of all construction work, with the Department of Defence being the single largest client.

A strict requirement for public accountability is the main characteristic of public clients. This is necessary since they spend taxpayers' money. However, the past practices of procuring projects using the traditional method of delivery based on full documentation and awarding a contract to the lowest bidder, selected through an open tender system, have long been replaced in Australia with more flexible tendering and procurement policies. Bids are now evaluated in terms of their technical and commercial merits rather than just on price. Apart from employing the traditional method of delivery, public clients also procure projects using alternative delivery methods such as the 'design and construct' and construction management methods. Furthermore, they commit to projects using the public private partnership (PPP) schemes that rely on private enterprise to fund, develop and operate major infrastructure projects. Partnering and alliance schemes have also been successfully employed on public projects (Uher & Davenport 2002).

Clients may develop construction projects for their own use as occupiers, for example private homes or factories; as a form of investment in the case of projects such as office buildings, hotels or apartments; or for public use such as schools, hospitals, airports, power stations, roads, telecommunication networks, water supply and so on.

Clients initiate projects with specific needs and expectations in mind. These are expressed as project objectives. Among these, the three most commonly expressed objectives in accomplishing a project are:

- meeting the cost budget
- keeping to the schedule
- achieving the required quality standards.

The matter of objectives will be discussed in more detail in Chapter 5.

Most construction clients raise funds through borrowings, but other funding models are also available. Public clients in particular are able to take advantage of securing funding from a private enterprise through the PPP scheme for infrastructure projects. Private finance initiative (PFI) is a specific type of PPP that enables a public client to rely on a private promoter to finance, build and operate a project for a concessional period. Funding of PFI projects is achieved through a mixture of debt and equity finance, shares, bonds and even lotteries (Walker & Smith 1995). Often a public client may be required to provide a loan to the promoter or a guarantee of income while operating the completed project.

A project manager

The client commonly appoints a project manager as early in a project's life as possible, preferably at the conceptual stage. The client then assigns overall responsibility and accountability (known as single-point responsibility) to the project manager for achieving successful outcomes, matched with appropriate empowerment.

The project manager's role is to manage a project and achieve its objectives. Since this can only be achieved by effective teamwork, the real role of the project manager is to provide leadership to a project team.

Leading a project team and co-ordinating and integrating its activities are thus the essential tasks of a project manager. They are also difficult tasks considering the highly competitive nature of the construction industry, entrenched as it is in a win–lose culture, a lack of personal trust, a high degree of contractual and industrial conflict, and often limited resources.

The project manager must also have detailed knowledge of activities taking place at each stage of a project lifecycle. He is expected to be a competent administrator with the ability to plan, organise and control. But his most important task is that of leadership. As such, the project manager must be an interdependent person who is self-aware, principle-centred and able to stimulate and inspire people. He is therefore expected to have maturity of judgment and the ability to communicate with integrity and clarity. He must also be a good listener and must show enthusiasm for tasks ahead. Delegating responsibility to team members, motivating them and empowering them are essential abilities in an effective project manager. To cope with a range of external influences, he also needs to display political astuteness and strong skills in negotiation, tactics and conflict management. The leadership qualities of project managers are summarised by the Chartered Institute of Building (CIOB 1988) as:

- ability to lead, motivate, command respect
- wisdom, knowledge, experience, self-confidence
- powers of delegation
- ability to prepare for a task
- enthusiasm
- clarity of purpose
- creation of opportunities
- example and personality
- responsibility and authority
- creation of an atmosphere of compatibility
- communication and co-operation
- fairness and impartiality.

Although a project manager is expected to have knowledge and skills across a wide spectrum of project management competencies, ideally he should not be directly responsible for any operational activities such as design, tendering, construction, cost control,

The Context of Project Management

scheduling or risk management. Specific operational expertise is provided by individual members of a team. A project manager's task is to lead a team in much the same way as a conductor leads an orchestra: by inspiring the individual musicians and synchronising the sounds of different musical instruments to make fine music.

Most textbooks on project management provide a comprehensive list of important or desirable attributes and competencies of a project manager. Those defined by PMBOK (1996) and referred to in Chapter 1 are most commonly quoted. A much broader checklist of attributes and competencies is provided by the Construction Industry Development Agency (CIDA 1993: 8–9):

- knowledge of project management concepts, theories and principles
- skill in the art of project management and the use of project management tools
- technical knowledge in the field of application
- experience in the practice of project management including the project management functions, phases of the project lifecycle, risk management, alternative delivery mechanisms
- a comprehensive vision required for project integration, which involves:
  □ comprehending all the technical, professional and human relationships in the project
  □ combining the separate components and participants in a complete, coherent, functional and productive whole
  □ managing internal technical interfaces
  □ co-ordinating across functions, for example cost with quality, cost with schedule, and design with functionality
  □ managing inputs of technical specialists and making sound judgments where their advice conflicts with one another or with other project parameters
  □ balancing contractual responsibilities with capabilities
  □ balancing devolution of risk with maintenance of control
- understanding of the commercial, organisational, industrial, legal, social, political and natural environment in which the project is embedded
- ability to manage the interfaces between the project and its environment, for example management of approval processes, management of industrial relations, and contract negotiation and management
- personal attributes including:
  □ ability to lead a multidisciplinary team
  □ communication skills
  □ interpersonal skills
  □ ability to think logically, comprehend technical matters and find creative solutions to problems
  □ energy, determination and endurance
  □ resilience and flexibility.

The client rewards a project manager by the payment of a fee, which is commonly expressed as a lump sum or a percentage of the final payment for the project. Apart from his overall responsibility for meeting project objectives, however, a project manager may also be required to assume the risk associated with meeting such objectives. In such a case, his fee may be linked to performance of a project. An emerging trend is for a project manager to assume more and more risk. In a recent case, a government client in New South Wales called for an expression of interest in the provision of project management services for a large hospital contract. Bidding project management firms were required, among other things, to assume the responsibility for the design and construction of the project. This represents a dramatic departure from the normal scope of a project manager's work. Whether this is an isolated case or the forerunner of a new trend in project management, in which a project manager may assume single-point responsibility for the entire project, only time will tell.

A financier

Most clients, no matter how large or small, obtain funds for new projects from financiers such as banks or finance companies. Securing finance is often a complex task that requires a considerable investment of time. Since a financier wants to recover the investment in full, it will attempt to reduce its exposure to risk by a range of reactive actions. For example, it may limit the amount of funds lent out or impose substantial collaterals. It may even insist on a low-risk construction and procurement strategy in the form of the traditional method of delivery based on full design and documentation, and competitive tendering. A financier may also insist on becoming a project team member to have a more direct access to project information and possibly to influence decision-making in order to safeguard its investment.

A better approach would be to invite a financier to become a project team member and involve it in both strategic and short-term decision-making. The financier would then have an opportunity to proactively contribute towards integrated decision-making, which would not only reduce its exposure to risk but might also improve the overall performance of the project.

A legal consultant

The legal consultant's role is to advise the client on the best contract strategy for the project at the conceptual stage of the project life cycle. This commonly involves drafting conditions of the main contract or amending standard conditions of contract, and allocating risk so that the client's exposure to risk is minimal. The services of the legal consultant may also be needed at the construction stage to help the client resolve contractual disputes involving client and contractor.

A design leader

On building projects the client normally engages an architect as the design leader, while on engineering projects the design leader is usually a structural engineer. The following discussion refers to the role of a design leader from an architect's perspective.

The architect is responsible for developing an architectural design of a new project and assembling the related design documentation. But depending on how a project is delivered, the architect may assume additional responsibilities. For example, projects delivered by the traditional method require the architect to take on the responsibilities of a contract superintendent at the construction stage. This involves administering the main contract between the client and the contractor. When a project is delivered by either the construction or the project management method, the architect's role is fundamentally that of a designer only. The role of a superintendent is taken over by the contractor or project manager (Uher & Davenport 2002).

The architect is commonly engaged by the client towards the end of the conceptual stage to assist in developing a brief. This may mean preparing sketch designs of alternative project strategies and their costing. It may also mean providing input on issues such as a project organisation design, budgets and appropriate methods for selecting a contractor.

In the design stage, the architect's main responsibility is to:

- develop schematic, preliminary and final designs
- design to a cost
- participate in value management studies of the design
- co-ordinate design activities of other designers, namely the structural engineer and services engineers
- secure development and building approvals
- prepare tender documentation in the form of working drawings and specification.

At the pre-construction stage, the architect's main role is to maintain production of the contract documentation, while at the construction stage the architect will be responsible for:

- revising contract documentation
- assisting in resolving design-related problems
- assisting in controlling quality of construction.

When a project manager leads the project team, the architect's main role at the commissioning stage is to assess the compliance of the completed project with the contract documentation.

THE ARCHITECT'S FEE

For a range of services that the architect provides, the architect traditionally receives a percentage fee calculated on the final contract price. Although some clients now engage architects on a fixed-fee basis, the use of percentage fees continues to be widespread. How the architect arrives at a particular percentage figure is often unclear and the client may well wonder whether or not such a fee appropriately reflects the volume of work. More importantly, the client may also ask whether or not a percentage fee provides an incentive to the architect to perform, considering that the architect benefits from project cost overruns. A prudent client may well ask why a percentage fee is used in the first place on projects of medium to low risk instead of a properly estimated fixed fee. There is no reason why an experienced architect should not be able to estimate with reasonable accuracy the volume of work required and price it accordingly. If the main contractor and subcontractors are required to enter into fixed-price contracts and assume the responsibility for their work and the mistakes they make, so should consultants such as architects. The architect should be accountable for errors in design and documentation, which should not be paid for from a project contingency. However, when imposed on high-risk projects, fixed-price contracts may lower the quality of the design and documentation if the designer has underestimated the true cost of the design and is unable to recover extra costs under the contract.

A main contractor

The client commonly selects the main contractors through a bidding process and awards the main contract on a lump-sum basis. The contractor's responsibility is to build the project according to the contract documentation within the required cost and time budgets, and to the specified standards. The execution of the main contract is administered by the project manager or, in the absence of a project manager, by the design consultant. The main contractor's responsibility is to build a project, but most of the work is performed by subcontractors. The important role of the main contractor is to co-ordinate the work of subcontractors. Other duties of the main contractor are to:

- plan, co-ordinate and control production
- organise resources, both human and physical
- co-ordinate subcontractors
- ensure compliance with the contract documentation
- deliver the project on time, within the cost budget and to the specified standards
- ensure safety of the workforce and the general public
- ensure compliance with various laws and regulations.

Subcontractors

Subcontracting is a way of subletting portions of the main contract to specialised firms in various building trades. The main contractor lets subcontracts and administers them. Subcontracting is popular and widespread, particularly in the building sector of the industry.

A cost consultant

On most projects, a quantity surveyor is engaged by the client to prepare a cost plan at the conceptual stage and a bill of quantities at the design stage of the project lifecycle. The quantity surveyor's role at the construction stage is to check the validity of progress and variation claims made by the main contractor.

Other consultants

Depending on the complexity of the project and the type of delivery method, the project team may require the services of other consultants such as a programming consultant, an IT specialist, an environmental engineer, a real estate consultant, and so on.

An end user of the project

Most industrial, commercial and public construction projects are undertaken by clients on a profit or non-profit basis for use by other organisations commonly known as end users. A project such as a new hospital in Sydney is developed and paid for by the Department of Health (the NSW Government) but operated by the Area Health Authority. Similarly, a new faculty building at the University of New South Wales is developed by the university for use by the Faculty of the Built Environment. These two examples of projects are occupied and operated by the end users. The success of these projects will be measured in terms of how satisfied the end users are with the layout of the project, its functionality and performance, the quality of services, types and quality of fixtures and fitout, and so on. The end users should be active members of the team, particularly at the design stage.

CONCLUSION

In this chapter, the broader context of project management was explored. A number of external factors that adversely influence the development of construction projects were identified and discussed, most notably physical, social, political, legal, economic and environmental factors. The key stages of the project lifecycle were defined as concept, design, tender, pre-construction, construction and commissioning. Specific project management activities were defined for each such stage. The remainder of the chapter focused on the composition of a construction project team and on developing teamwork. The roles of the key project team members were defined, particularly those of the client, architect and project manager.

EXERCISES

1   Why is it important to effectively manage the conceptual and design stages of a project lifecycle? What are the consequences of poor management?
2   What specific issues would you as project manager need to address to ensure the development of a team synergy in terms of:

- respect
- trust
- openness?

3   Who would you prefer for a post as project manager for construction of a new airport terminal building in Sydney and why:

- Person A: a civil engineer with 15 years' construction experience, expert in construction technology, materials handling, structural design and contract administration, or
- Person B: a commerce graduate with ten years' business and HRM experience, and four years' experience working with the property developer?

CHAPTER 3

ORGANISATION STRUCTURE

INTRODUCTION

Mintzberg (1979) defined organising as the process by which managers

- identify what works needs to be done to accomplish organisational goals
- divide work between units and individuals
- co-ordinate their efforts so that those goals can be attained.

One of the most important functions of a project manager is being able to design an organisational structure that is appropriate to the task to be performed. It follows that an understanding of the issue of structure and the impact this has on organisational effectiveness is one of the most basic skills needed in project management. Unfortunately, it is also one of the most neglected. The purpose of this chapter is to consider the many dimensions of organisational structure and their appropriateness in different project situations. It will also consider the changing work practices in the construction industry and the significant impact this has had on the task of organisational design.

ORGANISATIONAL STRUCTURE

An organisational structure should

- describe the assignment of tasks to individuals (division of labour)
- designate formal reporting relationships, information channels and levels, and lines of authority

- group individuals into departments and departments into the organisation as a whole.

It defines the way in which work is divided and co-ordinated among organisational units and it is often depicted as a formal organisational chart such as that illustrated in Figure 3.1.

Figure 3.1 A typical construction project organisational chart

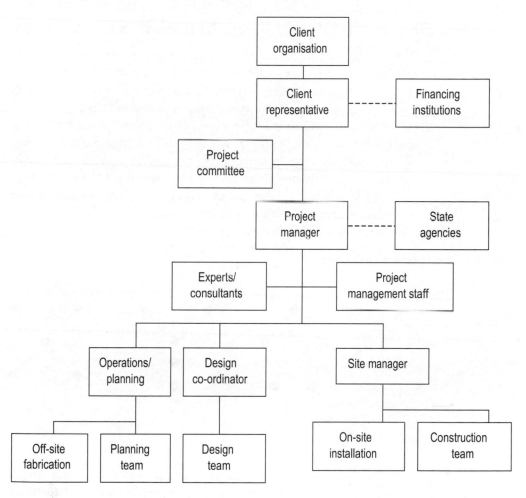

Such organisational charts are often accompanied by organisational manuals, which define in more detail the nature of people's duties and responsibilities. These manuals also contain standard operating procedures, which describe how and when these

functions are to be performed. In essence, the organisational chart represents an attempt by management to impose a communication and reporting system which constrains the independence of individuals in their day-to-day actions, avoids chaos and ensures efficient communication flow by keeping information channels free from irrelevant information (Galbraith 1973). The desired effect is to make the organisation efficient and fast in response to problems that may threaten the attainment of its goals.

THE IMPORTANCE OF ORGANISATIONAL STRUCTURE

The structure of an organisation is an important determinant of its efficiency, and the first to discover this were Leavitt (1951) and Shaw (1954). They investigated experimentally the relationship between the problem-solving efficiency of small groups and their communication structure. Leavitt artificially created small groups of people and gave them a simple problem to solve which required the pooling of information. The members of each group were physically separated and only permitted to use predetermined two-way communication channels. The group communication patterns were investigated and are illustrated in Figure 3.2.

Figure 3.2 Leavitt's experimental patterns

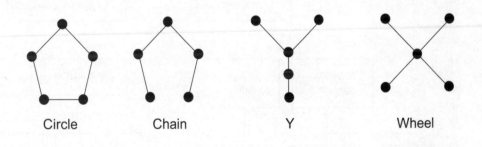

| Circle | Chain | Y | Wheel |

Within a predetermined communication structure, each group was given a number of problems to solve, enabling Leavitt to study the development of group efficiency in each pattern. Efficiency was measured by the amount of time required for solution, the quality of solution being considered irrelevant because of the artificial simplicity of the problem. Through observations, interviews and analysis of communications, Leavitt found significant differences in problem-solving efficiency between the patterns. He found that organisational efficiency developed most rapidly in the 'chain' followed by the 'Y', the 'wheel' and finally the 'circle'. In terms of leadership, the circle was seen as active, leaderless, unorganised and erratic but yet enjoyed by its members. In complete contrast, the wheel was less active, had a distinct leader, was well and stably organised, was less erratic but was unsatisfying to its members. Leavitt saw the concept of centrality as vital to an explanation of the results. In Leavitt's (1951: 38) view, 'the most central position in a pattern is the position closest to all other positions', distance being measured by 'the number of communicative links which must be utilized to get, by the shortest route, from one position to another'. Leavitt concluded that where centrality and hence interdependence are evenly distributed, there will be no leader, high activity, slow problem-solving but high satisfaction.

Shaw (1954) refined Leavitt's work by discovering that there was no one best organisational structure and that the most appropriate structure depended on the nature of the task. Shaw investigated how task complexity affected relative group efficiencies. Strikingly, he found that as tasks became more complex, the relative efficiency of various patterns completely reversed: the circle became the fastest and the wheel the slowest. In contrast, all other findings relating to satisfaction and leadership were in agreement with Leavitt's original conclusions. The explanation was that as problems became more complex, the central person in the wheel became overloaded with information and peripheral people became less willing to accept solutions offered by the central person. Since this early work, it has been discovered that in general, centralised wheel-type structures are more suitable for routine tasks with mechanised technology, carried out in stable environments by people who dislike autonomy. In contrast, decentralised organisations are more suitable for non-routine, non-mechanised tasks, carried out by professionals in an unstable environment. Construction activity approximates to the latter rather than the former and in this sense decentralised structures would seem more suitable.

FORMAL AND INFORMAL STRUCTURE

The organisational charts that adorn the walls of construction company offices and the pages of its accounts are an idealistic representation of a formally prescribed structure that may bear little resemblance to what actually exists. Like any imposed structure, the degree to which it is reflected in practice depends on the way it is monitored and on people's understanding of and receptivity to it. Systems are only as good as the people that use them, and there are numerous examples of organisations with sophisticated and expensive systems being undermined by the unpredictable behaviour of their employees. The Challenger space shuttle disaster was a tragic example of such an event in one of the world's most technically sophisticated organisations. Of course, some element of formal structure will exist in all organisations, but the reality is more likely to be defined by patterns of friendship and common interests than anything prescribed by managers. This is particularly so during a crisis, where people have a tendency to bypass formal procedures because these are inherently inflexible and cannot hope to cover all eventualities. In this sense they may become counterproductive by slowing the response to crises (Mintzberg 1976; Bennett 1991; Sagan 1991).

Thus it is evident that the informal system may be particularly dominant in organisations that demand high creativity and face much uncertainty and change in their markets and production systems. In such circumstances, spontaneity and the professional judgments of individuals become more important. This is the case in construction companies, which are concerned with the production of unique products in highly uncertain environments. It is the same in construction projects where managers must realise that the imposition of voluminous and prescriptive formal contracts can actually reduce a manager's control rather than increase it, as people covertly bypass the procedures set down. Loosemore (1996) has reported numerous instances of project participants bypassing procedures because of their restrictive rather than helpful nature, and he has also monitored the attempts of managers to reimpose the system and manage the resulting chaos. In organisations that face great uncertainty, there is real justification for managers to allow their workforce considerable autonomy and to work more closely with the informal system to accomplish their tasks. Managers must recognise that the organisation has a life of its own, that operatives often know from their past experience what is best for accomplishing a task and will find a way through the labyrinth of formal procedures to do so. In this way, the informal organisation has the potential to be both constructive and destructive, and an effective manager will monitor it to ensure that it has a constructive impact.

THE DIVISION OF LABOUR

A division of labour refers to the process of dividing a large task into smaller tasks, enough for individuals or groups to actually accomplish. The association of individuals with specific tasks encourages them to become specialists and therefore highly proficient. In this way, the division of labour and specialisation go hand in hand. The phrase 'division of labour' was first coined by Adam Smith in his famous *Wealth of Nations* of 1776 where he described the workings of a pin factory. The productivity gains experienced during the Industrial Revolution were largely a consequence of its widespread application. According to Adam Smith,

> a workman not educated to do this business, could scarce, perhaps with the utmost industry, make one pin in a day. But in the way in which this business is now carried on, not only is the whole work a peculiar trade, but it is divided into a number of branches, of which the greater part likewise are peculiar trades ... and the important business of making a pin is divided into about eighteen distinct operations ... I have seen a small factory, where ten men only were employed [who] could when they exerted themselves make among them about twelve pounds of pins in a day. There are in a pound upwards of four thousand pins.

The division of labour has become particularly evident and acute in the construction industry in the last 30 years with the growth of subcontracting. It is a trend that has been driven by a number of forces such as economic conditions, government policies and technological developments. Government policies around the world have increasingly been guided by the 'small is beautiful' philosophy. Increasingly volatile and competitive economic conditions mean that firms have rationalised and cannot afford to carry excessive administrative costs. Technological developments have demanded increasingly sophisticated knowledge and skills from those involved in their application. However, while there are many advantages to be gained from this division of labour or specialisation, the process does introduce problems of control because it means fragmenting the organisation into smaller parts and creating a larger number of potentially problematical interfaces to manage. The effective management of communication becomes a much more critical issue in such organisations, horizontally between working units and vertically from management downward. Indeed, with the growth of subcontracting in the construction process, this has become a major problem, as have problems of quality control and motivation (NEDO 1988; Gray & Flanagan 1989).

DIFFERENT WAYS OF DIVIDING LABOUR

At the most basic individual level, specialisation can be talked of in terms of line and staff. Line employees are those who work at the coalface, having direct responsibility for production. In a construction company these would be staff such as site managers, estimators, buyers, planners and so on. In contrast, staff employees are those acting in an indirect supportive capacity and would include people such as accountants, lawyers and personnel managers. Their task is to provide advice and support to the line managers. Both line and staff employees carry out specialised work, and line and staff functions should be designed to complement each other. Typically, line people have strong practical knowledge and staff people are more educated in an intellectual and theoretical sense. In poorly designed organisations, line and staff are often in conflict, particularly when staff employees consider themselves superior to line employees. Special attention must be paid to this potentially problematical interface.

Whereas the division of labour at individual level results in line and staff divisions, the division of labour at unit level results in departmental divisions. Departmentalisation is the grouping of activities by sub-units within the organisation, these sub-units being called departments. Departments in turn further divide the unit of work to be done among its individual members. The departmental structure does not replace the line and staff structure but is overlaid on it. Departmentalisation can be based on several criteria: function (skill), purpose (product), or geography.

Departmentalisation by function

Functional departmentalisation is a method of grouping people together who perform similar tasks on the basis of similar expertise or knowledge. For example, in a construction company one would expect to find a buying department, an estimating section, a planning department and so on. The advantage of having such a structure is that the centre of excellence and therefore the source of specialist advice is clearly identifiable within an organisation. Furthermore, the bringing of like minds together can have a synergistic effect through increased focus and specialisation, which can aid in the development of innovations and solutions. On the other hand, specialisation to this degree may cause a unit to become inwardly focused on its own activities, seeing them as a goal in themselves rather than as contributing to the organisation's wider goals. In this way communication problems can arise. Indeed, these problems can be exacerbated by the tendency for such departments to develop unique languages that prevent them conversing efficiently with other specialist sections. In fact there is a danger that an organisation can become characterised by a number of distinct cliques with little communication between them. A further concern is that with such specialisation, a unit is likely to develop a very narrow view of problems, which might prevent the organisation as a whole from adapting rapidly to market developments. An example of departmentalisation by function is illustrated in Figure 3.3.

Figure 3.3 Departmentalisation by function

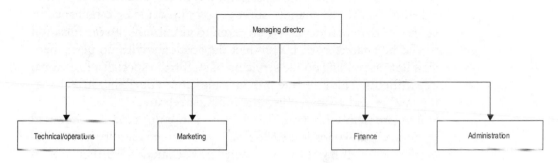

Departmentalisation by purpose

Departmentalisation by purpose is the grouping of people by the products they produce. For example, in a construction company one may find housing, commercial and civil engineering sections. This is illustrated in Figure 3.4.

Figure 3.4 Departmentalisation by purpose

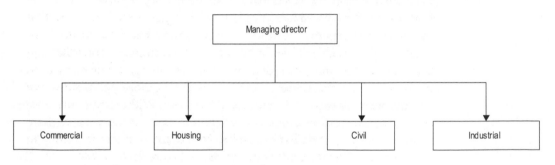

Organisation Structure

The problems associated with this structure stem from the competition for resources that may develop between different sections. There is a particular danger of this if section managers are empire-builders because the power struggles that develop between departmental heads can become an end in themselves and divert attention and energy from the main organisational task. Furthermore, there may develop an inefficient overlap of expertise between the different sections. For example, the housing and civil engineering departments may have their own dedicated quantity surveyors. The extent to which this is justified must be addressed in terms of the overlap of their skills. If they are very similar there may be some justification for having a mix of departmentalisation by function and purpose, in that a distinct functional quantity surveyor section could be established to service both departments. When taken to its extreme, an organisation would be characterised by product departments with no permanent staff but just relying on the servicing of staff from specialist functional departments. This structure provides maximum flexibility in a changing market and is often called the matrix structure.

The problem with a matrix structure is the difficulty encountered by product divisions in developing any sense of identity and teamwork since their membership is constantly changing. Furthermore, it can be a stressful system for those involved since people allocated to specific product sections find themselves with two bosses (the product section boss and the functional section boss); conflicts of interest can arise and leadership can become confused. Despite these problems, matrix structures have become popular, particularly in industries such as the university sector where academic departments service various courses throughout the university.

They are also popular in the aerospace industry where project-based teams are assembled to produce specific aircraft by order. Construction is also project-based and this structure is commonly found in construction companies. As projects are won by product-based departments, they are serviced by functional departments. So, for example, if the housing division won a tender they would request people from each of the functional department heads. The beauty of this system is its flexibility: it can provide each job with the exact amounts of functional specialism that it requires — if of course the specialists are not being used on other jobs. In this sense, labour supply can be closely tailored to the specific needs of each project that is undertaken. Without this structure, each product division would have to hold its own labour and there would be duplication of effort across a company and less flexibility in its provision. This structure is illustrated in Figure 3.5.

Figure 3.5 Matrix structure within a construction company

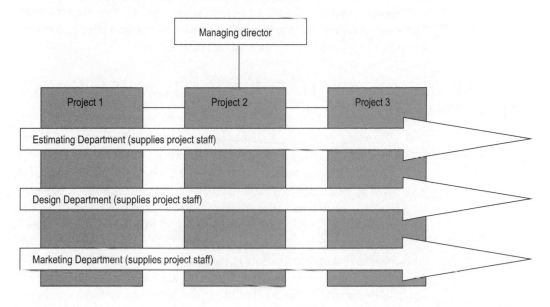

Geographical departmentalisation

Geographical departmentalisation is division by country. This structure is normally found in larger, well-established companies and arises from a desire to penetrate international markets. Its advantage is that it provides more direct access to these markets and thereby minimises costs. Furthermore, by using local labour, it can provide greater cultural sensitivity, which may be important in securing work and closer proximity to markets, which is in turn important in predicting and responding to market changes. Finally, a company wishing to become international in size may have no choice but to establish an overseas division with a local workforce since this may be a statutory requirement of an overseas government.

Customer departmentalisation

With the growth of partnering in the construction industry, some construction companies are establishing divisions that are dedicated to particular customers. This is only feasible with customers who build regularly and can supply enough work to keep a department viable and busy. Such clients may include supermarket chains, fast-food chains like McDonalds, or banks. The advantage of this arrangement for the customer is a custom-made, dedicated service which can routinely supply standard-made buildings in unusually fast times and with low costs. The advantage to the construction company is a long-term and stable relationship with a client who can provide consistency of workload into the future.

THE EVOLUTIONARY NATURE OF ORGANISATIONAL STRUCTURE

It is important to realise that organisational structure is not a static phenomenon and that a company's structure is constantly evolving as it grows or contracts and its markets change. The beginning of a company is normally characterised by an owner-manager system, where a few employees work very closely together. The structure is highly centralised, very simple, informal and fluid. The small size of the organisation ensures that there is little need for formal structure and rules and communication is largely by informal means. Finally, there is little departmentalisation of any kind since staff tend to be multiskilled and have to be flexible in fitting into any functional role required of them.

As a company grows, the owner inevitably becomes detached from day-to-day activities and is forced to focus on the task of management rather than on operational tasks. He is no longer able to run the company in a hands-on fashion and is forced to decentralise some of his decision-making responsibilities. In this sense, growth produces the need for greater attention to management as a specialist skill, and the first signs appear of differentiation between line and staff functions. With the taking on of ever more sophisticated projects, there also develops the need to create functional divisions. Eventually, as the number of projects increase, justification grows for grouping them into types and creating new departments on this basis to allow a greater customer orientation to develop in those areas. There is therefore a tendency as organisations grow to develop from highly formal and centralised structures into informal decentralised structures.

THE GROWTH OF TEMPORARY EMPLOYMENT

One important trend within the structure of organisations, which is resulting in a new division of labour, is the growth of temporary employment.

Organisational theory contains many warnings about the human impact of the seemingly unstoppable industrial trends towards technological innovation, mechanisation, specialisation, rationalisation and standardisation in the interests of ever-greater efficiency. For example, Hoxie (1915: 129) long ago argued that such trends 'inevitably lead to the constant breakdown of the established crafts and craftsmanship and the constant elimination of skill in the sense of narrowing craft knowledge and workmanship ... If fully realized, any man who walks the street would be a practical competitor for almost any workman's job'. Braveman (1974) wrote about the degradation of work and expressed grave pessimism about the decline of community and the threats to social harmony posed by such trends. But it is Handy (1994) who is credited with specifically foreseeing the destruction of prevailing patterns of work and the significant changes towards temporary employment which have taken place within Western labour markets during the 1980s and 1990s.

Handy (1994: ix) noted that during the 1970s

> the tradition of a man going out to work to support by himself a family at home became a statistical rarity; by the end of the decade only 14 percent of households fitted this stereotype. 'Long-term unemployment', 'youth unemployment' and 'redundancy' became familiar words, words which increasingly infected all social groups. Jobs began to be a scarce commodity, and 'work' started to mean other things besides the conventional full-time job ... The old patterns were breaking down; new patterns were forming.

In particular, he pointed to the collapse of work itself through the rise of part-time employment, contractualisation, temporary employment, odd jobs and the trend towards people holding a portfolio of jobs in a number of organisations, in parallel or at different times in their life.

As Sheldrake (1996) points out, the context of Handy's predictions was provided by the emergence of historically high levels of unemployment, particularly in the United Kingdom but also in other industrialised nations such as the United States and West Germany. This had largely been brought about by an unsustainable postwar boom, technological advances, increased competition associated with the erosion of international trade boundaries, a reorientation of government policies away from the pursuit of full employment towards the control of inflation, and the popularity of managerial trends such as rationalisation, re-engineering and downsizing. More specifically, Handy's predictions of the emergence of temporary employment was stimulated by right-wing government policies designed to break down the restrictive practices and industrial strife that had become associated with the 1970s and the growth of mega-enterprises and the associated union power.

In accordance with the highly influential thinking of management gurus such as Schumacher (1993), smallness in business was increasingly embraced by government and by the private business sector as the way forward in diminishing union power and thereby industrial strife and encouraging the flexibility and creativity essential in an increasingly competitive and dynamic business environment. In following this doctrine, governments created a system of incentives to encourage the formation of small businesses and broke up nationalised industries into smaller, more specialised production units. Private companies also began to decentralise their activities and downsize by outsourcing services that traditionally had been provided internally.

In essence, during the 1980s the structure of manufacturing and service industries underwent a fundamental change to a more fragmented, responsive but also vulnerable structure. While this was mainly a Western phenomenon, in recent years it has transpired that the developing Asian economies may be experiencing the same forces of change and at an accelerated rate. Their position appears to reflect that of the West during the 1980s in that they appear to be emerging rapidly from a period of economic growth and relatively full employment into an era that will be characterised by more rigorous fiscal control and higher unemployment. It may be only a short time before they experience the changes in labour markets that have been experienced in the West as a result of similar events.

TEMPORARY EMPLOYMENT IN THE CONSTRUCTION INDUSTRY

The way in which the construction process is traditionally organised ensures that construction projects are one of the most extreme manifestations of temporary employment organisations. This has become a well-known characteristic of the construction industry and has led Cherns and Bryant (1984) to use the term 'temporary multi-organisations' in referring to construction project organisation. While the increasing popularity of partnering and alternative methods of procurement is diluting this characteristic, it remains the case that the majority of construction project teams are temporary in nature. While construction industry participants have a long-term familiarity with this ever-changing personnel environment on construction projects, they have become used to relying on permanence of employment in the organisations they represent. This is particularly so for managerial personnel since the subcontracting out of traditional craft-based tasks at operational level has, over the last decade or so, become an accepted aspect of a construction firm's activities. It is not difficult to find statistics showing a dramatic growth in the subcontracting of these specialist construction services, which were once the domain of a directly employed workforce (Flanagan & Norman 1993). In recent years, however, in the search for even greater flexibility and efficiency, the concept of temporary employment has been expanded to include consultancy, self-employment, short-term contracts, job-sharing, overtime, restricted hours contracts, part-time work, casual work and fixed-term contracts. It is a trend from which managerial personnel cannot escape, in that it facilitates the outsourcing of their traditionally protected administrative roles. In this sense, it represents a strong force for rapid expansion of the service sector in the construction industry and the redefinition of the role of hitherto large multifaceted contracting organisations to primarily managerial roles in co-ordinating the diverse range of specialist firms that provide the activities they themselves offloaded. This is a trend that has been experienced in other industries, and there is no reason to assume that the construction industry will escape it.

TEMPORARY EMPLOYMENT AND FLEXIBILITY

While there have been many political reasons for the growth of temporary employment, the private sector's enthusiasm was largely founded on the greater flexibility and thereby responsiveness and efficiency it provided in an increasingly competitive and dynamic business environment. Flexibility refers to the ability to respond to new markets and to facilitate 'peak and valley' scheduling (Pollert 1991). The notion of organisational flexibility has received growing attention in recent years and it appears that in seeking to achieve it, organisations are restructuring in a way that more clearly differentiates between 'core workers' in permanent employment and 'peripheral workers' in temporary employment (Atkinson 1984).

Atkinson has developed a useful typology of different forms of flexibility provided by different parts of organisations:

- functional flexibility: the ability to redeploy employees through a multiskilled core workforce
- operational flexibility: the ability to increase or decrease the number of employees through a peripheral workforce
- financial flexibility: the ability to control the labour, plant and material costs of production, largely through core workforce control of product design and manipulation of the peripheral workforce.

In this sense, temporary employment is mainly of value in providing operational flexibility, though it also gives both functional and financial flexibility in an indirect sense. The value of Atkinson's typology is that it emphasises the need for other mechanisms to ensure complete organisational flexibility and warns against over-reliance on temporary employment as the main means of doing so. Complete flexibility can only be achieved by the development of a multiskilled core workforce and control over product design in association with temporary employment policies.

The consequences of temporary employment growth

The growth of temporary employment will inevitably lead to a reduction in the average size of contracting organisations by the displacement of non-core activities to peripheral status and the reduction in the number of middle management levels. As Pollert (1991) points out, one of the major consequences of this trend towards more dynamic and less distinct organisations will be the breakdown of interpersonal relations, problems in communication, a reduction in loyalty, motivation and teamwork, a loss of organisational identity and managerial control, and, ironically, ultimately a reduction in organisational flexibility. That means that in the long term the trend towards temporary employment may be self-defeating in creating more managerial problems than it solves. This is leading to the realisation that different strategies may be required for the management of core and peripheral groups, but it would seem that few companies have made any progress in developing and applying them (McAndrew 1993).

OVERCOMING THE PROBLEMS ASSOCIATED WITH THE DIVISION OF LABOUR

The productivity gains from the division of labour have been enormous. But as we have seen, the division of labour creates as many problem as it solves. These include problems of communication and co-ordination within an increasingly fragmented organisation. There are also human problems associated with the tedium and repetition such a system involves. For example, we have known for some time that when carried to extremes, the specialisation of work leaves employees with jobs so repetitive and meaningless that they become bored, dissatisfied, unmotivated and unproductive (Herzberg 1974). If this potential problem is not addressed by managers, then increased productivity can be offset by the slowdown such dissatisfaction produces. A number of techniques have been developed to tackle these problems.

Human problems

Job enlargement is a technique used to increase the number of tasks in a job and is based on the premise that people get increased satisfaction from greater variety. It is a form of horizontal expansion because tasks of the same nature are being combined. Job rotation is another technique used to expand the variety of tasks whereby employees move between different jobs, which allows them to have a change of environment and learn new skills. This can be done horizontally (between similar jobs at the same organisational level) or vertically (between different jobs at different organisational levels). Finally, there is job enrichment, a technique where employees are given greater autonomy in the management of their own jobs. For example, people may be given the chance to dictate work pace and work methods and participate in decision-making that affects their future.

Herzberg argues that enrichment is a more productive method because it associates people more closely with their jobs, whereas job rotation or enlargement merely adds more tedious jobs to the tedious task they had before. In the construction industry, the extent to which these techniques have been used is unknown. It is likely that tedium is not as great a problem in this industry as it is in factory line production industries since people are given some degree of variety in their different projects. Furthermore, construction activity is by its very nature a creative activity associated with the production of highly variable unique and one-off products. In any event, it is likely that the very strong divisions that have historically developed between the professions would prevent any job rotation or enlargement taking place. Having said this, the development of the multidisciplinary practice is one phenomenon that must contribute to increased workforce satisfaction, though it has not been widespread. At a project level, job enrichment is taking place to some extent with the development of new procurement systems that seek to involve constructors in decision-making activities, which were traditionally the province of the consultants.

Co-ordination problems

While organisations have to be split up into parts in the name of efficiency, those parts are interdependent and cannot work in isolation. A successful manager will be aware of these interdependencies and should develop a structure that will facilitate rather than obstruct them. There are two basic types of co-ordination mechanisms: individual and structural.

INDIVIDUAL MECHANISMS

One useful mechanism is the introduction of regular meetings between group representatives. There is also the option of using a liaison person, who is a member of a number of groups, acting as a point of contact between them by facilitating information flow. Then it is necessary to have liaisons between liaisons until a single point of contact is reached. In other words, co-ordinating mechanisms are hierarchical and need to be carefully planned. For example, liaison people may meet regularly in a committee of some kind to address ongoing issues. Task forces are also a useful co-ordinating mechanism based around a particular project. After they have achieved their objective, they are disbanded and the members return to their original departments.

STRUCTURAL MECHANISMS

These are connections between groups (chains of command) which are built into the formal organisational structure. That is, there is a requirement of one group that any decisions made will be communicated to another group. It is important to consider the appropriate span of control (that is, the number of groups or individuals reporting to one person) when designing reporting mechanisms such as these. For example, it is considered extremely difficult for one person to co-ordinate the work of ten groups and a span of control of around four is seen as about right. But it is also recognised that the appropriate span of control depends on a range of factors such as:

- competence of the workers
- competence of the supervisor
- nature of the task being supervised: complex tasks, non-routine tasks demanding smaller spans of control
- similarity of tasks performed by the groups: high similarity enabling larger spans of control
- location of work activity: widespread locations demanding lower spans of control
- function of the supervisor: if the supervisor merely receives information and assimilates it and passes it on, then higher spans are permitted, compared to when a supervisor has to assist groups in some way, such as in their decision-making.

It is important to point out that small spans of control result in tall and thin structures (that is, a large number of hierarchical levels), whereas large spans of control result in short and broad hierarchical structures. Therefore, organisations with highly qualified professional workers who like autonomy and can be trusted to make their own decisions on a small number of repetitive, simple and routine tasks carried out in one location will justify shorter, broader structures (since spans of control can be high) than taller, thinner ones. It is difficult to generalise about the structure of construction companies since they increasingly resemble consultancies more than traditional construction firms. That is, they employ largely professional-type people, though their work is generally of the non-routine kind apart from firms that specialise in certain low-technology markets such as housing. Similarly, in a project sense, the workforce is largely professional in nature, but the span of control will also depend on the complexity of the project. In this sense, each project needs to be treated differently. In general, however, it would seem that the imposition of rigid hierarchical structures and reporting chains by complex and legalistic contracts of employment may result in inappropriate reporting systems in some projects. It would seem that universal approaches to the imposition of organisational structure are misguided and this is one reason why different procurement systems have evolved and why some are more suitable to complex projects than others.

The different types of co-ordinating mechanisms available to managers are illustrated in Table 3.1.

TABLE 3.1 TYPES OF CO-ORDINATING MECHANISMS

Useful in mechanistic organisations (see Table 3.2)	Rules and procedures: identifying what actions are appropriate in different situations. They are standards to guide managers and employees.
	Hierarchical referral: when rules and procedures cannot co-ordinate separate units, co-ordination problems have to refer upward to a common supervisor. The hierarchy then acts as an integrator.
	Planning: what needs to be accomplished determines how people relate to each other. Planning sets objectives and defines what is to be accomplished and how people should interact to achieve this.
	Contact between managers: managers from separate units can work together to co-ordinate the work of their respective units.
	Liaisons: specific people are made responsible for providing co-ordination and communication between different units.
	Task forces: special work-groups are created from different units for the purpose of completing a specific task within a certain time limit.
	Teams: permanent task forces responsible for co-ordinating the work of several units over a long period and numerous projects.
Useful in organic organisations	Matrix organisation: matrix organisations give employees status in two or more groups simultaneously, while co-ordination is achieved through the formal matrix organisation.

CENTRALISATION VERSUS DECENTRALISATION

It is tempting to associate centralisation with co-ordination, in that the more information is focused onto a particular source the more focused is the control achieved. However, as we have said, there is a limit to information convergence: beyond it, information overload and problems result which can lead to a reduction in co-ordination. Therefore, in organisations dealing with complex and non-routine tasks, there is justification for decentralising decision-making in an attempt to stop certain information travelling beyond a certain point. That is, the decentralising decision-making power is an effective information-filtering system which prevents information overload and actually aids co-ordination. Of course, many managers do not like to decentralise their responsibilities because they fear losing control. Paradoxically, however, the likely result is an increase in control.

Many writers have argued that in an increasingly turbulent business world, decentralised structures will become much more common, particularly because decision-making is nearer the problems and therefore swifter. In this sense, decentralised structures are more responsive to change, more flexible and more able to deal with complexity due to a lower likelihood of information overload.

DESIGNING AN APPROPRIATE ORGANISATION

Burns and Stalker (1961) laid the foundations for the contingency school of thought. They were the first researchers to investigate the concept that organisations interact with their environment and are more successful when their structure is compatible with it. They developed a framework of two extreme types of organisation: organic and mechanistic, the latter being more flexible and suited to stable environments and the former more suited to dynamic environments. The characteristics of these two types of organisations are shown in Table 3.2.

Later, Daft (1983) introduced the idea of environmental complexity, which relates to the number of factors to be taken into account within that environment and their interdependencies. He produced the model in Figure 3.6 to illustrate the relationship between complexity, rate of change and structure.

Woodward (1958) also developed the contingency approach to show that internal as well as external characteristics should be taken into account. For example, the size of a company influences organisational design as much as its external environment. That is,

TABLE 3.2 MECHANISTIC AND ORGANIC ORGANISATIONS

MECHANISTIC	ORGANIC
Highly specialised tasks.	Individuals contribute as appropriate to company goals.
Separated tasks.	Tasks adapt to company's current situation.
Co-ordination by hierarchical control and formal rules and procedures.	Absence of formal rules and procedures.
Precise prescriptions of roles, relationships and responsibilities.	Individuals have autonomy to decide their own actions and co-ordinate themselves.
Top-down formal communication.	Informal lateral communication that can go up and down the organisation.
Insistence on loyalty and obedience.	Commitment to company goals valued over loyalty and obedience.
Use of coercive power to control subordinates.	
Centralised and hierarchical power structure.	Knowledge created anywhere establishes its own centre of authority.
Motivation by rewards and punishment against clearly defined and rigid goals set by senior managers.	Network structure with expectations to serve the common good.
	Decentralised power structure.
	Motivation by trusting people, providing autonomy and satisfying people's needs.
	Goals set by the workforce under the supervision of senior managers.

its structure must suit its size. In general, as size increases, structures become more departmentalised, hierarchical, formalised, have more managers and require more co-ordinating mechanisms. That is, they become more mechanistic. This clearly presents a danger to growing firms in rapidly changing environments, though Microsoft is living testimony to the fact that these rules do not always apply. The key to Microsoft's success lies in leadership and culture, issues that will be examined at pages 65–68 and in Chapter 4.

In construction project terms, there is clearly a danger that large projects will become slow at responding to their changing environment and vulnerable to crises. One way of controlling this tendency at company level is to form loosely coupled systems, that is, to split the company into semi-autonomous units that are almost independent in their actions but operate within the larger company

Figure 3.6 Daft's (1983) model

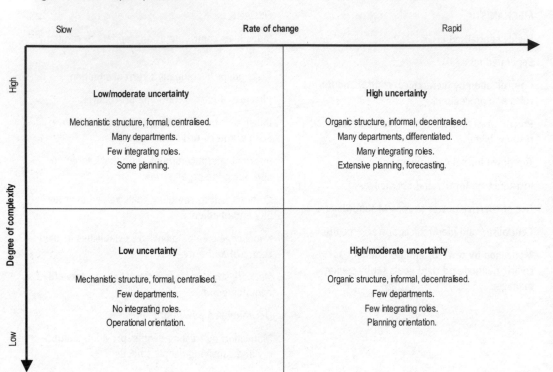

framework. Splitting a project into separate phases would achieve the same aim. Woodward also pointed to the importance of production technology as an influencing factor on structure. Technology refers to the degree of mechanisation in production, construction projects being very low tech. Woodward classified four types of technologies: unit production, small batch, mass production and process production. Unit production is a process in which one person works on one product from beginning to end. Small batch production is concerned with the production of small numbers of prototype units. Mass production is a process where numerous workers work on individual portions of a product. Process production is where several workers monitor machinery where the production process is largely automated. In essence, Woodward found that high-technology industries of mass production were better suited to mechanistic structures, whereas the organic structure was more suited to unit production. Where does the construction process fall in this typology? The answer is somewhere in between mass and small batch production, meaning that it justifies an organic structure.

DIFFERENTIATION AND INTEGRATION

Lawrence and Lorsch (1967) investigated the justification for having different structures among the units that comprise an organisation and referred to this property as differentiation. They found that the most successful companies were sensitive to the fact that different departments faced very different environments and undertook different tasks and therefore demanded different structures. They realised that to impose one generic structure across an organisation was counterproductive. However, the other key ingredient within these successful organisations was the existence of strong integrating mechanisms to bring these differentiated units together so that they worked as a united whole.

ORGANISATIONAL CULTURE

So far, the harder side of organisational design, which is concerned with structural issues, has been discussed. But there is a softer side to organisational culture. Culture refers to the shared values and beliefs that the members of an organisation hold about the way they should act, relate to each other and do their jobs. It is a crucial component which must be managed if an organisation is to be successful, but it is a challenging one because of its intangible nature. Cultural issues are important because they represent the values that underpin action, and in this sense the culture of an organisation must be appropriate if it is to support strategy. For example, it is no use having a decentralised structure to encourage innovation if the atmosphere within the organisation is not conducive to this — that is, if people are not encouraged to see the importance of being creative and innovative.

Before going on to discuss the issue of organisational culture, it is important to point out that the concept of culture can be defined at organisational, industry and national level, with all levels being relevant to organisations in the construction industry.

National culture

From a national perspective, an interest in cultural variability started to emerge in the 1950s, but the meaning of the word 'culture' quickly became confused. For example, Kroeber and Kluckhohn (1954: 24) collected over 300 definitions and produced their own definition, which was 'the total way of life of a people, the social legacy an individual acquires from his group'. Since this early work, many more definitions have been produced, with the programming model appearing to gain widespread acceptance (Fisher 1988; Hofstede 1992; Victor 1992; Hoecklin 1994; Deresky 1997). This model conceives culture as a programmed pattern of perceiving and reasoning, which begins at birth and is learned gradually through lifelong immersion in a culture and through a socialisation process. It is now generally accepted that the culture of a society is its shared values, understandings, assumptions and goals, learned from earlier generations and passed to future generations. It results in common attitudes, codes of conduct and expectations that guide behaviour. It is also widely accepted that cultural variables interfere with the process of communication by influencing the mindsets of the individuals concerned. Hofstede (1992) has arguably made the largest contribution to our understanding of cultural differences. He developed a typology that categorised differences by four criteria: masculinity (the degree to which male and female roles are separated and to which assertiveness exists); power-distance (the degree to which power-differences are respected); uncertainty avoidance (the degree to which uncertainty is avoided and controlled); and individualism (the degree of teamwork).

Construction industry culture

It is increasingly being recognised that different industries have their own unique cultures. Despite a growing awareness of organisational culture within the construction industry as a result of reports such as those by Gyles (1992) and Latham (1994), little is known about the influence of the construction industry's culture. This is because little is known about the specific characteristics of that culture and how it differs between countries. At present, the cultural understanding of most construction industries will have been crudely constructed from a wide range of existing literature that was never intended to be used in this way. For example, if one applies Hofstede's (1983) model of cultural variables to the UK construction industry, it is possible to form a preliminary picture of that industry's culture. Shrivastava and Fryer (1991), Gale (1992) and Bagilhole and colleagues (1995) argue that it is dominated by masculine values to the extent that they have become institutionalised. NEDO (1983, 1988), Gyles (1992) and Latham (1994) suggest that it is highly individualistic (selfish) and confrontational. Barnes (1989) and Uff (1995) suggest a high power-distance (a measure of acceptance of power inequalities) and Bennett (1991) suggests a high level of uncertainty avoidance (the degree to which people are threatened by ambiguity). While little is known about the cultural characteristics of individual countries' construction industries, even less is known about the cultural differences that differentiate them. Such comparisons are crucial because the culture of any construction industry can only be properly understood by placing it on a continuum with other countries' construction industries. Rowlinson and Root (1996) have attempted to do this by using Hofstede's model to investigate the cultural differences between the UK and Hong Kong construction industries. Their key findings are that in Hong Kong there is a higher power-distance but a similar level of uncertainty avoidance. In terms of masculinity, Hong Kong was lower than the United Kingdom, reflecting a greater sense of harmony rather than confrontation. Finally, the UK industry was characterised by higher individualism than the Hong Kong industry, where more of a collectivist culture predominated. However, such comparisons are rare in a construction context and for the time being discussions of culture must largely draw on the significant amount of research accumulated at national level. This has substantially deepened our understanding of cultural variability between people of different nationalities and is of great relevance and value to any person operating in a cross-cultural context.

Organisational culture

Therefore any organisation's culture will be influenced to some extent by the national culture of its members and by the culture of the industry it operates within. It is also important to appreciate that an organisation's departments may also have different cultures and that these will have to be integrated so that they collectively contribute to the organisation's goals.

One way of understanding organisational culture is to analyse an organisation's symbols, rituals and ideologies. Symbols include the name and logo that convey an organisation's image, the physical environment such as office layout, its decor and the general atmosphere within an organisation, which is reflected in people's behaviour, energy and modes of dress. The importance of symbols to organisational success is reflected in the recent and very controversial decision by British Airways to repaint its planes at the cost of millions of pounds at the same time as announcing significant redundancies and wage freezes among its flight crew staff. Rituals refer to customary and traditional actions within an organisation such as being at work until at least 7 pm, celebrating annual events and farewell parties and so on. Ideologies are principles that guide an organisation in its decision-making, such as having pride in taking care of customers or employees. It is fast being recognised in an increasingly competitive construction industry that attention to such issues will make the difference between getting work or not. Clients are finding it more and more difficult to differentiate on the basis of price alone and indeed are looking for alternative criteria on which to do so.

MANAGING CULTURE

A company's strategy defines its culture and therefore culture can only be changed with commitment from the top of an organisation. Top managers must set an example and ensure that people are educated about the new vision for the organisation and the new desired ways of operating and behaving. This can only be achieved successfully by involving them in its development and by rewarding people for adopting desired norms of behaviour and actions. By definition therefore, changing culture is a gradual process which takes time.

CONCLUSION

This chapter has reviewed the substantial literature surrounding the issue of organisational design. The problem with this issue is that it encompasses most of the organisational literature. The main point to emerge, however, is that there is no one best way to manage in all situations. Instead, design should be a function of the task to be performed in terms of its complexity, the conditions under which it has to be performed, and the people performing it. Organisational success is dependent on a thoughtful approach to organisational design, which incorporates both hard and soft elements. Managers should also recognise that any system is only as good as the people within it and that the key to success is getting good people and encouraging them to believe in the organisation's goals and means of attaining them.

EXERCISES

1. How can an organisation's structure affect its effectiveness?
2. What factors determine whether you use an organic or mechanistic structure?
2. How is the growth of temporary employment affecting modern organisations?
3. Why is it so important to manage an organisation's culture as well as its structure?

CHAPTER 4

MANAGING CULTURAL DIVERSITY

INTRODUCTION

Arguably, the most profound change in the business environment since the Industrial Revolution is being brought about by the process of globalisation. Trade boundaries are dropping, international markets are opening, companies are expanding internationally, and it is increasingly likely that managers will have to work within a culturally diverse setting. Indeed, it has been estimated that nearly all of the world's population growth in the next ten years will occur in non-indigenous, ethnic minority groups. For some countries, cultural diversity is more of an issue than others. For example, Australia has the world's second largest foreign-born workforce with approximately 30 per cent of its working population having been born in another country (DIMA 1998). Since 1945, almost 6 million people from over 150 countries have arrived in Australia as new settlers and they have been largely responsible for its population growth from approximately 7 million in 1945 to nearly 20 million in 2003. Approximately 70 per cent of these migrants originated from non-English-speaking countries and it is estimated that 16 per cent of the Australian workforce speak a language other than English at home (ABS 1999). Since the construction industry in Australia employs approximately 7 per cent of the working population and has a relatively large number of unskilled, menial and manual jobs, it tends to absorb a large proportion of migrant workers into its workforce. The situation is the same in many other developed countries. For example, in 1998 foreign workers made up 81.2 per cent of the total construction workforce in Singapore (CIDB 1998). Furthermore,

America, European countries and Middle Eastern countries are experiencing ever greater influxes of foreign migrants into their construction workforces. For example, in Saudi Arabia, which is also typical of most Arabian Gulf countries, the number of registered foreign labourers employed in the construction industry is approximately 30 per cent of the workforce (MLSA 1998). Similarly, recent concerns have been raised in the UK construction industry about the influx of illegal immigrant workers from Eastern European countries, a trend that is likely to continue as Europe's refugee problem continues to worsen (Glackin 1999). The purpose of this chapter is to explore the challenges and opportunities these trends hold for project managers in the construction industry.

THE CONSEQUENCES OF MISMANAGING DIVERSITY

The consequences of mismanaging cultural diversity are serious and include increase stress among the workforce, confusion, frustration and conflict which translates into lower morale, productivity, quality problems and higher accident rates (Allen 1976: Migliorino et al. 1994; Deresky 1997). Conversely, cultural diversity can be a significant asset to an organisation if it is managed well and many of the world's most successful companies have shown that it can be used to competitive advantage (Shaw 1995). However, such companies are in the minority and the construction industry has been singled out as having a particularly poor record in this area (Cavill 1999). It is not difficult to find evidence of this. For example, in Australia, O'Rourke (1998) wrote of increasing numbers of complaints to union officials by foreign workers of being called insulting names, being forced to eat in separate locations and having poorer career prospects. In the UK, Singh (2000) reported that 39 per cent of ethnic minority staff in the construction industry suffer harassment and in Singapore, communication problems with these workers was ranked the fifth most important issue in increasing productivity by construction companies (CIDB 1998). More recently, Loosemore and Chau (2001) found that there was a significant level of racial discrimination and harassment in the Australian construction industry. More disturbingly, there was minimal infrastructure to reduce this problem or manage it effectively when it occurred. In would seem that many constructions companies may be breaking the law every day and are unaware of the potential consequences of allowing racism to be expressed on their projects.

It would seem that discrimination and racism is common in the

construction industry and that its poor public image of equal opportunities for ethnic minority groups and women is justified. Indeed,
few companies have formal equal opportunity policies and programs
and ethnic minorities are under-represented at senior managerial levels. It is clear that the construction industry could benefit from a
more enlightened approach to managing cultural diversity. It must
be seen as one of the greatest challenges that will face managers in
the future, not least because many countries see immigration as the
only practical solution to halting declines in natural population
growth (AFR 2000).

NATIONAL CULTURE: WHAT IS IT AND
HOW DOES IT VARY?

We used the word 'culture' numerous times in the previous chapter
from an organisational perspective but have not yet defined it in a
national sense. This is an important omission because an understanding of national culture and the way that it varies within the
workforce is crucial in the effective management of multicultural
organisations.

The notion of culture is a fuzzy, broad and rather abstract one. But
it is generally accepted that national culture is the shared values
(beliefs about good and bad) that typify a society and lie beneath its
behaviour, art, architecture, styles of dress, language, food, codes of
conduct, expectations, mannerisms etc. and which are gradually
learnt from past generations through socialisation, passed onto future
generations and adapted by them. Essentially, culture is the differences that we experience and see when we visit a new country.

The best known research in this area has been undertaken by a
Dutch researcher, Hofstede (1980), who sees culture as 'the software
or collective programming of the mind that guides your behaviour
by determining the lens through which you see the world and the
way that you interpret information'. Hofstede and many other psychologists believe that we are all programmed by the societies we
live in to behave in certain ways and to expect others to behave in
certain ways. Of course, within every society there are exceptions to
the rule and it is dangerous to stereotype everyone from a particular
culture into the same mould. Furthermore, there is evidence all
around us that cultures are converging to some extent. For example,
Japan experienced a twentyfold increase in drug-related arrests
during the 1980s and multinationals such as McDonalds and Coke
are now as recognisable in China as they are in America. However,

cultures change very slowly because they relate to our innermost beliefs. A vivid illustration of this was the failure of the Cultural Revolution in China during the 1960s and 1970s to break down traditional cultural values of respect for experience, social position and age. Also consider the Hindu island of Bali, which exists within a state that has the world's largest Islamic population. Indeed, most of the world's conflicts are the result of one culture not being prepared to lose its identity in the face of imposition by another conflicting culture. However, while cultural differences are a common cause of conflict, the value of Hofstede's research is in illustrating that cultural differences are surmountable if people are able to understand them and to adapt their behaviour accordingly. To this end, Hofstede has produced the most rigorous framework for classifying national culture, and one of the most important aspects of culture from a managerial perspective is the way in which people relate to authority. Hofstede referred to this as a culture's 'power-distance'.

Power-distance

Power-distance relates to the distance a subordinate feels from their superiors and the extent to which the opinions and decisions of senior people are accepted and respected. In essence, it is a reflection of the degree to which inequality is accepted as a fact of life in a society and it has significant implications for managers. In particular, a person from a high power-distance culture would not expect to be involved in decisions but to be instructed autocratically what to do. The opposite would be true of someone from a low power-distance culture. The problems in a managerial setting are easy to imagine. For example, in a high power-distance culture (where there is relatively high respect for authority) a manager would not expect the participation of a subordinate in decision-making processes, which might cause resentment for a subordinate from a low power-distance culture. Conversely, a manager from a low power-distance culture could become frustrated with the unwillingness of a subordinate from a high power-distance culture to contribute to decisions. Hofstede found the variations between the power-distances of different countries. These are given in Table 4.1.

It is important to point out that Hofstede's work is now over 20 years old and, due to globalisation, is possibly somewhat out of date. There have also been criticisms of the representativeness of his sample and the validity of making generalisations to the wider population. Nevertheless, Hofstede's work is still of value to indicate the differences that exist between different cultures in the world. For

TABLE 4.1 POWER-DISTANCES OF DIFFERENT COUNTRIES

High power-distance

Malaysia	104	South Korea	60
Guatemala	95	Iran	58
Panama	95	Taiwan	58
Philippines	94	Spain	57
Mexico	81	Pakistan	55
Venezuela	81	Japan	54
Arab countries	80	Italy	50
Ecuador	78	Argentina	49
Indonesia	78	South Africa	49
India	77	Jamaica	45
West Africa	77	USA	40
Yugoslavia	76	Canada	39
Singapore	74	Netherlands	38
Brazil	69	Australia	36
France	68	Costa Rica	35
Hong Kong	68	Germany (West)	35
Colombia	67	Great Britain	35
Salvador	66	Switzerland	34
Turkey	66	Finland	33
Belgium	65	Norway	31
East Africa	64	Sweden	31
Peru	64	Ireland	28
Thailand	64	New Zealand	22
Chile	63	Denmark	18
Portugal	63	Israel	13
Uruguay	61	Austria	11
Greece	60	*Low power-distance*	

Managing Cultural Diversity

example, it is noticeable that colonised countries feature relatively high on the power-distance scale and so do less developed nations. The latter could mean that high power-distance is bad for business or alternatively that greater prosperity produces a fall in power-distance. The presence of countries like Hong Kong and Singapore in the high power-distance sector supports the latter argument. It is, however, important not to see these figures as indicating good or bad because this itself would be a culturally bound assumption.

Individualism versus collectivism

The next indicator of culture that Hofstede said was important in a managerial setting was individualism versus collectivism. In an individualistic culture the emphasis is on self-interest, self-reliance, independence, freedom and getting ahead, and the ideal is to become a good leader. In contrast, collectivist cultures emphasise the importance of belonging and the ideal is to be a good member (of a family or group). In this sense, individualism verses collectivism is a reflection of the level of trust in others, the need for the security of a group and the need to have personal relationships with those that you work with. A good illustration of the difference between the United Kingdom (a highly individualistic culture) and Japan (a highly collectivist culture) are two common proverbs used in both countries. In the United Kingdom it is common to hear 'every man for himself' but in Japan it is common to hear 'the nail that stands up will be hammered down'.

Hofstede's ranking of different countries on the individualism verses collectivism scale is given in Table 4.2.

Table 4.2 indicates that in the United States and Northern Europe, business concerns and family ties are kept separate but that in most of Africa, the Middle East and much of Asia and Latin America, personal relationships and family ties are of central importance to business. In these latter cultures, people derive most of their work satisfaction from being part of a group and the management decision-making systems that have evolved reflect this. For example, consider the traditional Japanese Ringi system where before a decision is made, a written proposal is made by a group and is informally passed around for consultation. Amendments are then made from feedback and a final report is prepared under the group's name rather than an individual's (to maintain humility), which is then simply rubber-stamped by decision-makers because it has already been widely seen and discussed. In contrast, consider the Western way of doing things which often involves preparing a report in secret, inserting one's

TABLE 4.2 INDIVIDUALISM VERSUS COLLECTIVISM OF DIFFERENT COUNTRIES

Individualistic

USA	91	Turkey	37
Australia	90	Uruguay	36
Great Britain	89	Greece	35
Canada	80	Philippines	32
Netherlands	80	Mexico	30
New Zealand	79	East Africa	27
Italy	76	Portugal	27
Belgium	75	Yugoslavia	27
Denmark	74	Malaysia	26
Sweden	71	Hong Kong	25
France	71	Chile	23
Ireland	70	Singapore,	20
Norway	69	Thailand	20
Switzerland	68	West Africa	20
Germany (West)	67	Salvador	19
South Africa	65	South Korea	18
Finland	63	Taiwan	17
Austria	55	Peru	16
Israel	54	Costa Rica	15
Spain	51	Indonesia	14
India	48	Pakistan	14
Japan	46	Colombia	13
Argentina	46	Venezuela	12
Iran	41	Panama	11
Jamaica	39	Ecuador	8
Brazil	38	Guatemala	6
Arab countries	38	*Collectivistic*	

Managing Cultural Diversity

name in capitals on its cover, submitting it to the boss in a surprise attack and getting it 'bounced back' to have amendments made to it.

It is noticeable that religion tends to play a greater role in collectivist cultures than in individualistic cultures. For example, in the Islamic and Hindu cultures of North Africa, the Middle East and South Asia, religion is a way of life and spiritual beliefs strongly influence managerial actions and organisational processes. City Bank experienced problems when they recently entered the Middle Eastern market and began lending to local businesses. They failed to realise that many Muslims consider insurance cover a challenge to Allah's will and City Bank lent a considerable amount of uninsured money before realising this.

Masculinity versus femininity

The next dimension of culture that Hofstede identified as being important in a managerial setting was masculinity versus femininity. This does not directly refer to issues of gender but distinguishes between the hard and soft approach to management and the extent to which male and female roles must be maintained in society. For example, in a masculine culture men and women have strictly defined roles in society, the latter being afforded little power and authority. In such cultures people also tend to be highly task-oriented and management styles are assertive and supported by strictly defined and inflexible rules, regulations and procedures. People are best motivated by being given strong and clear chances for promotion, high earnings and individual recognition. In contrast, people from feminine cultures are more people-oriented and see the roles between men and women more flexibly. There is a greater sense of equality in such cultures and people are motivated more by prestige than by money alone and by having a good relationship with the people around them. This clearly has important implications for managers and it is not difficult to foresee problems arising when a woman from a highly feminine culture has to work with a subordinate man from a highly masculine culture. In the first instance the man is very unlikely to accept orders from the woman and would feel very uncomfortable with the informal managerial style she would adopt.

Hofstede's rankings of countries on the masculinity–femininity scale is given in Table 4.3.

Notice that Sweden has the lowest masculinity. This is reflected in the way that managers operate in this country, with companies like Volvo having both men and women equally represented at all levels

TABLE 4.3 RANKING OF COUNTRIES BY MASCULINITY

Masculinity

Japan	95	Singapore	49
Austria	79	Israel	47
Venezuela	73	Indonesia	46
Italy	70	West Africa	46
Switzerland	70	Turkey	45
Mexico	69	Taiwan	45
Ireland	68	Panama	44
Jamaica	68	France	43
Great Britain	66	Iran	43
Germany (West)	66	Spain	42
Philippines	64	Peru	42
Colombia	64	East Africa	41
Ecuador	63	Salvador	40
South Africa	63	South Korea	39
USA	62	Uruguay	38
Australia	61	Guatemala	37
New Zealand	58	Thailand	34
Greece	57	Portugal	31
Hong Kong	57	Chile	28
Argentina	56	Finland	26
India	56	Costa Rica	21
Belgium	54	Yugoslavia	21
Arab countries	53	Denmark	16
Canada	52	Netherlands	14
Malaysia	50	Norway	8
Pakistan	50	Sweden	5
Brazil	49	*Femininity*	

Managing Cultural Diversity

of the organisation. In contrast, in the United Kingdom and Japan, women are largely under-represented in the workforce, particularly at senior levels, and the traditional roles of men and women are maintained. The Indian construction industry is a good illustration of women's roles within highly masculine cultures since here it is the women who labour on the sites and the men hold the managerial positions. As managers, we must resist judging this as right or wrong but instead focus on how we adjust our managerial styles to accommodate the differences between ourselves and the culture we are in contact with.

Uncertainty avoidance

The final dimension of culture identified by Hofstede as relevant in a managerial context is uncertainty avoidance. This measures the extent to which people feel comfortable with uncertainty and ambiguity. In high uncertainty-avoidance cultures, people dislike uncertainty and are reluctant to take risks or break the rules, even if it is in the best interests of the organisation to do so. Furthermore, people tend to stay with organisations for a long time, often feel nervous and insecure in work, are very orderly and precise, and do not challenge the status quo. For example, the Japanese Ringi system of decision-making is essentially a way of reducing uncertainty, whereas the American way of doing business is to shoot first and ask questions later. Once again, the implications for managers are not difficult to imagine. For example, someone from a high uncertainty-avoiding culture would feel very insecure working with a manager from a low uncertainty-avoiding culture who would tend to create a very informal environment, make fast decisions and bend the rules to get jobs done.

Hofstede's ranking of countries by uncertainty avoidance is shown in Table 4.4.

TROMPENAAR'S WORK ON CULTURE

Another Dutch researcher, Trompenaar (1993), has taken Hofstede's research further and discovered three extra dimensions of culture which are important in a managerial setting.

Neutral versus affective relationships

This concerns the way that different cultures express emotions. At one extreme you have the reserved British who traditionally hide their feelings and at the other end of the spectrum you have the more openly emotional South Americans. In contrast to the latter, the

TABLE 4.4 RANKING OF COUNTRIES BY UNCERTAINTY AVOIDANCE

Uncertainty avoiding

Greece	112	Ecuador	67
Portugal	104	Germany (West)	65
Guatemala	101	Thailand	64
Uruguay	100	Iran	59
Belgium	94	Finland	59
Salvador	94	Switzerland	58
Japan	92	West Africa	54
Yugoslavia	88	Netherlands	53
Peru	87	East Africa	52
Argentina	86	Australia	51
Chile	86	Norway	50
Costa Rica	86	South Africa	49
France	86	New Zealand	49
Panama	86	Canada	48
Spain	86	Indonesia	48
South Korea	85	USA	46
Turkey	85	Philippines	44
Mexico	82	India	40
Israel	81	Malaysia	36
Colombia	80	Great Britain	35
Venezuela	76	Ireland	35
Brazil	76	Hong Kong	29
Italy	75	Sweden	29
Pakistan	70	Denmark	23
Austria	70	Jamaica	13
Taiwan	69	Singapore	8
Arab countries	68	*Uncertainty tolerating*	

Managing Cultural Diversity

British believe that it is a sign of weakness to express your emotions too intensely. One can easily imagine emotions interfering with the messages being transferred between people from these two cultures in a management setting. Trompenaar's findings are illustrated below and they show that some countries are markedly different in this respect.

The potential for misunderstanding between neutral and affective cultures is significant. For example, affective people may find it difficult to detect when a neutral person is aggrieved since they would expect to see this expressed openly in emotional terms. Indeed, an affective person would find it very frustrating to work with someone who hides their emotions and will be unlikely to give the clues to their emotional state. In contrast, the emotional displays of an affective person could accidentally insult a neutral person, who would interpret them as being far stronger than they really are. The potential for an accidental escalation of a dispute in such an insensitive relationship is quite serious.

Interestingly, the expression of emotion has been linked to one's gender and social class. One has to only think of the rigid, unemotional stereotyped view of the upper classes in the United Kingdom as an illustration of this — the aristocracy being sent to boarding schools to learn how to suppress their emotions. In highly masculine countries gender is also an important determinant of who can acceptably express emotions openly. For example, in the United States it is generally unforgivable for a man to weep openly in a business situation, whereas a woman would be frowned upon but probably forgiven. Conversely, if a woman expressed extreme anger and aggression in a meeting this would probably be seen as unacceptably masculine. The opposite is true in Arab countries, where men are acceptably emotional and weep easily whereas women are expected to be controlled, calm and practical.

Specific versus diffuse relationships

Trompenaar's second dimension of culture is specific versus diffuse relationships. This refers to the balance between family and working life, to people's sense of private and public space, and to the degree to which people feel comfortable sharing their feelings with others who are outside their immediate family. We all have various levels of personality from a more public to a more private level and the ease of penetrating these layers varies between cultures, diffuse cultures being easier than specific cultures. People from specific cultures tend to keep their private life separate from work and to guard it closely,

Figure 4.1 Neutral and affective relationships

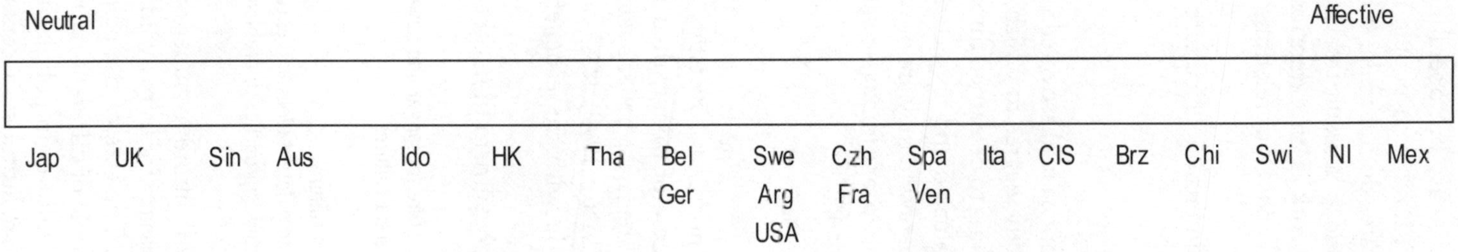

Neutral   Affective

Jap | UK | Sin | Aus | Ido | HK | Tha | Bel | Swe | Czh | Spa | Ita | CIS | Brz | Chi | Swi | Nl | Mex
Ger | Arg | Fra | Ven
USA

Arg Argentina   Jpn Japan   Aus Austria   Mex Mexico   Bel Belgium   Nl Netherlands   Brz Brazil   Sin Singapore   Chi China   Spa Spain
CIS Former Soviet Union   Swe Sweden   Czh Czechoslovakia   Swi Switzerland   Fra France   Tha Thailand   Ger Germany   Ven Venezuela
HK Hong Kong   UK United Kingdom   Ido Indonesia   USA United States   Ita Italy

Figure 4.2 Specific and diffuse relationships

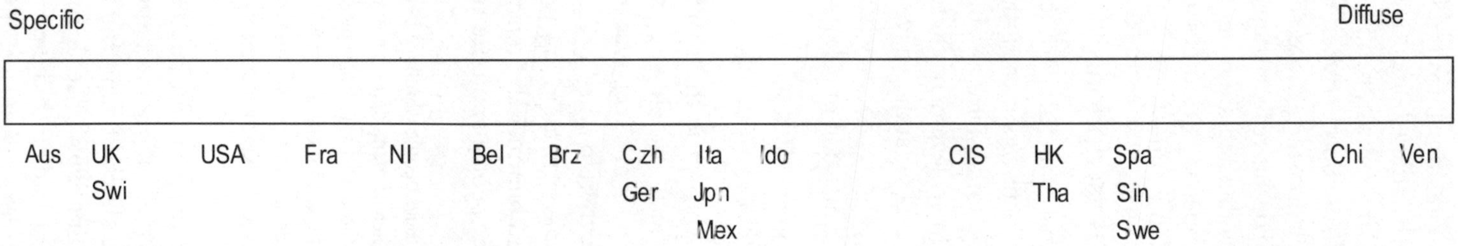

Specific   Diffuse

Aus | USA | Fra | Nl | Bel | Brz | Czh | Ita | Ido | CIS | HK | Spa | Chi | Ven
Swi | | | | Ger | Jpn | Tha | Sin
| | | | | Mex | | Swe

Figure 4.3 Achievement versus ascription

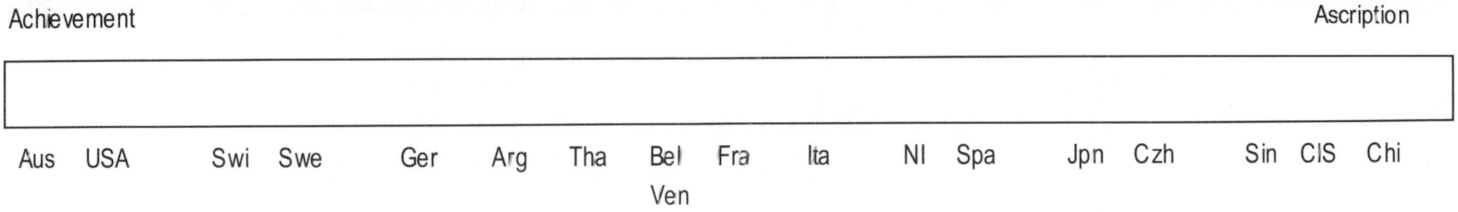

Achievement   Ascription

Aus | USA | Swi | Swe | Ger | Arg | Tha | Bel | Fra | Ita | Nl | Spa | Jpn | Czh | Sin | CIS | Chi
Ven

working relationships tending to be open and easy but operating at a superficial level since it is hard to get to know such people intimately. In contrast, people from diffuse cultures do not make such a stark distinction between work and home and as a result treat business relationships more seriously because work is an extension of home. For this reason they often seem initially cool, but relationships once established are more meaningful.

Trompenaar's findings relation to this dimension of culture are shown in Figure 4.2.

Achievement versus ascription

The final dimension of culture that Trompenaar discovered was achievement versus ascription. This relates to the degree to which cultures see social status as a birthright or something to be earned through merit. In achievement cultures people are evaluated and respected on the basis of what they have achieved rather than who they are or what school or university they have attended. Trompenaar's findings relating to this dimension of culture are given in Figure 4.3.

CULTURE SHOCK

Ethnocentrism is the natural tendency for people to overstate the importance of their own culture in the scheme of things; their cultural values are deeply seated within their innermost belief systems and are therefore highly resistant to change (Victor 1992). It is understandable that anything which does coincide with these beliefs is seen as abnormal and rejected as inferior.

Some people are more susceptible to ethnocentrism than others:

- older people who have been immersed in their culture for longer and have not had as many opportunities to be exposed to different cultures
- those who have not travelled
- those who come from very strict and well-defined cultures
- uneducated people who have not been exposed to different ideas and encouraged to think about the value of different views and approaches to problems.

The tendency for people to see their own culture as superior to others is a defence mechanism which is necessary for us to avoid the culture shock of adjusting to other cultures. The size of the culture shock one experiences in interacting with people of different cultures varies according to the differences between cultures in politics, religion, security, language, customs, climate, food, attitudes, acceptable behaviour and so on. The stages of culture shock we experience when we interact with people of another culture are as follows.

1. Spectator stage: an exciting, interesting phase where people stand back and observe different ways of doing things and differences in people's behaviour. As a manager one cannot stay in this phase for long since people have to be managed.
2. First shock: a miscalculated decision is made or one is made without regard to cultural differences and the response is not what was expected. In this situation a person has three options:

    • Fight it: force through one's own ways of doing things.
    • Go native: over-identify with the culture, which means 'bending backwards' to accommodate its practices. This is likely to result in more problems because one does not understand the culture deeply, because people do not expect someone to do this, because it might be seen as a sign of weakness, and because one's culture is suited to one's environment and adopting another culture may be inappropriate.
    • Assimilation/adaptation: through trial and error one discovers what works and what does not and one learns to avoid simple mistakes.

But these three options only produce superficial changes in a manager's behaviour and only permit operation in a limited range of situations. To manage effectively in the longer term, one must build a deeper understanding of cultural differences and the means of transcending them. It is the next round of failed experiments that tend to be more serious than the first and that lead to the second culture shock.

3. Second culture shock: This phase involves some serious thinking on the manager's part, particularly in relation to what is acceptable behaviour in their own culture but what is not so in other cultures. Once again, one can either fight it or adapt. If one can adapt, one eventually arrives at a deep working knowledge of a culture which enables one to relax, set more realistic expectations and targets, communicate more effectively, and build meaningful relationships.

RACISM IN THE CONSTRUCTION INDUSTRY

Earlier in this chapter, we provided evidence of racism in the construction industries of countries such as Australia, Singapore and the United Kingdom. We also highlighted the negative implications of racism for organisations. In order to control racism, it is important for managers to understand its psychological base, how to recognise it, and how to deal with it effectively.

The psychological basis of racism

Racism refers to pervasive and systematic assumptions about the inherent superiority of certain groups and the inferiority of others based on differences in race, culture, ethnicity, national origin and descent (Thompson 1997). It is rooted in two forms of human behaviour which are central to people's cognitive processing: ethnocentrism and stereotyping. Ethnocentrism refers to the natural human tendency to see one's own culture as superior to others and stereotyping refers to people's tendency to categorise others into distinct social groups and to arbitrarily generalise about their distinguishing traits, normally in a negative way (Brislin 1993; Sawin 1995). This arbitrary social categorisation or stratification is usually based on salient and physically identifiable features such as age, colour, appearance, disability, gender, race, ethnicity and social status. Furthermore, in many countries it is based on learned fact. For example, Thompson (1997) points out that in the United Kingdom, the link between race and class became strongly established after World War II as an increasing population of black migrants became over-represented at lower levels in Britain's socio-economic class system. Soon black people became seen as the cause of poverty rather than victims of it. Furthermore, the social problems experienced by black people were translated by their white 'hosts' into matters of personal failing, weakness or inadequacy. The final step was to blame black people for the problems experienced by white people, which led to the emergence of racism on a much wider scale than ever before.

Both ethnocentrism and stereotyping are coping mechanisms that perform an adaptive function for people in novel, complex and uncertain situations by reducing the world to a more manageable number of categories. They are also valuable 'sense-making' mechanisms to explain the actions and behaviours of 'out-group' members, particularly when things go wrong or are threatening. In these circumstances, people try to preserve a sense of self-esteem and personal power by blaming groups that are 'different' and in a 'minority' (Altmeyer 1988). In particular, they help people to orient themselves to the realities of intergroup life and legitimise actions that would otherwise be socially unacceptable. For example, Hunter et al. (1991) investigated how violent acts were justified by Protestant and Catholic students in Northern Ireland and found that each group made 'external attributions' (for example, retaliation and fear of attack) for their own group's violence and 'dispositional attributions' (for example, aggressiveness and psychopathy) for out-group violence. Stereotypes and ethnocentrism produced social explanations of behaviour which were supportive and self-serving of in-groups but derogatory to out-groups. When reinforced by the power of the group, this generated a dangerous cycle of self-fulfilling prophecies which legitimised otherwise anti-social actions and encouraged members to act in a way that was consistent with in-group expectations.

Thus individuals employ a variety of defence mechanisms to protect themselves from unacceptable internal impulses and threatening environmental circumstances. In these situations, negative internal impulses, personal internal conflicts and inadequacies are projected in an arbitrary way onto minority groups who are perceived as different, threatening and inferior (Eagly & Chaiken 1993). Interestingly, Azar (1997) found this behaviour to be more likely in 'powerful' people. Furthermore, such behaviour is often reinforced by the mass media, which play an increasingly important role in moulding and reinforcing attitudes towards other nationalities (Oskamp 1988; Eagly & Chaiken 1993). As Wood (1997) argues, in an increasingly busy world most people rely on the media for their information and to be newsworthy, social comparisons tend to be oversimplified, exaggerated and caricatured. In other words, the media shape, validate, legitimise and perpetuate value judgments and understandings of how different cultural groups vary, playing a powerful role in determining how they value themselves and others.

The scientific basis of racism

So prejudice has an inherent cognitive component which performs an important adaptive function for people and is a natural and inevitable response to the increasingly uncertain and culturally diverse world in which they find themselves. Indeed, so natural is this behaviour that recent research has found that racist behaviour can often be activated automatically without conscious awareness, even among people who embrace egalitarian beliefs (Devine 1989).

In the early 20th century racism had a spurious scientific justification from biological determinists who argued that behavioural, social and economic differences between human groups (races, classes and sexes) arose from inborn biological distinctions. These social Darwinist theories were a vulgarisation of the natural selection theories of Herbert Spencer and Charles Darwin and, later, led to the now discredited science of eugenics (the science of improving population by controlled breeding). These theories were developed during the 18th and 19th centuries and lent intellectual justification to imperialist expansions of many European countries. Indeed, the emergence of psychology during this period also contributed to these theories through research which sought to use measurements of physical characteristics such as skull size as proof of the superiority of the European (male) brain (Hannaford 1997). These ideas have now been discredited by genetics, which proves that the concept of race has no basis in biology and that racial categories are largely arbitrary. For example, it has been found that there is more genetic variation within most racial groups than between them (Phinney 1996). Genetics has shown that people are far less variable than was once thought, and it would seem that externally visible physical features such as skin colour only has relevance to an individual's worth when a society arbitrarily loads it with differential social value. Nevertheless, 19th-century notions of biological superiority have proved to be resilient in some sectors of society where it remains a highly salient political and social construct. One such instance appears to be the construction industry, which has a particularly poor reputation for representing and protecting the interests of minority groups.

Expressions of racism

Together, the dual processes of ethnocentrism and stereotyping result in racist behaviour. This manifests itself in the form of discrimination or harassment, where one group is treated (physically, emotionally and/or economically) unfavourably, unjustly and unequally compared with another and therefore feels oppressed, devalued and intimidated (Hollingsworth et al. 1988). Most developed countries now have laws which criminalise such behaviour and differentiate between four main types of discrimination:

- direct discrimination
- indirect discrimination
- harassment
- victimisation (RDA 1975).

Direct discrimination occurs when a person receives less favourable treatment than another person in the same position. An example would be an employer paying an Australian worker more than an Asian worker for the same job or not providing the latter with the same training, opportunities or entitlements. Indirect discrimination occurs when policies or actions that appear to be fair have an adverse affect on a higher proportion of people in one racial or ethnic group. An example would be a policy of only promoting people who can pass a test in English at a certain level. If set too high, this could discriminate against those of a non-English-speaking background, especially if this level of competency is not necessary for the work in question and is unreasonable in the circumstances. Harassment is offensive or abusive behaviour or public acts based on race, colour, ethnicity etc. which are reasonably likely to offend, insult, humiliate or intimidate those groups. Examples are racist comments written on walls, or racist jokes circulated by e-mail, radio or any form of publication. Victimisation includes threats against employees based upon race or ethnicity, which intimidate employees and adversely affect their ability to operate effectively.

While legislation to punish racism is widely welcomed, Nesdale (1997) questions whether it has significantly reduced levels of discrimination in society. He argues that contemporary expressions of racism have simply become more covert than in the past and therefore less easy to detect. Old-fashioned (often called blatant) racism based on segregation and physical abuse has now been replaced by a more symbolic and subtle variant known as modern racism. This is covertly framed in terms of values and ideologies rather than expressed in overt dislikes and aggressive behaviour, and manifests itself in limited access to resources and entitlements. Recent research suggests that while people may increasingly support the principles of racial equality, they still maintain a set of beliefs and attitudes about how minorities should behave and act in their dominant society (Pettigrew & Meertens 1995). It seems that acts of racism are now increasingly rationalised on a non-racial basis. A good example of this is provided by Wetherell and Potter (1992), who studied the language that white New Zealanders used to talk about their Maori neighbours. They found that the overwhelming majority of their white middle-class respondents used language which legitimised the existing social order of inequality in a way that avoided their being labelled as racist. This research has been replicated in other countries to support the view that contemporary racism is more subtle, obscure and ambivalent than in previous generations (Augoustinos & Sale 1997).

Dealing with racism

In countries such as Australia and Britain, the initial response to racism was the development of assimilationist policies (Rowse 1993; Thompson 1997). These encouraged and often forced ethnic minorities to integrate into mainstream white society so that they did not attract hostility by being different. The idea was to treat everyone the same and to ignore cultural differences. However, it was a 'colour-blind' approach which was still grounded in the ideology of racial superiority of white over black. Inevitably, the benchmark standard of behaviour applied was that of dominant whites, which meant that black people were expected to become white in all but skin colour. It was an approach that ignored the different needs of ethnic minority groups, the positive attributes of different cultures and the need of people to feel cultural belongingness.

In contrast, contemporary anti-racist policies dismiss the oppression and cultural imperialism of assimilationism and recognise the cultural pluralism of multicultural societies. The emphasis here is not on minimising differences between blacks and whites and on 'sweeping ethnic differences under the carpet' but on valuing cultural diversity and on recognising and allowing for the differences between people from different cultures. Indeed, the validity of the term 'ethnic' to refer to minority groups is being questioned because ultimately everyone has an ethnic background. Importantly, modern anti-racist policies also recognise the structural basis of racism and tackle it by seeking to redress inequalities and imbalances in power between different ethnic groups. 'Affirmative action' policies are a good example of this 'empowerment' approach. Contrary to common belief, affirmative action is not about discriminating in favour of a particular minority group and disfavouring others but is about condoning and ensuring equity in society. This involves recognising the accumulation of disadvantage that ethnic minorities have suffered as a result of racism and developing policies to help overcome the problems this causes them.

Corporate racism

Although the above arguments have tackled racism from an individual perspective, every person lives within an interdependent societal and institutional environment. The issue of racism must be considered at all three levels if it is to be fully understood. For example, official government policy statements and laws on discrimination influence company policies and conversely, company statements on immigration affect individual perceptions of out-groups. Similarly, intergroup conflicts can spur government action, as can the feelings of individual voters.

Just as individual expressions of racism have become less blatant, so have government and organisational policies in many countries. For example, Vasta (1993) argues that although the Australian government has officially abandoned its segregatory 'White Australia' policy and moved to one of multiculturalism, there are many policies that prevent the full integration of minority groups into Australian society. The same applies at an institutional level, where the widespread absence of operationalised equal opportunity policies ensures that minority groups suffer relatively high levels of work-related stress, accidents, harassment and discrimination. By failing to act, governments, managers and individuals condone and perpetuate racism, and research indicates that the result is division, conflict, low morale, high labour turnover rates, low productivity, high absenteeism, stress-related illnesses, costly legal suits and a poor public image (Hoecklin 1994; Deresky 1997). Indeed, directors of companies are vicariously liable for the racist actions of their employees if they cannot prove that all reasonable preventive steps had been taken such as developing, communicating and implementing equal opportunity policies, providing training and establishing effective monitoring and complaints procedures. Conversely, the productive, economic and creative benefits of a harmonious and well-managed multicultural workforce are enormous (Napoli 1998).

INTERCULTURAL COMMUNICATION

Research has shown that on average, project managers spend 70 per cent of their time communicating in one form or another. It follows that being able to communicate effectively is crucial for effective project management. But the barriers to doing so are particularly great in an international context. Indeed, Victor (1992) ranked communication difficulties as the number one problem in international business affairs.

What is communication?

Communication is something more than the transfer of information between two or more people. Rather, it is about the transfer of meaning, since the purpose of communicating is to transfer an idea from one person's mind to another's and for it to be perceived in the same way. This involves a series of discrete steps: encoding, choosing a medium, decoding and providing feedback.

Encoding

The communication process begins with a sender who has an idea that they wish to transmit to someone else. In order to do this, they have to first encode it, which involves converting it into symbols which a receiver can interpret, for example words, drawings, body language, equations, numbers. The choice of symbols is crucial to enable the message to be interpreted correctly, since some receivers may be able to interpret some symbols more clearly than others. For example, when dealing with someone who cannot speak your language you are forced to rely on non-verbal symbols. British airways made a famous error when they named their executive lounges rendezvous lounges without knowing that this means 'a house of loose women' in Portuguese! Even when dealing with people of the same nationality one's communication skills in writing, speaking and acting are critically important. Indeed, AT&T spend US$6 million a year teaching their workers minimal proficiency skills in reading and writing in an attempt to improve communications within its organisation (Victor 1992).

In general terms, it can be said that the art of writing is in thrift. That is, keep it as simple as you possibly can and use as few words as you possibly can to communicate a particular message. However, having said this, there may be situations in an intercultural context where following this rule may be inappropriate. For example, compared to the sequential thought processes of Australians, Arabs tend to think in loops and build a lot of repetition into their conversations. You would have to do the same when communicating with them even if appears inefficient to you.

A useful model to help people think about the likely response to the language used in their communications is given in Table 4.5. The idea behind this model is that communications that are perceived to be evaluative, contrived, controlling, unsympathetic, condescending and dogmatic will be perceived negatively and will induce a defensive behavioural response and vice versa. This has important implications for the way that one communicates at any time but particularly during a construction project dispute since it indicates how disputes can unintentionally escalate simply as a result of the language used in communications.

TABLE 4.5 ASPECTS OF COMMUNICATION CONTENT THAT INDUCE DIFFERENT RESPONSES

INDUCES DEFENSIVENESS	INDUCES SUPPORTIVENESS
Evaluative	Descriptive
Controlling	Collaborative
Contrived	Spontaneous
Unsympathetic	Empathic
Condescending	Equal
Dogmatic	Provisional

SOURCE Gibb 1984

Language problems are arguably the greatest threat to the encoding process, and with approximately 2800 languages spoken in the world it is impossible to master them all. While English is the international language of business, spoken as the first language in eight of the world's leading trading nations, complacency is unwise and there are some simple rules that can minimise misunderstandings with people whose first language is not English. These are as follows:

- Avoid slang and colloquialisms.
- Keep language simple.
- Speak slowly.
- Rephrase frequently.
- Build redundancy into messages.
- Back verbal messages up with written messages.
- Play on words which are similar in different languages.
- Don't tell jokes.
- Don't try and be subtle but be direct, forthright and clear.
- Get to know the people you are dealing with and learn a few words of their language. This breaks down barriers and opens communication channels.
- Use a translator, but be careful, because the use of a translator is a skill that needs learning and practice. The choice of translator is also an important consideration.

Choosing an appropriate communication medium

Once encoded, the message has to be transferred via a medium (e-mail, fax, letter, telephone) to the receiver and the choice of medium is also important. For example, there is little point using e-mail if you are dealing with an organisation or a country that does not have equally well developed e-mail capabilities.

Essentially, communication mediums can be classed as one-way (fax, letter) or two-way (meeting, telephone), and the consideration of which to use is important from an intercultural perspective. For example, two-way communication may be less effective in high power-distance cultures because the degree of participation and interaction it induces is likely to be small. In contrast, two-way methods may be more effective when dealing with high-context cultures (cultures that place more emphasis on the way something is said than the content of what is said) because they facilitate the building of relationships before doing business. For example, Arabs see letters as a greater sign of formality than westerners do and may become suspicious of their use. In contrast, an American's relatively high task orientation and lesser reliance on personal relationships would make them prefer dealing in one-way methods as often as possible. Another factor that may influence the use of two-way or one-way mediums is the potential for loss of face caused by the subject of the communication. For example, if one was having to communicate blame for something, the use of one-way mediums, which are more private than two-way communications, would be important, particularly in cultures where loss of face is especially humiliating. The Japanese culture is a good example of this. However, with such people, the use of two-way mediums may be more appropriate when blame is not an issue because they tend to withhold information about potential problems and meetings are the best way of detecting this. Other practical considerations in deciding to use one-way or two-way mediums of communication are those of cost and time, since two-way mediums are more expensive and time-consuming than one-way mediums.

If one decides to use two-way mediums of communication this has important ramifications for the encoding process because one's greater physical exposure in the communication setting increases the chances of misunderstandings occurring in one's non-verbal signals (body language). This is particularly so in high-context cultures where more emphasis is placed on how something is said than on what is said. High-context cultures approach business in a more leisurely way, the establishment of personal relationships being a prerequisite to successful business communications. They also place relatively small emphasis on the written word and prefer to work informally and make agreements on verbal promises rather than written contracts. In essence, the attention to detail of people from high-context cultures is low but their attention to intention is high.

Research has shown that non-verbal messages can account for between 65 and 87 per cent of interpreted communication, especially with people from high-context cultures. The symbols and codes which represent non-verbal communication can generally be categorised into six types:

- kinesics
- proxemics
- paralanguage
- object language
- time
- physical appearance.

KINESICS

This refers to body movement, posture, facial expressions and eye contact, and while research has shown that most people across different cultures can detect basic emotions such as anger, sadness, fear and happiness, the detection of more subtle emotions can be far more problematical. For example, Japanese people tend to feel uncomfortable with the intensity of Americans' eye-to-eye contact because they are taught to bow their head as a sign of respect to others. Conversely, the depth of a bow communicates different messages to the recipient that Americans may not be aware of. Indeed, Americans may perceive bowing as a sign of weakness.

Eye contact is particularly important in high-context cultures and when dealing with people who have relatively undeveloped language skills because such people rely more heavily on non-verbal cues. In this sense, wearing sunglasses when talking to foreign workers, as is the common practice in Australia, may not be the most effective way of communicating with them.

Other examples of differences in body language between cultures are:

- Arabs being insulted by the use of the left hand for shaking hands
- tapping the nose, which has also been found to have six different meanings in various countries
- folding one's arms across the chest, which in many countries is a sign of defensiveness but in Fiji is a sign of respect for the person you are communicating with.

PROXEMICS

This refers to people's use of proximity and space and their level of tactility in communications. Arabs, South Americans, southern Europeans and eastern Europeans are generally high-contact cultures and stand relatively close in conversation, often touching each other. In contrast, people from the United Kingdom, North America, northern Europe and Australia come from a low-contact culture and have a strong sense of personal space, seeing it as a form of territory to be protected and feeling uncomfortable when it is invaded.

The issue of touch (haptics) is an important one because responses to it are not culturally uniform. For example, when shaking hands in the United Kingdom, one would expect to treat a stranger less enthusiastically than a friend. Too enthusiastic a handshake too early in a relationship would make someone feel uncomfortable or may imply a desire to manipulate or control. In contrast, the Spanish have no such sensitivities and have a tradition of shaking hands more enthusiastically early in a relationship. Indeed, a Frenchman may hug and kiss you on first meeting.

Smell is also an important issue which is related to proxemics; the role of smell in different cultures varies considerably. For example, in the United Kingdom, Australia and the United States the presence of body odours can be offensive, but in some countries natural body smells are considered inoffensive and fewer attempts are made to mask them with perfumes. On construction sites, where people from many different cultures work and eat in close proximity, it is surprising how much bearing obscure issues like smell can have on relationships in the workplace — not that it is easy to manage.

PARALANGUAGE

This relates to how something is said, such as the rate of speech, tone and inflection of the voice. Strangely, one of the strongest forms of paralanguage is silence, which in some cultures conveys enormously strong messages. For example, studies of American conversations have found that they contain a relatively small amount of silence compared to conversations held in countries such as Japan. Americans openly admit that when they interact with Japanese people, they find the conversation too slow and the spaces between ideas and words uncomfortably long and frustrating. In response, Americans tend to fill in the gaps left by the Japanese, which in turn offends and frustrates the Japanese.

As another example of the importance of voice usage, it has been found that Western cultures perceive a fast and high voice to be a sign of weakness, whereas a slower, deeper, calm voice implies a greater sense of control, dignity and power. There are no such stereotyped views about voice tone held in Arab countries, and loudness indicates strength to a greater extent than tone or speed. To Australians, the loudness of Arabic conversations may be overly intrusive, forceful and offensive.

OBJECT LANGUAGE

This refers to how we communicate through material artefacts such as clothes, cars or cosmetics. Appearance is a very powerful communicator, particularly of social status. The origins of these associations lie in the history of warfare where different levels of army command wore different uniforms – more seniority being associated with more ornate and styled clothing. However, the messages communicated by different appearances vary considerably between cultures. For example, the British are renowned for their conservative and formal dress, particularly in a work context. The British tend to have definite uniforms for work and play, while in Denmark and Australia there is far more informality in dress and the distinction between dressing for work and play is less sharp. Often this is related to the practicalities of climate, cold climates being associated with more formality than warmer climates.

TIME

Different cultures condition people to see time differently. In some cultures such as Australia, the United Kingdom and the United States, time is of the essence, life is planned in great detail, and punctuality is very important. In other cultures (generally warmer climates) time is less important and communications are far less orderly and time-structured. The differences in attitudes towards time are most vividly experienced in the different approaches to meetings, where in time-oriented cultures there are precise agendas that have to be adhered to and any delay is considered a waste of time. In other cultures, scheduled time is far less important and time is seldom considered wasted. These different attitudes towards time have huge implications for project managers who are working to tight time schedules, since in non-time-oriented cultures delays are not seen as a problem, particularly in the short term, and may therefore go unreported. This behaviour could be very frustrating and difficult to understand from the opposite cultural perspective and may be a significant source of potential conflict.

PHYSICAL APPEARANCE

Research has shown that people's physical appearance has a major influence on the way that people judge them. Apparently we all judge books by their cover to some extent. Since different nationalities vary considerably in their physical make-up, it is likely that this would be an important factor in intercultural communication. For example, Swedes are generally tall, blond and slim, Germans are a fairly large people and Southeast Asians tend to be relatively small. The importance of physical appearance as a measure of one's social worth has become increasingly important in society and it appears that the world is being swept by a preoccupation with physical appearance, which has been referred to as 'body-mania'. This has been largely fuelled by the emergence of the mass media, which constantly expose us to images of perfect people and convince us that it is possible for us to look the same.

Research has found that the following physical characteristics have an important influence on the way we judge people:

- Attractiveness: attractive people tend to be given positive behavioural attributes and are generally seen as being more competent and successful. Unattractive people tend to be seen as introverted.
- Weight: overweight people tend to be held in low esteem and seen as lazy. On the other hand they are also seen as warm-hearted, calm and sociable.
- Fitness: a good physique is now perceived to be a triangular shape and tends to be associated with an outgoing and strong personality. On the negative side, extreme fitness is associated with low intellect.
- Height: tall people tend to be held in higher esteem than short people, being seen as more powerful and forceful. Height has been found to relate directly to professional success in a number of studies. However, excessive height, which produces thinness, is generally seen as undesirable.
- Facial beauty: there have been strong correlations drawn between facial beauty, talent and social success. However, the barbie doll, blonde bombshell image has been found to be associated with a low intellect.
- Stature: the way a person holds themselves produces strong messages relating to power and social and professional success. People with slumped shoulders and curved backs tend to be held in lower esteem than those who hold themselves upright and hold back their shoulders.
- Hair: female brunettes receive the most favourable ratings, followed by blondes and then redheads. Brunettes are seen as ambitious, sincere, strong and predictable, blondes are seen as weak, simple and beautiful, and redheads are seen as cold, tense and excitable.
- Eyes: round eyes are associated with alertness, whereas closed eyes are associated with untrustworthiness. Bulging eyes are associated with excitable, impulsive, unpredictable people.
- Facial structure: people with wrinkles in the corners of their eyes and upturned mouths are perceived as more positive and friendly and as having a good sense of humour. Thin-lipped people are seen as being more conscientious than thick-lipped people, who are seen as lazy and unresponsive. (Deresky 1997; Victor 1992).

Decoding

When the formulated message finally gets to the receiver, as much as 40 per cent of the original message may have been lost due to cultural distortions, and it has been shown that a further 30 per cent can be lost at the point of reception due to cultural differences as the message is decoded (Victor 1992). Earlier we said that cultures are the lenses through which we see the world and that these lenses naturally determine the way messages are perceived and interpreted. The greatest distortions in decoding arise from people's natural tendency to use standard lenses to view different cultures — that is, their tendency to stereotype. This tendency is particularly strong in the construction industry because the constantly changing nature of project teams prevents people from building relationships and gives them no other choice than to base a relationship on preconceived impressions of similar people in the past. The implications of stereo-typing for communication efficiency are serious since they cause people to see what they expect to see rather than what is actually there. That is, their lenses filter out information which does not con-form to their expectations, resulting in a self-fulfilling prophecy. The frightening thing about stereotypes is that they operate subcon-sciously, which means that people don't realise their influence. But the most worrying aspect of stereotypes is that they are almost always simplistic, negative, judgmental and destructive, producing a distorted view of reality that works against true communication and co-operation — and therefore good business. Stereotypes are also contagious and represent the foundation of racism and discrimina-tion around the world.

Essentially, research tells us that there are three main types of people in terms of their tendency to stereotype:

- those who are openly biased and prejudiced and show it in their overt actions
- those who believe that they are not prejudiced and try to act that way but show covert signs of bias
- those who believe they are not prejudiced and show no signs of bias in their overt and covert actions.

Research has found that the people who are most prone to rely on national stereotypes are those who:

- have not travelled or had the chance to get to know people of other nationalities (stereotyping is a product of ignorance)
- those who have had strong positive or negative experiences of particular national groups (stereotypes are shaped by our past experiences)
- those who are under time pressure and do not have the time to explore real cultural differences and think about the implications of their actions (stereotyping is the easy and quick way of judging people)
- those under stress, with low self-esteem or who feel vulnerable and threatened (stereotypes are a defence mechanism and are nearly always judgmental and negative towards other groups).

Feedback

Feedback is essential to the communication process because it gives the sender information about how the receiver perceived the message. In this sense, the process of communication should not be seen as a one-off event but as a continuous process where sender and receiver repeatedly share and exchange information over time in order to converge on a mutual understanding of each other's intentions and ideas. In some cultures, however, particularly those of a high power-distance, feedback may not be forthcoming and it is essential that in these circumstances special efforts are made to facilitate its provision.

PROGRAMS FOR MANAGING CULTURAL DIVERSITY

Research in the construction industry by Loosemore and Lee (2001) suggests that the construction industry has a long way to go in sensitising itself to the challenges of cultural diversity. Their research in Australia and Singapore indicates that most companies deal with cultural diversity as a potential problem rather than a potential opportunity. The limited measures to deal with a multicultural workforce focus on training people to cope with problems such as discrimination rather than on how to harness the productive potential of different cultures. Although the problems of communicating with different language groups are significant for most supervisors, initiatives for tackling the issue are ad hoc, individualistic and inconsistently applied from site to site. Company-wide policies are almost non-existent. In both Australia and Singapore, communications with different language groups appear to be left to chance and the most common strategy used by individuals to overcome problems associated with language differences in the workplace is to work through 'cultural gatekeepers' who have no formal support or training. This informal reliance on such people in the construction industry is in stark contrast to best-practice companies in other industries, which provide training and support infrastructures to improve the language skills of those have difficulty communicating. Here the aim is to encourage a greater sense of collective responsibility for better intercultural communication in the workplace. However, if construction companies were to move down this path, they would be wise to heed the advice of Migliorino and colleagues (1994). They found that most training schemes focus on those who do not speak the indigenous language, ignoring the need to provide indigenous managers with a basic vocabulary in the array of languages they have to deal with. Furthermore, they discovered that language training tends to be too general and not improve technical work-related communications in the workplace. Finally, is it rarely evaluated effectively or linked to rewards. Indeed, many language courses are voluntary, being provided after work hours, making it a low priority that adds to workloads and competes with family responsibilities. Thus if greater training opportunities are afforded, it is important that people have the motivation and learning skills to benefit from it and that they are 'family-friendly'. Furthermore, operatives in both Australia and Singapore are likely to need pre-training in basic literacy, numeracy and learning skills before embarking on any training program.

It is also important to appreciate that self-awareness is just as important as cultural awareness in achieving competency in intercultural communication. Simply learning a language is not enough. An important part of improving intercultural communications is giving people a better understanding of their own culture as well as those with which they interact. This may not be achievable in a classroom and may mean providing opportunities for people to socialise with different cultures so that they can understand the values, beliefs, rituals, expectations and superstitions that underpin them. Finally, having trained one's supervisory and operative workforce to be more culturally astute, it is important to nurture and sustain these relationships. For example, supervisors could be encouraged to specialise in a particular culture and their future responsibilities could reflect this. Allocating supervisory responsibilities on construction projects according to culture as well as professional expertise would certainly be a major innovation in the management of human resources in the construction industry.

As well as providing better training in diversity issues, companies can do much at policy level to prevent problems of discrimination. The consequences of not doing so are social, criminal and economic and manifest themselves at both the individual and corporate level. First, companies should review their equal opportunity policies to ensure that they are up-to-date, comprehensive, easy to understand, openly communicated and properly implemented. This may mean translating them into several different languages so that they can be communicated to various stakeholder groups. Summaries should also be distributed to every employee and displayed in the workplace. Such policies should set out grievance procedures in details and make clear everyone's collective roles and responsibilities in the equal opportunity environment. Only from a collective position is it possible to break down racist structures and cultures and institutionalised practices.

In addition to keeping managers, supervisors and employees informed of their rights and responsibilities, it is within individual attitudes that the biggest contribution towards eliminating racism can be made. Only by acting at this level can a genuine anti-discriminatory culture be developed. However, distilling the principles of anti-racist attitudes and behaviour is by no means an easy task. People must be made aware of the reasons underlying the behaviour of people from other cultures, and they must be made aware of their own stereotyped perceptions of other racial groups and how this influences their behaviour. Making people conscious of the way they behave and the implications of this for themselves and others is widely accepted as the most effective intervention strategy to break down prejudice in organisations.

This should not be done by stimulating feelings of guilt but through positive educational awareness programs in partnership with minority groups. For example, a site induction process can be used to communicate and reinforce equal opportunity policies, to increase racial tolerance on site, and to establish expectations of behaviour. Educators therefore have a crucial role in setting the context for cultural change, in helping people understand the reasons and implications of racist behaviour, and in developing strategies to combat it. Role models of different cultural backgrounds who are outstanding achievers can also be used to champion the cause and increase the perceived status of their respective groups. Finally, regular meetings and consultative committees (with a very senior person as chair) can be formed to discuss equal opportunity issues and to reinforce exemplary behaviour. Senior management commitment is crucial to ensure that anti-racist policies are premised on more than just good intentions but are underpinned by a genuine commitment to tackle some difficult, confronting and painful issues. In recent years there has been considerable rhetoric about anti-racism and the notion of equal opportunity in general and there is a danger that anti-racism sentiment and policies stay as rhetoric.

CONCLUSION

This chapter has demonstrated the potential for misunderstandings to occur between people of different nationalities in construction projects. Cultural differences have serious implications for the way negotiations are conducted, decisions are made, human resources are managed and the way that construction project managers motivate and lead their workforce. Indeed, cultural differences affect every aspect of the managerial function. But the chapter has also demonstrated the complexity of considerations that have to be made by managers and workers. It is impossible to understand and accommodate all cultures and the most important thing is to be aware that differences exist and to be sensitive to them.

EXERCISES

1   Why is it increasingly important to be able to understand and manage diversity in the workplace?
2   How do national cultures vary?
3   What is the cause of racism and how can it be effectively managed?

Managing Cultural Diversity

PART 2

MANAGING A PROJECT

CHAPTER 5

MANAGING THE PROJECT CONCEPT

INTRODUCTION

The concept stage of the project lifecycle defines the extent of the project work and the end product. It is the first and by far the most important stage since the effectiveness of its execution determines the ultimate project outcome. The work associated with the concept stage has the greatest ability to influence the cost, time and quality of the project. For example, a carefully defined scope, clear objectives, and appropriately formulated design and construction strategies should stimulate a smooth and trouble-free execution. Most projects that overrun on time and cost are those that have been poorly conceived (Bromilow 1969, 1970, 1971; Levido et al. 1981). A graph of the impact of the concept stage on project outcomes, particularly the cost, is given in Figure 5.1.

The aim of the concept stage is to prepare a blueprint of important decisions necessary for the development of design and construction strategies. In doing so it seeks answers to a series of questions such as:

- What does the client need and want?
- What resources are available (funds, time, land, people)?
- What standard of performance is required?
- What external risk factors are likely to impact on the project?
- Who will make the key decisions?
- Who is the end user?
- What does the end user need and want?

A client initiates a construction project for one of a number of reasons: to satisfy its accommodation needs, to make money out of

Figure 5.1 The impact of decisions in the concept stage on project outcomes

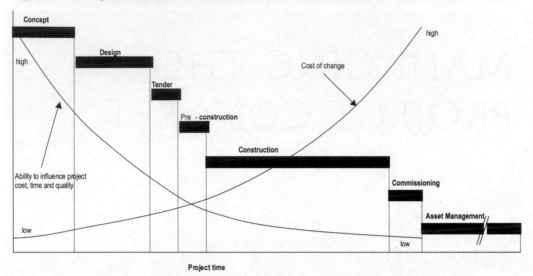

a new development, or to build some form of public utility such as roads, schools, hospitals or power stations. An experienced client may be able to perform the work involved in the concept stage, but on larger projects a client commonly seeks the services of a professional project manager. Together, they interpret the client's needs and develop appropriate strategies. Depending on the type of project, its size and the degree of complexity involved, the client may seek the assistance of other specialist consultants such as a lawyer, risk manager, designer, quantity surveyor or financial expert. In the context of this book, the application of project management is viewed as a task that an independent professional project manager is best placed to perform across each stage of a project lifecycle.

Figure 5.2 illustrates a series of important tasks that are performed at the concept stage. These will now be examined in detail.

APPOINT A PROJECT MANAGER

The role and appointment of a project manager have already been discussed in detail in Chapter 2. The timing of the appointment is the remaining issue worth examining. The client may appoint a project manager immediately after deciding to initiate a project or the client may prefer to do it sometime later. It is in the client's interest to have a project manager on board as early in a project's life as possible in order to capitalise on his or her expertise in formulating early decisions.

Figure 5.2 Activities at the concept stage of a project lifecycle

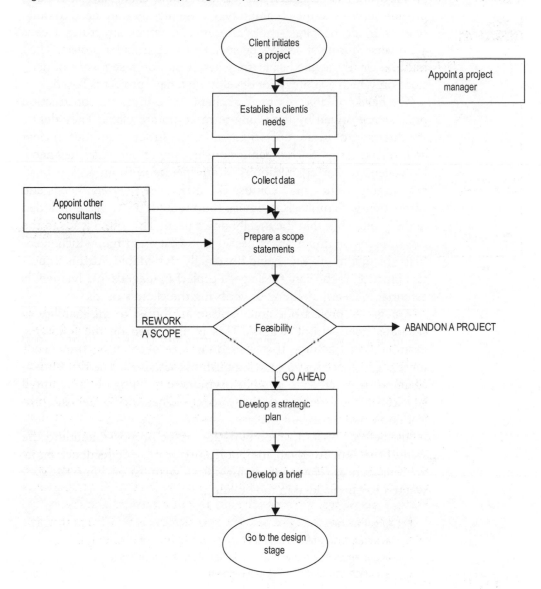

ESTABLISH A CLIENT'S NEEDS

A client initiates a construction project to meet a specific need. A small client, such as a house owner, may need to build a larger house for the expanding family, while a large organisation such as a bank seeks an attractive investment opportunity, such as an office building,

Managing the Project Concept

for the accumulated shareholders' funds. Public clients are required to provide services to the community. Consequently they need to build schools, roads, hospitals, gaols, airports and similar structures. Needs are diverse but must always be specific to a particular project. They must be clearly defined since they set in motion a sequence of decisions on which a strategy for developing a new project is based.

The needs of large commercial clients who invest in construction projects are driven by their long-term corporate plans. They define the direction of future operational and investment activities in clear and concise terms. Such organisations usually know what they need. However, smaller organisations often operate without a clear business strategy. They may confuse the difference between needs and wants by over-emphasising issues they would like to have included in the project but that they really don't need. This often leads to the loss of focus on important issues such as cost and time, which eventually results in projects being financially non-viable. At this important junction in the concept stage, a project manager's involvement is essential in clearly defining or verifying the client's needs.

A negative project outcome is often attributed to the inability to separate 'needs' from 'wants'. This is illustrated on the following example. Let's assume that a medium-size accounting firm needs more office space for its growing business activities. The firm's management has decided to build its own office building of the required size in the preferred location of Sydney. It has formulated the project's objectives and budgets.

Just before the completion of design of the proposed building, the staff of the firm are given an opportunity to provide feedback on its appropriateness. The feedback includes a wish list of things the staff wanted to be included in the building:

- larger offices
- a better quality of office furniture
- better amenities in the form of additional shower facilities
- a squash court located in the basement.

To keep the staff happy, the management has decided to include all of these items in the design. The additional costs are deemed acceptable and the project is given a green light. However, a serious delay occurred in the construction stage as a result of design changes arising from the decision to increase the size of offices and to build a squash court. Consequently, the project was delivered late and well over its budget, making it financially non-viable. The firm may have no other alternative but to sell the building and retain the existing office space.

The knowledge of a client's real as opposed to perceived needs is a condition precedent for formulating specific project objectives and an appropriate business strategy for achieving them. It may well be that a client's need to achieve a maximum return on invested funds may not be satisfied by investing, for example, in an office building. Another investment alternative may better meet the client's needs and objectives.

COLLECT DATA

Now that a client's needs have been defined, the next task is to generate and collect specific information from a client or a client's organisation in support of the defined needs. Such information will be wide-ranging and may include:

- the nature of a client's business and the mode of operation
- a client's mission statement and objectives
- a client's competitors and the market share
- a client's financial position
- a client's organisation strengths and weaknesses
- a client's organisation culture and structure
- a clear description of the proposed project, its intended function, and the desired technical and operational requirements
- the client's preliminary budget (the maximum amount of money the client is willing to invest)
- a time-frame for development
- the desired quality standards
- a client's historical data from previously developed projects (if available)
- the likely sources of funds
- a client's organisation, decision-making, and communication processes
- a brief summary of key decision-makers in a client's organisation and a process of communication between them
- the name of a decision-maker in charge of a project and his or her level of authority
- the location of a preferred site and site conditions
- a summary of potential constraints on a project such as financial, time, staffing, technical, operational or economic, political, social and environmental
- a summary of relevant laws, regulations and codes relevant to a project.

After becoming familiar with the client's organisation, its needs and objectives, a project manager will attempt to verify the need for a project through market research. Of particular interest is the level of demand for the type of project the client intends to develop. Other issues that may be unearthed are the client's market position, its reputation and its competitors.

It goes without saying that more information derived at this point in time will make decision-making in the concept stage more effective.

APPOINT OTHER CONSULTANTS

The size of a project, its type and complexity, and the extent of a client's and a project manager's expertise will generally dictate what other consultants may be needed in the concept stage. For example, a design consultant may assist in assessing preliminary structural or architectural design schemes of a proposed project. A quantity surveyor's input may be invaluable in formulating an overall project cost budget and assessing the feasibility of alternative development schemes. Since construction projects are exposed to risk, a risk manager is commonly required to advise on the likelihood and the consequence of risk, and on appropriate treatment strategies.

A financial expert may be needed to assist in identifying sources of funds for a project and helping to raise such funds. A lawyer's presence at the concept stage will ensure that a proposed project complies with various acts, codes and regulations. A lawyer may also provide advice on risk allocation in contracts, on the appropriateness of a preferred method of project delivery and on an option for a contract price, and may also assist in developing conditions of a contract for the engagement of consultants. A lawyer's assistance in fund-raising may also be invaluable.

An environmental expert may assist in ensuring that a proposed project meets environmental laws and regulations. Some projects, particularly those that are likely to be controversial, may benefit from expertise provided by a professional lobbyist, whose task will be to gain support for a project from organisations and community groups that are likely to express some concern.

PREPARE A SCOPE STATEMENT

A 'scope' is a written statement that defines the extent of work (based on the expression of the client's needs), formulates key project objectives of cost, time, quality and function, provides the necessary supportive data, formulates a preliminary development strategy from the identified alternative schemes and establishes a preliminary cost budget. It forms a basis for future project decision-making.

PMBOK (1996) defines a scope statement in much broader terms, by focusing on what is to be developed, why and also how it is to be built. Although there is no consensus in the literature on either the definition of a scope statement or a logical sequence of steps in the concept stage, a view taken in this book is that a scope statement is fundamentally concerned with what work the client wants to accomplish and why, while the issue of how it is part of a strategic plan will be examined later in this chapter.

A project manager, together with the client and any other consultants, will formulate a scope. Since the client's needs have already been defined and the necessary supporting information collected, the reason for a project and the extent of work to be undertaken are already known.

Project objectives

The next step is to formulate project objectives. The client's objectives for a project may not necessarily be compatible with those of individual team members. To ensure a successful outcome, every attempt should be made to develop project objectives common to each team member, including the client.

Cost, time and quality are the most common examples of project objectives. Their achievement will determine whether or not a project has been successful. The total amount of money that the client is willing to invest sets a cost objective criterion, while the overall amount of time available to complete a project establishes a time objective criterion. Quality objective criteria are usually those defined in design documentation and in a project quality assurance plan.

Depending on a project type and a client's particular need, other objectives such as function, utility or aesthetics may need to be satisfied. These, however, cannot be measured using quantifiable criteria. Instead, some form of qualitative assessment will need to be applied.

Inexperienced clients often mistakenly believe that the objectives of cost, time and quality can be maintained in equilibrium throughout the entire project lifecycle. This would be possible only if they were independent of each other. Instead, they are closely correlated. For example, a project constrained by tight cost and time budgets may have its quality standards eroded. This is because it may become necessary to lower standards in order to pay for cost or time overruns. Similarly, a project that is expected to be built to the highest quality standards, such as the Sydney Opera House, would require cost and time budgets to be substantial to accommodate unforeseen quality-related problems. The Australian Parliament

House in Canberra serves as an example of a project that was constrained by a very short and perhaps unrealistic schedule. The project was completed on time but at the expense of a huge cost overrun which almost doubled its original budget.

Once objectives have been defined, they must carefully be prioritised in order of their importance. This information is particularly important for a design consultant to know in designing a project. A project manager will also rely on this information in developing plans and controlling progress.

Projects built under the concept of partnering are often associated with 'softer' or largely qualitative objectives such as:

- achieving open communication
- maintaining teamwork
- achieving the best safety standards
- developing best practice
- providing the best possible service to the community
- having fun.

These are commonly assessed using a subjective form of evaluation specifically design for partnering projects. For more information, see CII (1989), USACE (1990), Uher (1994), AGC (1995).

The formulation of objectives may be preceded by the development of a project mission. A mission statement defines the philosophy of a project organisation, what a project team wants to achieve, and the values or principles on which the work of a team is based. The development of a mission statement is more commonly associated with partnering and strategic alliance projects. But non-partnering projects may also benefit from a mission statement; examples of a real-life mission statement associated with construction projects are given below.

Example 1 of a project mission statement: We are committed to deliver safe and improved assets through the provision of innovative maintenance services and to maximise the value for all stakeholders through integrity, best practice, openness, trust and knowledge sharing.

Example 2 of a project mission statement: We, the A Team, will work in an open and trustworthy manner. We are committed to the timely achievement of project goals through decisiveness and respect for the project stakeholders. In the process of project delivery, we seek to be honest and ethical, sharing the knowledge and understanding gained from working together in an enjoyable partnership.

Example 3 of a project mission statement: Our focus is on people through leadership, trust, honesty, caring and commitment to individuals and the team to develop open communication, account-ability and courage. We will achieve excellence in safety, quality, performance, teamwork and professionalism, delivering a bench-mark project with the personal satisfaction of all involved.

Setting a time-frame for a project

Financially successful construction projects are those that are not only completed within a cost budget but also within a defined time-frame. Delays in completion not only increase the cost of a project cost but also delay its commissioning, the impact of which is either an extra cost to the client or a loss of income. It is therefore neces-sary for a project manager and the client to establish an overall time-frame for a project. This is best achieved by determining a date on which the client expects a project to be completed. Working back-wards from that date, a project manager will apportion the available time to each stage of a project lifecycle. If the time available is found to be insufficient, the client may be agreeable to providing extra time by extending the date of completion. However, when a project must be completed by a specified date, which is common to large retail projects such as shopping centres, a project manager would need to fast-track a project by overlapping some of its stages, particularly design, tendering, pre-construction and construction. This decision would almost certainly influence the design of a project organisation structure and the method of delivery. These issues will briefly be dis-cussed later in this chapter.

Formulation of a development scheme

The next step in developing a project scope is to formulate two or more alternative development schemes that would meet the client's stipulated needs and objectives. Enough information is now avail-able to define performance parameters that a proposed project is required to meet, a project size, and its main structural and architec-tural features. For example, assume that the client's requirement for a coal-loading facility is to store and handle 40 000 m^3 of coal per day. This performance parameter defines the size of the facility and the technical specification of its handling equipment. Since the pro-posed facility could be built in either steel or concrete, a project team will consider the implications of each structural scheme. It will also examine the impact of such structural schemes on the overall archi-tecture of the project.

Another example is a factory building, which could be designed either as a steel or concrete frame or as a load-bearing wall structure. In the latter case, walls could either be prefabricated or built in situ. A team may discard one or more of these structural concepts at this point and test the financial viability of the remaining schemes by a process of feasibility Project feasibility will be discussed in detail later in this chapter. Clearly, formulating more than one development scheme is beneficial in finding the best solution.

A project organisation structure

With preliminary decisions made on what is to be built and how, and in what time-frame, an important task for a project manager is to design an organisation structure best suited for a project. A project organisation structure defines the roles and responsibilities of its members, how they are contractually linked, and how they communicate with one another. Establishing open lines of communication and empowering an organisation's members are essential requirements. This issue has already been discussed in detail in Chapter 3.

Design of a project organisation structure is then overlaid with an appropriate delivery method that will closely match stipulated contractual and communication links. The choice is between the traditional method, which is characterised by the involvement of a client, a design consultant and a contractor, or one of many non-traditional methods, in which a project manager commonly assumes leadership. The detailed information on different types of delivery methods can be found in Uher and Davenport (2002). Selecting an appropriate organisation structure and a method of delivery in the very early stage of a project life may seem premature, but it is necessary for subsequent decisions, particularly those affecting a project design.

Availability of resources

One of the important tasks of managing the concept stage is to identify important resources needed for a project and to assess their availability.

The client may have already purchased or leased land for a proposed project. If not, a suitable site must be found. A number of factors will influence site selection, particularly the type of project, its size, and the actual location. The aim is to find a building site with the greatest propensity for future capital gains since it would significantly improve the overall project value.

The project manager should ascertain whether or not the client has already secured necessary funds for a project. If no funds have been secured, the project manager together with the client will need to take the necessary steps towards doing this. There is no point moving a project forward without a guarantee that funds will be available.

Other resources needed for a project, such as people, materials and plant, are often regarded as being readily available when needed. This, however, is a risky assumption to make since available resources may be limited in quantity or quality. It is therefore necessary for a project manager to ascertain the availability of the important human and physical resources.

Preliminary cost budget

The amount of money the client is willing to invest in a project is a sum that must cover all the costs incurred throughout all the stages of a project lifecycle, from inception to completion. The client articulates to a project team the maximum amount of funds available and whether or not it includes a contingency for risks and for things that are generally unknown at that particular time. For example, let's assume that the client wants to invest $100 million to build a project. Let's further assume that this is the maximum amount of money that the client is willing to spend. It means that this sum includes a contingency. The client expects a project to be designed and built to this sum, which represents a total project cost budget.

Establishing a total project cost in the concept stage is a difficult task since there is not enough specific information, such as the project's design, to arrive at an accurate estimate. A project will also be exposed to a wide range of risks, which may increase its cost. It is therefore necessary to quantify the risk and inaccuracies in cost estimating, and express them as a contingency sum. The total project cost is then a sum of a net project cost plus a contingency.

Assume that for the above hypothetical project its net cost was estimated as $88 million and the contingency as $15 million. Clearly, the total cost budget is exceeded by $3 million. If the client can provide no additional funds, the proposed development has to be reviewed in order to lower the amount of the net cost or reduce the exposure to risk. But if the net cost is estimated as $80 million and the contingency as $15 million, the client may use a surplus of $5 million to either increase the scope of a project or increase the amount of the contingency.

While difficult, estimation of a project cost at the concept stage is vital for assessing its financial viability. Because only sketchy and imprecise information is available, a cost estimate will probably be inaccurate. Consequently, an estimate of a contingency sum is likely to be conservative. For these reasons, a project cost budget is commonly regarded as preliminary only at this stage.

It is the responsibility of a project manager to ensure that a project is conceived, designed and built within a total project cost budget. It is important for all team members to recognise that the client owns a contingency sum and that its expenditure must be authorised by a project manager or the client.

One of the outcomes of scoping is the identification of alternative development schemes. These must now be costed to ensure that they could be built within a total cost budget. A quantity surveyor will most commonly perform this task. A cost estimate will comprise a cost incurred in each stage of a project lifecycle.

Costs incurred at the concept stage are fairly easy to compile. They comprise consultants' fees, the cost of land on which the project will be built, the cost of borrowings, and the client's direct costs and overheads.

The main component of the design stage cost is a fee paid to a design consultant. If other consultants such as a project manager, quantity surveyor or value manager are involved, fees paid to them would also need to be included as costs. Additional costs commonly incurred in the design stage are those such as statutory charges and the client's direct and overhead costs.

The design stage cost is more difficult to estimate in the absence of a project's design. Although a project scope statement has already defined one or more possible development schemes, the extent of the design work cannot be fully established. Design uniqueness and the degree of complexity involved have a direct bearing on how much work is needed to produce design and tender documentation. The volume of design work will be reflected in the amount of fees paid to a designer and other consultants.

The main component of the cost at the tendering stage is fees paid to designers and other consultants. Other costs are those related to printing of multiple copies of tender documentation, and direct and overhead costs of the client. Estimation of these costs is a fairly simple task.

The most difficult task at the conception stage is to estimate the cost of construction in the absence of a design. At best, only a preliminary construction cost may be estimated using either a single price-rate estimating method or an elemental cost-planning method. Volume or area methods are examples of a single price-rate technique. They enable a construction cost to be estimated from the volume or the area of a proposed project, which is then multiplied by a historical price rate derived from past similar projects. The problem is that a margin of error in estimating of around 30 per cent is uncommon for these single price-rate estimating techniques.

If the main characteristics of design work are known, it may be more appropriate to use an elemental cost-planning technique for estimating a cost of construction. Cost planning allows a project to be divided into elements and sub-elements, which are then priced using the standard price data. The Royal Institute of Chartered Surveyors (RICS) and the Australian Institute of Quantity Surveyors (AIQS) have developed a standard approach to cost planning based on the definition of standard elements and sub-elements (see for example Ferry & Brandon 1991). Using cost planning, it is possible to build up a reasonably accurate estimate with only sketchy information, with the margin of error around 15–20 per cent.

Other costs incurred at the construction stage are preliminary costs, statutory charges, fees of consultants, and the client's direct and overhead costs.

If the client intends to occupy a completed project, a cost of the commissioning stage is likely to comprise the cost of refurbishment, the client's relocation expenses, and the cost of maintenance contracts. If a completed project is to be leased, incurred costs are likely to be those related to leasing and legal charges, the cost of maintenance contracts and the cost of advertising. If, however, the client intends to sell a completed project, the cost of commissioning may include a real estate agent's commission and a stamp duty fee. In all of the above cases the cost of commissioning would also include fees of consultants and the client's direct and overhead costs. Commissioning costs are easy to estimate and have a high degree of accuracy.

When combined, the costs of individual stages of a project lifecycle provide an estimate of a net project cost. The remaining component a total project cost that needs to be estimated is a contingency sum for risks and 'unknowns'. It too should be estimated for each stage of a project lifecycle. The amount of contingency is likely to be very high at the concept stage, reflecting the lack of specific information about a project's design and construction. As a project progresses through its lifecycle, more information becomes available and uncertainty is progressively replaced with more specific decisions. Consequently, contingencies should get progressively smaller in later stages.

Responsibility for the management of a cost contingency commonly rests with a project manager. However, there should be no doubt among team members that a contingency belongs to the client. If a cost contingency is not needed, a project team has no right to spend it.

The work associated with developing a project scope is now almost complete. In the final step, a project manager will summarise issues and decisions made thus far into a written scope statement as a basis for subsequent decision-making.

FEASIBILITY ANALYSIS

Feasibility analysis is the process of assessing the financial viability of proposed development schemes. It is an attempt to determine if such schemes meet the defined project objectives. It is a process undertaken by a team composed of the client, a project manager and other expert consultants such as a financial expert, a risk manager or a quantity surveyor.

Feasibility analysis may be a single activity or a series of progressive appraisals of development schemes. For example, if the client has a preferred scheme in mind at the start of the concept stage, it may be appropriate to appraise it at that point and reappraise it later, when other alternatives have been considered.

A feasibility analysis commonly involves technical and economic assessment. Technical feasibility assesses different development schemes in terms of their performance, design features, practicality, good practice features, safety, demand for specialised resources and lifecycle performance. Some technical parameters such as performance are assessed quantitatively against defined project objectives, while others may be assessed qualitatively.

Economic feasibility is an assessment of the likely financial performance of proposed schemes and may use one or more of the following methods:

- return on investment
- payback period
- net present value
- internal rate of return
- cost–benefit analysis.

Each of these methods will now be briefly discussed.

Return on investment

Return on investment is a simple ratio of net profit after taxes to the total investment value. In combination with other ratios such as

liquidity, leverage and profitability, it serves as a useful insight ,
the strengths and weaknesses of investment options. A return c
investment ratio is a meaningless number in itself unless compared
with a similar ratio calculated for another investment option. For
example, a return of 5 per cent may seem low, but when compared
with the industry average, which may be 3.5 per cent, it would probably be seen as adequate. A return on investment ratio provides only
a fraction of the information needed for decision-making and must
be supplemented with other appraisal methods to provide a complete picture of investment viability.

Payback period

Payback is a measure of the time required to recover the initial
investment. The underlying assumption is that investment with the
shortest payback period is preferred. It is a popular appraisal method
due to its simplicity and ease of calculation. Its application is demonstrated on the following example. Assume that the client is considering three project development schemes, each requiring an
investment of $1 million. The net returns from the three equally
risky investments A, B and C, are given in Table 5.1. The payback
period for project A is ten years, for project B six years, and for project C four years. According to this analysis, investment in project C
would be preferred to investments in projects A and B.

TABLE 5.1 PAYBACK PERIODS FOR PROJECT INVESTMENT OPTIONS A, B AND C

YEAR	PROJECT A (INCOME)	PROJECT B (INCOME)	PROJECT C (INCOME)
1	100 000	50 000	200 000
2	100 000	100 000	200 000
3	100 000	200 000	300 000
4	100 000	250 000	300 000
5	100 000	300 000	200 000
6	100 000	300 000	100 000
7	100 000	300 000	50 000
8	100 000	350 000	50 000
9	100 000	350 000	50 000
10	100 000	350 000	50 000

Managing the Project Concept

Although easy to calculate, the payback method may lead to ~~~ng decisions. The most notable shortcoming is that the method ~~~nores income derived outside a payback period. This may happen when a project derives more income in later years. Project B in Table 5.1 demonstrates this point: it outperforms both project A and C from year 5 onward.

Discounted payback is an enhancement of the basic payback method. It takes into account the concept of the time value of money to future cash flows. The basic and discounted payback methods are regarded as inferior for investment appraisal (Flanagan et al. 1989).

Net present value

The net present value method (NPV) is a discounted cash flow technique for analysing investment opportunities. It is computed by first calculating present value (PV) of the expected net cash flow of an investment, discounted at the cost of capital, from which the initial cost outlay of a project is subtracted. If NPV is positive, the investment should be accepted; if negative, it should be rejected. This method overcomes the shortcomings of payback since it takes into account all the revenues from a project and accounts for the time value of money.

A simple example in Table 5.2 illustrates the use of NPV. Assume that the client is considering investing in two projects, A and B. Each requires an investment of $2000. Further assume that the client borrowed the funds at 6 per cent. Net returns from the two equally risky investments A and B over a period of five years are discounted at a 6 per cent rate. In this case, only option B is positive and therefore viable.

TABLE 5.2 NPV OF PROJECT INVESTMENT OPTIONS A AND B

YEAR	DISCOUNT RATE (6%)	PROJECT A (INCOME)	PROJECT A (PV)	PROJECT B (INCOME)	PROJECT B (PV)
1	0.94	600	564	350	329
2	0.89	500	445	450	401
3	0.84	500	420	600	504
4	0.79	300	237	650	514
5	0.75	300	225	700	525
PV			1891		2273
NPV			−109		+273

One major criticism of NPV is that decision-ma[ ]
cult to interpret the results (Flanagan et al. 1989). F[ ]
NPV of +273 represent poor, average or high profi[ ]
there is no simple answer to this question. Intern[ ]
(see below) is said to overcome this shortcomin[g] ....ss,
NPV is a useful appraisal method, particularly in ranking investment options.

Internal rate of return

Internal rate of return (IRR) is defined as 'the interest rate that
equates the present value of the expected future income to the cost
of the investment' (Weston & Brigham 1968: 133). It extends the
NPV method by trying to find the value of the interest rate that produces a zero NPV. If, for example, the IRR for a particular investment
is determined to be 5 per cent and the firm borrowed the funds at
5 per cent, the firm would come out exactly even on the transaction.
If, however, the IRR is 8 per cent, the investment would be profitable. Conversely, if it is 3 per cent, it would result in a loss.

The IRR cannot be calculated precisely but must be derived by trial
and error. For example, project A in Table 5.2 shows PV of $1891 for
the investment cost of $2000 using a discount rate of 6 per cent. To
find the value of the IRR, a discount rate must be lowered until PV of
the investment and the investment cost are equal to each other. This
will be achieved when a discount rate is 3 per cent. Hence the IRR for
investment A is 3 per cent. Since project B's PV is higher than the initial investment cost, a discount rate will be increased until PV of the
investment and the investment cost are equal to each other. Thus the
IRR for project B is approximately 11 per cent.

Cost–benefit analysis

Cost–benefit analysis attempts to relate the costs of a new investment to its projected financial benefits. If these exceed the costs,
then a project passes the test. The technique has been used extensively in the public sector on projects such as roads, airports and
power plants.

A difficulty may arise, however, in converting projected benefits
to money. In some cases the conversion may be fairly easy. For
example, the time of travel saved by building a new freeway can be
converted to a saving in the fuel cost and the reduced wear and tear
of car engines as a result of less frequent braking and accelerating,
and a lower incidence of overheating of engines. It may also be possible to cost the time saving achieved by each person travelling. But

Managing the Project Concept

in other cases it may be difficult to attach a value to non-quantifiable things such as the value of a human life. This poses a problem and renders the cost–benefit method vulnerable to manipulation. For example, Ferry and Brandon (1991: 13) refer to a case in 1988 where, for political purposes, the UK Department of Transport arbitrarily doubled the value of a human life in order to justify construction of a new road.

Feasibility assists the client in deciding whether or not to abandon a project altogether, or proceed with one of the alternative schemes or perhaps go back and rework the scope. The client's decision to abandon a project at this stage may be costly, but the cost that has already been incurred may be insignificant in comparison to the losses that may arise if a 'risky' project goes ahead.

The client may decide to rework the scope of a project in order to make the project financially viable. This may involve altering its objectives or their priorities, or modifying a development strategy.

If the client decides to proceed with a project, a project manager will then formulate a strategic plan for the selected development scheme. The completion of feasibility and the selection of a preferred development scheme represent a major decision point.

STRATEGIC PLANNING

In broader terms, strategic planning is 'a disciplined effort to produce fundamental decisions and actions that shape and guide what an organisation is, what it does, and why it does it' (Bryson & Alston 1996: 3). While the definition suggests that an organisation is an ongoing corporate entity, the concept of strategic planning is equally relevant to an organisation with a limited life span, such as a construction project organisation. It is in this latter context that the concept of strategic planning will be discussed.

A strategic plan is a written document that describes in detail strategies and specific processes that have been formulated for achieving successful project outcomes. These strategies are based on the client's needs, project objectives, and specific technical, operational and management requirements. Individual tasks in strategic planning are illustrated in Figure 5.3.

Let's now focus on how strategic planning is done. Teamwork and systematic analysis of project variables are prerequisites for formulating an effective strategic plan. Under the leadership of a project manager, the client and various expert consultants work together in seeking and clarifying answers to many questions, including:

Figure 5.3 Tasks associated with strategic planning

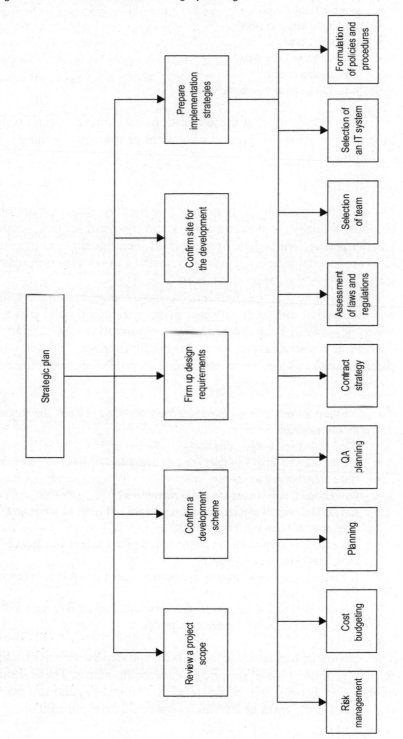

Managing the Project Concept

- what is to be done?
- why is it being done?
- what is the extent of work?
- what will it cost?
- how is it going to be done?
- who is going to do it?
- when is it going to be accomplished?

The answers to most of these questions have already been formulated in a project scope. So a starting point of strategic planning is to review the scope.

Review a project scope

The concept stage is, by its nature, highly fluid. As more information becomes available, a project manager and the team acquire a much broader perspective of development ideas. Because changes to initial ideas are inevitable, it is necessary to review a previously formulated scope and clarify its appropriateness.

It may well be that the client's needs have changed since the scope was prepared or that some unforeseen external factors have arisen. It is also possible that the defined project organisation structure and a method of project delivery are no longer appropriate. The following checklist serves as a guide in verifying the appropriateness and completeness of a project scope.

- Are the previously defined project objectives still valid and if not, how should they be modified?
- Is the project cost budget adequate?
- Are the required funds available? Are they adequate?
- Is the project period adequate?
- Are resources, both human and material, available?
- Has the likely impact of relevant laws, regulations and codes on the project been assessed? How will it be treated?
- Has a site been secured for this project? Has the impact of site conditions been considered? How will it be treated?
- Is a project organisation structure appropriate? Does it stimulate open communication? Does it empower people?
- Has the risk of contractual disputes been assessed? How will it be mitigated?
- Is the method of project delivery appropriate?

The outcome of this review process is an updated statement of scope with firm objectives and specific project requirements. These should be regarded as being 'set in concrete'. If changed beyond this point, the impact in the form of cost increases could be significant.

Confirm a development scheme

In the light of possible changes to the scope, a project team will be required to review a previously defined development scheme. If any modifications have been made, a revised scheme would need to be reappraised by a feasibility analysis to determine its appropriateness and viability.

Firm up design requirements

The technical requirements of the project that have already been stipulated will now be firmed up to ensure a clear and concise definition of design requirements in terms of:

- the physical scale of the project and its technical specification
- the shape of the project and aesthetic considerations
- performance requirements
- quality standards
- functional requirements
- services.

Confirm site for the development

The project manager will assess the suitability of available sites for the project and recommend the preferred site. The recommendation will take into account:

- evaluation of the most appropriate land use in functional and economical terms
- access to transportation (air, water, road and rail)
- access to utilities (water, sewer, electrical, gas, communication)
- soil conditions
- climate conditions.

Prepare implementation strategies

In the initial phase of strategic planning, a project team reviews and reappraises previously made decisions. In the next phase, it then formulates specific strategies or processes for the implementation of a project. These will now be briefly discussed.

RISK ASSESSMENT APPROACH

A risk manager will assist a project team in a detailed assessment of risk, both internal and external to a project. Initially, specific risks will be identified. Then their likely impact will be assessed and their likelihood and consequences prioritised. Finally, treatment strategies will be developed. This information will assist a team in formulating realistic schedules and budgets. It will also assist in a more efficient management of resources. The topic of risk management will be discussed in detail in Chapter 15.

COST BUDGETING

A preliminary cost budget developed earlier at the concept stage will now be firmed up and formulated as a final project budget with a separate contingency arrived at through risk assessment. Estimates of costs associated with accomplishing each stage of a project lifecycle will be also be firmed up and turned into cost budgets for individual stages, accompanied by their respective contingencies. A team will also prepare guidelines for the management of contingencies.

PLANNING

A team formulates an overall time schedule for a project, which shows work to be accomplished across the entire project lifecycle within the defined period. It is either a bar chart or a critical path schedule and shows major activities with only scant detail. Depending on the overall period of a project, the scale of the schedule is either in weeks or months. Apart from showing a logical sequence of major activities, the schedule also defines duration of each stage of the lifecycle. Furthermore, it should highlight the key decision points.

An overall project schedule is commonly supplemented with other plans, schedules and charts. For example, a plan of the work breakdown structure is prepared to show a top-down hierarchy of work arranged in an increasing amount of detail. Since resources may not always be available when needed, detailed schedules of important resources such as labour, materials and plant are also prepared. A chart of a project organisation structure assists team members in better understanding the nature of the organisation, its contractual links and its communication process.

QUALITY ASSURANCE PLANNING

A project manager and a team specify quality standards that a proposed project is required to meet. The quality requirements are compiled in the form of a quality assurance plan in accordance with relevant standards such as the AS 3900 or ISO 9000 series. For quality assurance planning, see Dawson and Palmer (1995), Gilmour and Hunt (1995) or Oakland and Sohal (1995).

CONTRACT STRATEGY

Since work associated with a project is executed under the terms of contracts that individual project team members enter into, and considering that many of such contracts are present, a clear and concise strategy for forming and executing contracts is needed. Its aim is to provide a framework for efficient administration of contracts in a timely manner, and for minimising or completely eliminating contractual conflict.

A project contract strategy is designed to promote:

- the use of fair and equitable conditions of contract
- equitable risk-sharing
- openness in communication as a means of preventing disputes and resolving them as soon as they occur
- the use of an IT system for effective and efficient contract administration
- the use of partnering where possible.

A project contract strategy describes in detail the type of delivery method selected and the specific roles and responsibilities of team participants. It also specifies whether a contract price will be formulated as either fixed or cost-plus. It may further comment on a tendering process and on how a contractor or contractors will be selected. The detailed information on contract strategy can be found in Uher and Davenport (2002).

ASSESSMENT OF LAWS, REGULATIONS AND APPROVALS

Various laws and regulations control the development of construction projects. They govern to a smaller or a larger extent work in each stage of a project lifecycle. Project participants must be familiar with such laws and regulations, and must comply with them, particularly in areas of:

- urban planning
- building codes
- health and safety
- taxation
- environment
- licensing
- industrial relations.

For the benefit of other project participants, the project manager prepares a summary of relevant laws and regulations that require strict compliance.

Managing the Project Concept

Some laws stipulate a formal process of applications and approvals to be followed in specific stages of a project lifecycle. For example, under the NSW Environmental Planning and Assessment Act 1979 (amended 1997) local councils are responsible for the administration of development and building applications. A project manager should then summarise approval requirements under various laws and regulations, and specify the extent of information needed for submission. Since the work associated with preparing applications for approval takes time, periods needed for approvals must be included in an overall project schedule. It also costs money, and such approval costs must be included in a project budget.

SELECTION OF TEAM MEMBERS

A project manager is responsible for managing a project, including its individual lifecycle stages and their interfaces. But the actual task of managing involves leading members of a project team who do the actual work. They are experts with the specific skills needed for the execution of a project. Successful projects are commonly characterised by having effective teamwork and synergy. Selection of project team members is therefore of the utmost importance. While price is almost always the most important selection criterion, it should not be the only one. The ability to work in a team and effectively communicate with one another is another criterion vitally important for achieving teamwork. An effective selection process of team members is an integral part of the management strategy.

It should be noted that a condition precedent for developing teamwork is to appoint the right project manager who has the necessary knowledge and skills. Selection of a project manager is entirely in the hands of the client. Only that project manager who has the ability to openly communicate, create and maintain teamwork, delegate authority and responsibility, and effectively administer a project should be selected; see Chapter 2.

SELECTION OF AN INFORMATION SYSTEM

Construction projects generate a large volume of information, which is then shared by a large number of participants in the project. One of the most daunting tasks of a project manager is to co-ordinate and integrate the activities of so many people. This is difficult to achieve without an effective information flow. It is therefore essential for a project manager, at least on large or complex projects, to install an appropriate information system for timely and effective flow of information.

It is the responsibility of a project manager to plan activities in each stage of a project lifecycle, engage required resources, and control progress in terms of cost, time, quality and use of resources. With modern IT technology, a project manager can significantly improve the effectiveness of planning and control activities. Employing sophisticated computer information systems such as Timberline, the process of generating planning information and tracking progress of a project can be made more efficient and effective.

FORMULATION OF POLICIES AND PROCEDURES

A project manager also formulates a protocol of a wide range of policies and procedures necessary for effective management of each stage of a project lifecycle.

If the client is satisfied that a proposed development scheme detailed in a strategic plan is feasible and appropriate, the client will require a project manager to prepare a brief for design.

DEVELOP A BRIEF

A brief is a detailed account of the client's requirements for a new construction project from which a design consultant develops a functional, aesthetically pleasing and economically viable design and prepares the necessary documentation. It sets out in clear terms the details of a project scope, including needs, purpose, cost, and timing and objectives. It also describes a preferred development scheme.

A brief is a vital document which must be complete, accurate and clear in its description of what is to be built, why and for what purpose, and how it is to be built. Information contained in a brief must be detailed enough to enable a design team to interpret it correctly and translate it into a design concept that would meet the required objectives and performance criteria. A brief that is incomplete, inaccurate, vague or contradictory may be misinterpreted by a design team, particularly its intent. The final design may not fully reflect the client's requirements.

The project manager formulates a brief in co-operation with the client and the design consultant. The design consultant's input will ensure clarity and completeness.

Barrett and Stanley (1999) developed a template of a brief document which is an extension of the work of Becker (1990). It contains a range of issues that should be included in a typical brief. A modified version of the template is given in Table 5.3.

TABLE 5.3 A TEMPLATE OF A BRIEF FOR A HIGH-RISE COMMERCIAL OFFICE BUILDING

PROJECT STAKEHOLDERS	DETAILS
Information on project stakeholders	
Empowerment	
Responsibility for decision-making	
PROJECT OBJECTIVES	
Client's needs and objectives	
PROJECT DESCRIPTION	
Detailed project description	
Project location	
Functional requirements	
Performance requirements	
Site conditions	
FIXED CONSTRAINTS	
Project time and cost budgets	
Design time and cost budgets	
Design contingency	
Specific decisions already made	
Specific conditions that must be satisfied	
PROJECT ORGANISATIONAL STRUCTURE	
Information on project organisation structure	
Method of delivery	
Option for contract price	
Communication links	
CONTRACT STRATEGY	
Conditions of the main contract	
Special conditions	
Who is responsible for contract administration?	
IMAGE EXPECTATIONS	
Appearance of the building and its interior	
TECHNICAL SPECIFICATION	
Area of the site	
Total area of the required floor space	
Services and their required performance	
Quality requirements	
Security	
Parking	
Information technology system	
OPERATIONAL SPECIFICATION	
Interior space requirements	
Maintenance needs and running costs	
OTHER ISSUES	

SOURCE Adapted from Barrett & Stanley 1999.

The decision to produce a brief signifies the major commitment on part of the client to proceed to the next stage of the project lifecycle, which is known as the design stage.

CONCLUSION

The concept stage is regarded as the most important period in the entire lifecycle of the project because the decisions made at that stage have a major impact on project cost and schedule. Effective management of the concept stage is therefore vitally important. It involves a systematic assessment of the client's needs, which leads to the formulation of a detailed project scope and alternative development schemes. These are then appraised through feasibility analysis, after which a strategic plan is formulated for a preferred development scheme. Finally, a brief for a design consultant is prepared. The project now moves to a new stage of its development, known as the design stage.

EXERCISES

1 Experienced clients commonly appoint a project manager just before the development of a scope statement. Do you think that they should employ a project manager right from the beginning of the concept stage? If you do, how would you persuade clients to do so?
2 What is the main purpose of developing a scope statement?
3 Are you familiar with your firm's mission statement, values and objectives?
4 Do you think it important to develop a mission statement for a project? How would you go about developing a project mission statement?
5 What is the main purpose of strategic planning?
6 What is the main function of a project brief? Who prepares it and for whom it is intended?

CHAPTER 6

MANAGING DESIGN

INTRODUCTION

Design in any field is a creative task that expresses the personal ideas of the designer. In the case of a construction project, a design attempts to meet the client's needs by integrating project objectives, as defined in the brief, with technical, performance, functional, aesthetic and regulatory requirements. Even though constrained by such objectives and requirements, a good designer frequently challenges existing approaches and concepts in an attempt to develop something new and unique. At the same time, the client expects the design of a project to represent a realistic and financially viable solution to a problem.

There are numerous definitions of design and even more arguments about which definition is more appropriate. Jones (1970: 11) expressed design in the following terms: 'Design is to initiate change in man-made things', while Lawson's (1990: 23) definition is more focused on outcomes: 'Design is the optimum solution to the sum of the true needs of a particular set of circumstances'. The design fraternity recognises that finding a single definition of design acceptable by everyone is unlikely. It agrees that 'searching is probably much more important than the finding' (Lawson 1990: 23).

The achievement of successful project outcomes very much depends on how well the project is conceived, designed and documented. Research has shown that the greatest opportunity for significant cost efficiencies lies in the conceptual and design stages of a project lifecycle (Bromilow 1970, 1971; Levido et al. 1981). Conversely, faults in design and documentation may substantially increase the cost of a project. This was briefly discussed in Chapter 5.

The purpose of the design stage is to develop a design concept together with the necessary documentation that meets the objectives of a project, ensures functionality, is technically competent and aesthetically pleasing, and meets the requirements of various laws and regulations.

Developing a design for a construction project is a highly complex task. Its successful accomplishment requires a disciplined approach to planning, a sharp focus on co-ordination and integration of design activities, and a tight control over the budget, the construction schedule and the quality of the finished structure (Morris 1994). Perhaps the most important and also difficult task is co-ordinating and integrating the activities of many different design consultants. The client expects the final design to be complete, accurate, buildable and cost-effective. This can only be achieved by effective management of the design stage.

The outcome of the design stage is a set of drawings and specifications, a bill of quantities (if required), and general conditions of contract. Together, these are referred to as the tender documents. A bidding contractor uses the tender documents for estimating the cost of construction. After the award of a contract to a contractor, they become contract documents from which the contractor builds the project.

This chapter presents a systematic approach to the management of the design stage, including selection of a design leader, formation of a design team, preparation of a design management plan, and finally development of schematic, preliminary and final designs and their documentation.

OUTLINE OF DESIGN AND DOCUMENTATION ACTIVITIES

The design stage begins after a project brief has been finalised and approved by the client and ends when tender documentation (working drawings, specifications, conditions of contract and a bill of quantities) has been completed. The work undertaken in the design stage involves:

- appointment of design consultants
- establishment of a design team
- development of a design management plan
- formulation of a schematic design
- formulation of a preliminary design
- development of a final design
- completion of specifications and a bill of quantities.

The logical sequence of activities performed at the design stage is illustrated in Figure 6.1. These will be the focus of discussion in this chapter.

Figure 6.1 A sequence of activities at the design stage

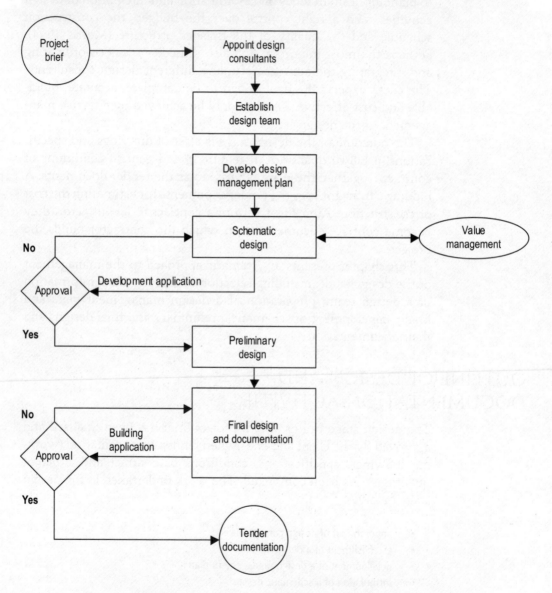

A project's type and size and the ⸎
design requirements determine wh⸎
engaged at the design stage. An archi⸎
document a simple house project, whi⸎
pital project may require the services ⸎
and other consultants such as:

- an architect
- a structural engineer
- services designers (electrical, hydraulics, mecha⸎ ...on)
- an environmental engineer
- an interior designer
- a landscape architect
- a quantity surveyor
- a value manager
- an IT specialist.

The client may choose to enter into a contract with one design consultant only — who then becomes the design leader — or with several. Depending on the type of project, the design leader could either be an architect or a structural engineer. An architect is the preferred choice for building projects, while a structural engineer usually becomes the design leader in civil engineering projects.

When the client contracts with one design consultant only, the client assigns overall responsibility for the production of design and documentation to that person or firm. The design leader then contracts with other design consultants for the supply of specialist design services. When projects are executed using the traditional method, the client may require the design leader to assume responsibility for the overall management of the design stage. Alternatively the client may assign this responsibility to a project manager. The main advantage to the client from having only one design contract is that there is then a single point of responsibility for design.

Contracting with several design consultants has some distinct advantages and disadvantages for the client. The main advantage is that the client has direct control over the contract fees of individual designers. This reduces the risk of 'bid shopping' (lowering a bid price by negotiation), a practice that a design leader might apply to other design consultants. It also gives individual designers much greater security of payment. On the negative side, the client has more design contracts to administer, and the task of co-ordinating and integrating the design activities of separate designers is much more arduous. This is when the project manager steps in.

Because effective management of the design stage is essential for achieving successful project outcomes, it is best left for a project manager to perform. The following discussion assumes that a project manager is in charge of the design stage.

At the start of the design stage, a project team has only two members: the client and the project manager. It will not be long before they are joined by other participants, the first of which will be the design leader.

The project manager has the overall responsibility for the completion of all design tasks, strictly in accordance with the brief and within the requirements of the project's objectives. Specifically, the project manager is responsible for:

- formulation of a statement of duties and responsibilities, and conditions of engagement of the design leader and the other consultants
- engagement of other consultants
- conduct of meetings, co-ordination and management of interfaces including:
  □ client/design team
  □ design team/project team
  □ design team/authorities
- development of a design management plan, its implementation, and control of progress
- management of the cost budget, design expenditures and contingencies
- management of all statutory approvals and permits
- review of progressive design reports
- implementation of value management appraisals
- progressive reporting to the client.

The project manager may delegate some of these responsibilities to the design leader as long as an adequate level of monitoring and progressive reporting is maintained.

APPOINTMENT OF THE DESIGN LEADER

Before appointing this design leader, the client and the project manager will agree on how closely this consultant will be involved in the project, particularly in the subsequent stages of the work. This decision should be aligned with those already made at the conceptual stage regarding the type of organisation structure and the method of project delivery.

The project manager then proceeds to formulate a statement of the design leader's duties and responsibilities and allocates an appropriate degree of authority. Of particular importance is a clear definition of the design leader's role in the co-ordination and integration of design tasks. Equally important is to specify a process of communication that the design leader will be required to follow, which includes reporting requirements and the format of a correspondence protocol. The project manager also determines an appropriate monetary limit to the professional indemnity insurance that the design leader would need to have. Furthermore, the project manager specifies what documention is required for the work, such as:

- drawings (architectural, engineering, services, interior and landscaping))
- specifications
- a bill of quantities
- a site investigation report
- schedules of materials and finishes.

The project manager also specifies quality standards for deliverables, such as:

- the size, quality and number of copies of prints of drawings
- the size and quality of models and perspective drawings
- whether drawings, specifications and a bill of quantities are to be produced electronically.

Finally, the project manager recommends to the client standard conditions of contract under which the design leader will operate, paying particular attention to:

- the determination of the fee (whether expressed as a fixed price or a percentage price)
- the terms of payment
- the mechanism for a fee escalation
- the payment of reimbursables.

Design consultants are commonly engaged under general conditions of contract prepared by their respective professional institutes. In Australia general conditions drafted by the Royal Australian Institute of Architects form the basis for the engagement of an architect. Increasingly, these conditions have been modified by clients to allocate more risk to the architect.

Common practice is for the client to engage a design consultant at the conceptual stage to assist with the preparation of the briefs. The engagement is short-term and is related to the provision of a specific service. If, at the end of briefing, the client is happy with the work, the client would normally award a new contract to this design consultant for the design and documentation of the project. Or the client may award a contract for design and documentation to another design consultant.

Engagement of a design leader is commonly based on direct negotiation, though select tendering has gained some popularity in recent years. But select tendering may be to the detriment of the client if the client places too much emphasis on price. Design contracts awarded on a price basis only may seriously erode the design and documentation quality, which in turn may give rise to claims for extras by the contractors.

There are many general criteria for appointment of design consultants. They would include:

- past experience with a design firm
- experience and reputation of a design firm and its staff
- capability and availability of a firm's staff
- the track record of current staff and principals of a firm
- a business approach by a firm
- creativity and innovation
- the use of technology such as CAD and IT
- the financial capacity of a firm
- the work on hand at present
- the ability to administer a project
- the level of fees
- a reputation for designing to a cost and meeting a schedule
- a reputation for quality.

After the design leader has been engaged, a project team, which now comprises the client, the project manager and the design leader, will select and progressively engage individual specialist design consultants. Other consultants such as a quantity surveyor, soil engineer, value manager or lawyer may also be engaged.

ESTABLISHING A DESIGN TEAM

With the required consultants on board, the design team is formed. It will operate as the project team for the duration of the design stage. The design leader heads the design team and has responsibility for accomplishing the required design tasks, such as co-ordination and integration of the design activities. The design leader reports to and liaises with the project manager, whose task is to manage the design stage and ensure completion of all design tasks in accordance with the brief and the approved schedule.

In managing the design stage, the project manager first formulates a design management plan and then ensures its implementation. Throughout the design stage, the project manager fosters and maintains good and effective teamwork between all the consultants.

DEVELOPING A DESIGN MANAGEMENT PLAN

Developing a sound plan of action is a prerequisite for achieving a successful outcome. The purpose of a design management plan is to set out clearly a strategy for achieving the best possible design and documentation that will meet the client's needs and the project's objectives, also to resolve any technical and co-ordination problems and to satisfy legal and environmental requirements. It is the project manager's responsibility to develop such a plan in collaboration with the design team (Chitkara 1998). Important steps in developing a design management plan will now be discussed.

Review of a project brief, a client's needs and project objectives

Members of the design team must first learn as much as possible about the project and its scope, and in particular about the client's needs and objectives. They must become familiar with the specific requirements defined in the brief. The project manager communicates the necessary information to the team and explains and clarifies decisions already made. In particular, the team needs to know what project objectives have priority. These will influence the development of a preliminary design concept and help to establish a realistic cost budget and time schedule. Let's assume that quality and aesthetics are the main objectives. The design leader would conceptualise them into an appropriate design scheme that would need to be unique, eye-catching, innovative or perhaps even daring. Such a design would almost certainly cost more and take longer to produce and build. Accordingly, a cost budget and a time-frame for design would need to be made more substantial.

Development of a design schedule

One of the main components of a design management plan is a schedule of the design work, clearly showing what is to be done, by whom, how long and when. The project manager and the design team develop this schedule through a series of meetings in which relevant issues are discussed in detail and decisions confirmed. They include (Chitkara 1998):

- breaking the design process down into elements from which a detailed list of activities is compiled
- identifying risks at the design stage, assessing their likely impact, and developing suitable mitigation strategies
- developing a preliminary schedule of design activities
- finalising a schedule in the form of a critical path program or bar chart
- allocating sufficient resources to the schedule
- obtaining commitment to the schedule from each design team member
- agreeing on regular updating of the schedule.

A completed design schedule is now ready to be used for organising work and controlling its progress. When presented as a critical path schedule, it gives information such as:

- the duration of design work, start and finish dates of each activity and specific milestones
- the identity of critical activities
- the amount of float, that is, the time by which non-critical activities may be delayed without delaying the completion of the whole schedule.

For information on critical path scheduling, see Uher (2003).

It is good practice to assign the names of team members to the activities for which they are responsible. This transparency about responsibilities improves understanding of who is responsible for what and provides a basis for effective co-ordination and integration of design tasks.

Firming up of a cost budget

A project cost budget was prepared at the conceptual stage, usually by the quantity surveyor. It was built up from estimates of costs of individual stages of a project lifecycle and an estimate of a contingency. A contingency sum is usually included to allow for unexpected items that will always arise as the project develops.

Estimating the cost of construction at the conceptual stage is always difficult because of the lack of specific design information. Estimates are generally produced using price-rate methods such as a per square metre rate, which unfortunately may introduce a substantial margin of error. A contingency allowance must be large enough to account for such an error. Although inaccurate, an estimate of the construction cost usually becomes the budget, which a design leader should not exceed in designing a project. It is the responsibility of the project manager to ensure that the project is designed to that budget. To meet this budget, the quantity surveyor is engaged to prepare periodic estimates of costs as more specific design information becomes available. The quantity surveyor costs the actual design at a particular point in its development and compares it to the budget. If the cost is over the budget, the design team will be required to modify the design to bring the cost back on budget. Unless specifically approved by the project manager, a design contingency should not be used to pay for an over-the-budget cost. If the cost is under the budget, the client may decide either to retain the savings or spend it in return for obtaining a higher design value, such as higher-quality finishes or faster lifts (Levido 1990).

The quantity surveyor prepares a final estimate of the cost of construction on completion of the design work and its associated documentation. If the cost is within the budget, a final estimate becomes 'the client's estimate' of the cost of construction, against which competing contractors' tenders are assessed. It may often become necessary for the project manager to authorise the use of some of the contingency at the design stage to pay for unforeseen events, such as client-initiated changes to the design. An unspent amount of contingency may be retained by the client or added to a contingency sum set aside for tendering and construction. If, however, a final estimate exceeds the cost budget and there is no contingency left, the client, on advice from the project manager, may either accept a higher cost, provided he has the capacity to pay for it, or he might order an appraisal of the design in order to bring the cost back on budget.

The ownership of a design contingency and its use may sometimes become a contentious issue since designers often erroneously believe that they own it. In fact the client owns a contingency and empowers the project manager to manage it. In managing a contingency allowance, the project manager applies a simple rule: a contingency can only be used to pay for the cost of risks, unforeseen events, and specific tasks that have not been fully defined in a brief. A design contingency should not be used to pay for design mistakes or over-the-budget costs unless specifically approved by the client.

Design co-ordination and integration

Design of a construction project is a creative process that brings together, depending on the size and complexity of a project, designers with varying expertise in fields such as architecture, structural engineering, building services, environmental engineering, and interior and landscape architecture. Since IT and security systems have become an integral part of design, experts in those fields may also be invited to join a design team.

Co-ordinating and integrating the activities of so many experts is probably the most difficult part of the design development. Here the project manager must provide effective leadership to the design team by promoting teamwork, maintaining open communication, and enforcing discipline in meeting scheduled target dates. Only then can the project manager and the design team be assured of the completeness and accuracy of the design and its documentation. To this end, the project manager will constitute weekly meetings with the team. Apart from providing an excellent platform for promoting openness and teamwork, these meetings are used to exchange information and to resolve design issues and problems. They also review progress of the work against the plan (CIOB 2002). Between meetings the project manager may need to discuss with the client any matter brought up at a meeting that impinges on the design brief.

Another challenge for the project manager and the design team is to cope with changes to the project scope. These commonly arise when the client's requirements change or when an unforeseen event not considered at the conceptual stage has emerged. Scope changes may also be symptomatic of ineffective management of the conceptual stage, indicated by an incomplete or vaguely framed brief. This risk should be reduced significantly by the presence of a professional project manager as leader of the project team.

Scope changes could add significantly to a project cost and schedule, and must therefore be kept to the absolute minimum. The project manager's responsibility is to make the client aware of the impact of scope changes on cost and schedule. This is a highly effective strategy that reduces the incidence of client-initiated changes. But no matter how hard the project manager may try to eliminate them, some scope changes are bound to occur, and a contingency allowance to cover them should be made in the budget.

Fast-tracking

A project may be developed by advancing it sequentially throughout its individual lifecycle stages. This is the most commonly used approach. When some of the stages, most commonly design, tendering and construction, are overlapped, a substantial saving in the overall project period can be achieved. This is commonly referred to as 'fast-tracking'. The aim is to reduce the overall development period by starting construction earlier, with design and documentation only partially completed. By shortening a period, the client is able to commission the project earlier, which usually has significant financial benefits (Uher & Davenport 2002).

Fast-tracking a project is, however, a rather risky proposition. Changes to or errors in the design of those parts of a project already under construction are likely to cause serious cost and time overruns. Managing a fast-tracking project requires a highly disciplined approach to the formulation of a brief, which must remain unchanged, and to the development of the design and documentation to a strict schedule.

Packaging the work

Closely associated with fast-tracking is 'packaging'. This is the process of breaking up the construction work into a number of reasonably large parts which are individually designed. The decision on fast-tracking and packaging is made at the conceptual stage and is defined in the project brief.

The aim of packaging is to design and document each segment or package of a project separately and build it separately. Once the design of a first package has been completed, it is let to a contractor who starts building it. When the design of the next package is completed, it will be let to another contractor, and so on.

Fast-tracking and packaging are integral features of the construction management and the project management methods of project delivery (see Uher & Davenport 2002).

How a project is broken up into packages largely depends on its size, type and complexity. It may also be influenced by the personal preference of the project manager. For example, a high-rise office building may be divided into six packages, each let sequentially to a separate contractor:

- package 1: ground works (demolition, excavation, underpinning, de-watering, temporary access roads)
- package 2: structure (footings, slabs, beams, columns, stairwells, lift shafts, external walls)
- package 3: services (electrical, mechanical, fire protection, hydraulics, IT)
- package 4: façade and roof
- package 5: internal walls
- package 6: internal finishes.

In this particular case, the packages are awarded to separate contractors in a predetermined order dictated by the logic of construction. The actual construction sequence adopted defines the parameters of a design schedule. In managing the design stage, the project manager's task is to adhere strictly to the design schedule to ensure delivery of each package on time.

The example above refers to packaging within a single structure. With six contractors operating within the same structure, it is simply impossible to accurately apportion responsibility for preliminaries, safety, industrial relations, and materials and personnel handling to any one contractor. Such responsibilities are therefore commonly assigned to the construction manager or the project manager.

Projects with several separate structures may be packaged differently. In such cases, each separate structure may become a package. For example, a new thermal power station project may be broken up into the following packages:

- package 1: the administrative building
- package 2: a structure housing thermal generators
- package 3: chimney stacks
- package 4: cooling towers
- package 5: the rail or road access for fuel and its storage
- package 6: the transformer and the power grid.

For the above project, the packages may not necessarily be let in any particular order and it is also possible that some of them will be let concurrently. Each package becomes a self-contained contract with a contractor responsible for preliminaries, safety, industrial relations and the like.

Packages are usually awarded to contractors by competitive tendering on a lump-sum basis. This means that the design of each package must be accompanied by full documentation. It is the project manager's responsibility to ensure that the tender documentation for each package is complete and accurate.

Design appraisal

The design and documentation of a project takes a considerable time and has a tendency to influence project outcomes. The importance of effective management of the design stage has already been emphasised.

Design is a creative work that requires designers to formulate the best possible solutions to a set of specific requirements. Apart from meeting technical and commercial requirements, designers are also required to be innovative in creating aesthetically pleasing and unique structures. In searching for new ideas, designers sometimes venture beyond the boundaries of their current knowledge and experience. Some are willing to take a risk in pursuing a unique design scheme, hoping to make it work. The danger is that such an approach, whether or not it achieves the desired outcome, may exceed the project budget. There are plenty of examples of projects with unique and highly innovative design features, whose high architectural value was achieved for a substantially higher cost. The Sydney Opera House is probably the best example of such a project.

To ensure that designers develop schemes that satisfy the client's needs and requirements and meet stipulated project objectives, yet at the same time search for innovative and aesthetically pleasing solutions, it is necessary to provide them with a tool that appraises and is capable of adding value to their design. Such a tool is the concept of 'value management'. Value management was defined by Connaughton and Green (1996: 6) as 'a structured approach to defining what value means to a client in meeting a perceived need by establishing a clear consensus about the project objectives and how they can be achieved'. The treatment of value management in any depth is outside the scope of this book; information on it can be found in Kavanagh and colleagues (1978: 235) and McGeorge and Palmer (1997: 11–52). Only the key elements of value management will now be briefly discussed.

Value management is a disciplined approach to appraising design to ensure that it meets project objectives and functions in the most economical way. It is conducted by a structured workshop facilitated by a professional value manager and involves members of the design team, the client, the project manager and the quantity surveyor.

A value management study is not aimed at appraising the effectiveness of designers; rather, it focuses on the effectiveness of design outputs. It is commonly performed at the schematic and preliminary design phases. The cost of staging a value management study is around 0.1–0.3 per cent of the construction cost, which is insignificant in comparison to the cost savings of around 10 per cent that such a study commonly achieves.

The fundamental approach of value management is to create a work breakdown structure or a value tree structure of a design problem by separating objectives into sub-objectives, which are further subdivided into relevant elements and sub-elements. Each objective, sub-objective, element and sub-element in a value tree is then weighted. Although the assessment process is subjective, quality decisions are reached through consensus between participants in a workshop by addressing the following questions (Kavanagh et al. 1978: 240):

- What is it?
- What does it do?
- What does it cost?
- What else would work?
- What does that cost?

Effectiveness of design options is best assessed using a design evaluation matrix (Tucker & Scarlett 1986), which comprises:

- criteria: these are elements to be measured
- weights: they determine relative importance of the criteria
- performance scale: it compares the measured value of the criterion to a benchmark value
- performance index: it determines performance.

A simple example (Figure 6.2) demonstrates the use of a design evaluation matrix. Design criteria are listed in the upper part of the matrix, and a performance scale from 0 to 10 located in the middle and the bottom part of the matrix gives performance scores, weights and values. Each criterion's value is obtained by multiplying the score by its weight. The sum of the values then gives the performance index. If, for example, the minimum acceptable performance score for a design option in Figure 6.2 is set at 6 for each design criterion, the minimum value of the performance index would be 510. The appraisal score of the design option in question is clearly well below its expected performance.

This method of design appraisal can also be used to compare different design options as well as to compare designs of different projects.

A process of value management also assists in overcoming a problem associated with lack of buildability. Designs are often criticised for neglecting buildability. This is particularly important for projects delivered by the traditional method, when contractors bid on full documentation. The manner in which the contractor builds the project is largely defined by the design. For example, if the structural design requires joints between concrete slabs and the service core to be monolithic, the contractor would be unable to construct the service core using either slip-forming or jump-forming, techniques that the contractor may prefer to use on the grounds of speed and superior overall productivity

Monitoring and control

The project manager maintains progress by closely monitoring and controlling design activities through regular meetings with members of the design team. Apart from assessing performance and progress, these meetings provide a forum for identifying issues and problems and developing appropriate solutions. The project manager also regularly reports to the client on progress achieved.

Figure 6.2 A design appraisal matrix

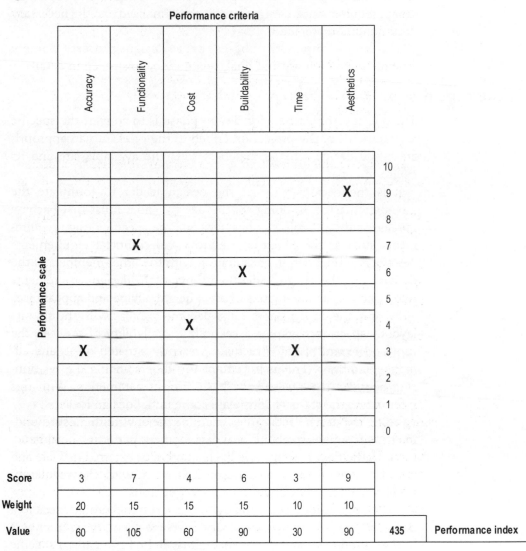

SOURCE Adapted from Tucker & Scarlett 1986.

DESIGN PROCESS

The process of design development starts by a detailed analysis of a brief, from which an initial design scheme is formulated. It progresses to a preliminary design phase where the scheme is firmed up and confirmed as the most appropriate solution to the stipulated design requirements. Design work is then finalised and the necessary documentation prepared.

The three important phases of a design process, namely schematic, preliminary and final, will now be examined in detail.

Schematic design

The objective of a schematic design phase is to convert the specific requirements of the project described in the brief into an appropriate initial design scheme that best suits the available site and its environment.

It is the responsibility of the design leader to formulate the schematic design. In doing so, the design leader must first become familiar with the project's objectives and the specific design requirements such as the functions, services and technical requirements described in the brief. In consultation with the other members of the design team, the design leader then converts the design requirements into basic project parameters of area, height, shape and appearance.

Second, the design leader must become familiar with the site: its layout, aspect, topography, accessibility, availability of services, the type and extent of any structures presently situated on it, and climatic conditions. This is important for understanding the relationship between the site and the basic project parameters. With this information, he starts to formulate some initial design ideas.

Third, the design leader must examine the environment surrounding the site and learn about local development patterns, social structures, traffic flow patterns and likely sources of external risk. He and his design team must also become familiar with the regulatory framework, including environmental regulations.

The design leader is now ready to convert initial design ideas to a design scheme. Relying on feedback from the client, other members of the design team and the project manager, he formulates a specific design scheme and prepares sketch drawings that show the basic outline of the project, its size, shape, height and orientation, as well as the position of the project on the site and access to the site. Throughout this phase, the project manager monitors and controls operations and ensures that project objectives are met.

It is now time to conduct a first value management study of the schematic design to verify that the proposed design scheme is indeed the best design concept for the project in terms of meeting requirements and objectives and providing the best value for the client.

The client's acceptance of the proposed design scheme represents a decision to proceed by lodging an application for the development approval (DA) with the local council.

The development application consists of a detailed description of the project's purpose and function, sketch drawings of the proposed design scheme, and any other information required by the local council such as an environmental impact statement (EIS). The EIS states the level of impact that the proposed project will have on the surrounding environment.

The local council determines if the development scheme in question meets all the urban planning laws and regulations and its own planning requirements. It also determines whether the proposed project's design is sympathetic to the local environment and verifies that the site is correctly zoned for the project.

The council will either accept or reject the development application. Acceptance is frequently made conditional on meeting specific requirements. For example, the client may be required to dedicate a part of the proposed project to the local community in the form of a child care facility, a library or just open parkland. The client may also be required, for example, to widen the road surrounding the project, dedicate a part of the site for parking, or even upgrade some community facilities such as parks and playgrounds. Needless to say, conditional approvals are likely to incur extra cost and may also extend the project schedule. Securing the DA may, depending on a council, take many weeks. Since this activity poses the risk of delays and associated extra costs, an appropriate contingency must be made for the likely impact of such a risk.

To mitigate this impact, the project manager and the design team should, before submitting a development application, obtain as much intelligence about the local council as possible, for example:

- details of the local planning regulations
- an analysis of the past DA decisions of the local council
- information on the consistency of the local council's DA decisions
- the speed of processing applications.

Whenever possible, the project manager and the design team should seek advice directly from the local council on how best to prepare a development application.

If the DA is rejected, the client may either abandon the project or consider revising its development strategy. The client may also consider appealing the decision of the local council to the Land and Environmental Court. Win or lose, a substantial delay and an extra cost are likely to occur.

Granting of the DA is a major milestone in the project development process. The design will now move towards its next preliminary phase.

Preliminary design

The aim of a preliminary design phase is to formalise the approved design scheme into a coherent concept that includes more detailed sketches of architectural, structural and services components. The design team appraises structural systems and prioritises construction techniques, selects materials and verifies performance requirements. The outcome of the preliminary phase is a spatial solution of design requirements and the development of an exterior design of the project. Because design elements now show more details, the quantity surveyor is able to prepare a more accurate estimate of the cost of construction.

It is now time for the project manager to engage the remaining design consultants and induct them into the design team. The project manager's role in the preliminary phase is to closely co-ordinate the information and activities of different design consultants and control performance in terms of cost, time, functionality and buildability.

Projects delivered using the traditional method are expected to be fully designed and documented before tender and construction. This, however, is rarely achieved in practice. In fact most projects go out for tender on partial documentation only. There are several reasons for this. Often the client delays the definition of some project requirements such as fixtures and fittings until the project is well into its construction. Sometimes the design stage period is insufficient for completion of design and documentation. In such cases the project manager and the design team will leave out of the design some parts of the project that are not needed in the initial period of construction. These will be designed and documented later. It may also be the client's deliberate decision to remove some parts of the project from the design in order to reduce the length of the design stage.

It is therefore necessary for the project manager and the design team to decide at the end of the preliminary design phase what parts of the project, if any, will be left out of design and documentation. For these omitted parts, the quantity surveyor will formulate estimates of cost in the form of a provisional sum (PS) to be inserted into the contract for each part that will be defined and formally priced later in the final design phase.

Another important issue to be considered at this point is the likely impact of the design on operating and maintenance costs of the completed project. The Sydney Opera House is an example of a project with large operating and maintenance costs that far exceed the original cost budget. Staging another value management study may help to fine-tune the design and ensure its low impact on operating and maintenance costs.

The approval of the preliminary design by the client moves the design process to its final and most critical phase — the development of final design and documentation.

Final design and documentation

With the preliminary work on design completed, the design will now be finalised across architectural, structural, services, interior and landscape tasks, and design documentation produced. This is the crucial phase of design development. Different design tasks must now be brought together, co-ordinated and integrated into a design with error-free documentation. The final design is expected to satisfy the project objectives and meet the required technical, functional and aesthetic requirements, and to comply with all the appropriate codes and regulations. The outcome of the design stage is a set of fully co-ordinated and integrated design documents that include:

- working drawings
- a schedule of materials and finishes
- specifications of materials and equipment
- a bill of quantities
- general conditions of contract.

Effective management of the final design phase is of the utmost importance considering the extent of work to be accomplished within a limited time-frame on large projects. Hundreds of working drawings will be produced, showing all the information that a contractor requires for construction of the project. There will also be thousands of pages of specifications, describing in detail types and qualities of materials and equipment used in the project. This information must be complete, unambiguous and accurate.

It is the responsibility of the project manager to manage the design stage effectively and efficiently. Apart from keeping to the schedule, an equally important task for the project manager is to implement a strategy for co-ordination and integration of design information to minimise the incidence of errors, omissions and ambiguities in the documentation. He must also keep the cost to its budget.

The general conditions of contract under which a contractor will build a project is another important document that must be prepared at the design stage. It defines the legal obligations of client and contractor, describes how the contract will be administered, and allocates risk. These general conditions must be closely aligned with a delivery method that has already been defined and agreed to at the conceptual stage. The preferred strategy is to use a suitable form of standard general conditions. Clauses in standard conditions of contract provide more certainty in meaning, since they will have been tested many times in courts of law. However, special conditions are often added for specific or unique circumstances. The preparation of conditions of contract is best left to a legal consultant, with whom the project manager will liaise closely.

The last document that is commonly produced is a bill of quantities. Although not used throughout the world, a bill of quantities is popular in most Commonwealth countries, including Australia. The production of this document depends on the completion of working drawings and specifications. If their production is delayed, the quantity surveyor would need to accelerate the rate of work in order to complete the bill of quantities on time. This is a common problem that often reduces accuracy of the bill.

With sufficient documentation available, the design team prepares an application for a building approval (BA) to the local council or to a private certifier.

The BA process ensures that the design of the proposed project complies with building codes and regulations, particularly in the areas of structural integrity and fire safety. The probability of failing to secure the BA is minimal, but approvals are rarely unconditional. Special conditions imposed may, for example, restrict hours of work on the site or limit deliveries of materials to specific times. A tree preservation order may even force a redesign of the project to accommodate existing trees on the site.

An application for the BA consists of:

- a formal letter of application
- architectural drawings
- structural drawings
- a specification
- any other information required by a local council or a private certifier.

Before the project moves to the next tendering stage, the quantity surveyor will prepare a final estimate of the cost of construction with clearly defined amounts of PC and PS items. This final cost estimate will be used as a benchmark for assessing tenders submitted by bidding contractors. The quantity surveyor will also assist the project manager in formulating the amount of contingency set aside for construction. This contingency must be able to accommodate additional costs arising from variations, cost escalations, time extensions, and rise and fall claims.

CONCLUSION

The development of design and documentation for a particular project is a co-ordinative and highly integrative process that translates the client's needs and objectives expressed in a project brief into a specific design scheme that satisfies functional, technical, aesthetic, economic and environmental requirements.

The project manager is responsible for the effective management of the design process. This requires the appointment of design consultants, the establishment of a design team, the development of a design management plan inclusive of a schedule and a cost budget, and the co-ordination and integration of design activities across the schematic, preliminary and final design phases. The project manager and the design team appraise the design using value management to ensure that it is of the highest quality, is functional and buildable, and represents best value for the client.

The outcome of the design stage is the production of quality design documentation. It will be used by bidding contractors to formulate the cost of construction in the next, tendering stage of a project lifecycle.

EXERCISES

1   Try to recall the last project that you have been involved in. Who was responsible for the management of the design stage? Was the design stage well managed? If not, what went wrong? Did the project suffer cost and time overruns as a result of a poorly managed and documented design stage?

2   What are the typical activities of the design stage? Do they vary for different options of project delivery?

3   Both the design leader and the contractor enter into a contract with the client to provide specific design and construction services respectively. But the design leader bears little or no risk when paid on a percentage fee basis, while the contractor, who is paid on a lump-sum basis, bears

the full brunt of the construction risks. Is this difference in the way the design leader and the contractor are engaged justified? If so, on what grounds?

4   What are the potential benefits to the client and the project team of engaging the design leader on a lump-sum fee basis? Should the design leader be responsible for the errors, omissions and ambiguities in the design documentation that commonly give rise to variations?

5   Give a list of issues for consideration in selecting a design leader.

6   Is it necessary for a design team to be managed? Who should be responsible for the management of a design team?

7   What are the main components of a design management plan?

8   What is fast-tracking? What is the relationship between fast-tracking and work packages?

9   Do you agree that it is necessary to appraise design using a concept of value management? How can value management improve the efficiency of design consultants?

10 What is the purpose of securing the development and building approvals? When are applications for such approvals made?

11 You are a project manager in charge of a design team. What specific strategy would you put in place to ensure that the design process is fully co-ordinated and integrated? Explain in detail.

CHAPTER 7

MANAGING TENDERING

INTRODUCTION

Completion of design documentation moves a project development process to the tendering stage. The purpose of tendering is to select a best tender price for a project from bidding contractors.

Depending on the extent of design documentation, a contractor may be required to submit a tender for a whole project or a part of it. The latter case refers to tendering for a 'package'. In both cases, a process of selecting a winning tenderer is the same.

Not all construction contracts are formed on the basis of tendering. Some are entered into by negotiation. The client may negotiate a contract directly with a contractor with whom the client prefers to work. In the client's opinion this contractor has the expertise, reputation for performance and the necessary resources to build a project successfully. Direct negotiation may also save the client a considerable amount of time that otherwise would have been spent on tendering. However, a contract price that has been negotiated directly between the client and the contractor is likely to be at a premium.

Another approach that clients sometimes use is to negotiate with two or three low tenderers for a price after the conclusion of tendering. While not illegal, this practice may be viewed as contravening the principles of competitive tendering. By playing one contractor against the other in the post-tender negotiation, the client is likely to get a more competitive price that will be lower than the tender prices. This practice is commonly referred to as 'bid shopping'. While delivering a lower price in the short term, bid shopping is undesirable to the client in the longer term. This is because the contractor will attempt to recover the profit lost by bid shopping, by applying the same bid-shopping practice to subcontractors or by 'working the contract'. The latter practice refers to exploring errors, omissions, discrepancies and ambiguities in the contract as a means of raising claims against the client. Neither of the above defence mechanisms benefits the client. By having their profitability lowered, subcontractors may respond by lowering the quality of work. Low profitability may even force subcontractors out of business. By 'working the contract' the contractor will increase the frequency of claims, which in turn will increase the cost of a project.

Contracts are most commonly formed by tendering. The contractor's tender represents an offer. If accepted by the client, a contract between the contractor and the client is formed, provided other legal principles such as intention, consideration and capacity of parties are satisfied.

Tendering is a competitive process. It brings together a number of bidders who compete against each other. Each bidder's objective is to win a job by offering as competitive a price as possible. Tendering is eloquently described by Flanagan (1986: 1) in the following terms:

> Competitive tendering confers undoubted benefits on all organisations in the building industry. It reconciles the potentially conflicting desire of the client for the most competitive price, and of contractors for the best attainable profit. In addition, the data generated by competitive tendering are useful to those individuals in the building industry who are involved in estimating future construction prices or forecasting future trends.

This chapter first reviews tender documentation, after which it examines a tender process that leads to the selection of a best bid. Note that the terms 'tender' and 'bid' are synonymous and are used indiscriminately throughout this chapter.

TENDER DOCUMENTATION

Documentation required for tendering commonly comprises:

- a notice to tenderers
- conditions of tendering and a form of tender
- general conditions of contract under which a contractor will build a project
- working drawings
- specification(s)
- a bill of quantities
- any other relevant documents.

The project manager manages the tender stage and recommends to the client a competent contractor with whom the client should enter into a contract. The first important task of the project manager is to finalise tender documentation (working drawings, specifications, a bill of quantities and general conditions of contract), most of which should already have been completed at the design stage. What remains for the contractor is to formulate conditions of tendering and issue a formal notice to tenderers. Let's now briefly examine individual tender documents.

Notice to tenderers

The project manager drafts a suitably worded text of a notice to tenderers. It informs them of an opportunity to either tender for a particular project, or express interest in tendering, or pre-qualify for future tendering. A notice may be in the form of a newspaper or internet advertisement, or a letter written to selected tenderers. If it is a newspaper advertisement, it is unlikely to include conditions of tendering. Bidding contractors would need to obtain these from the project manager on demand.

Conditions of tendering

Conditions of tendering specify operational and administrative requirements that bidding contractors must comply with. In drafting conditions of tendering, the project manager gives particular attention to the following issues:

THE TIME AND THE PLACE FOR THE SUBMISSION OF A TENDER

This information must clearly be defined since late submissions and submissions made to a different place will not be accepted.

The quality of submitted tenders is not only related to the completeness and accuracy of tender documentation but also to the length of time given to contractors for tendering. If insufficient time is given, contractors have limited time to closely scrutinise tender documentation and may be unable to uncover errors, omissions and discrepancies, which are almost always present. If uncorrected now, they are unlikely to be discovered until the project reaches the construction stage, when they become costly variations. A short period of tendering limits the extent of contractors' risk assessment. It may also prevent them developing and assessing alternative construction strategies.

There is no simple answer to how much time is appropriate and adequate for tendering. Some undesirable outcomes arising from a short period of tendering have already been discussed. Having more time allocated for tendering may also lead to problems. For example, an unscrupulous contractor might intensify 'bid shopping' of subcontractors or might even engage in some form of collusion with other bidders. Another issue to be considered is an increase in an overall project period and, since time is money, an increase in project cost. There are no hard and fast rules for determining an appropriate period of tendering. Such a decision will, in most cases, be influenced by the past experience of the project manager and other team members.

A decision on whether or not to accept late tenders is another contentious issue. As a general rule, late tenders should be rejected on the grounds of fairness and equity. However, in the absence of explicit rules governing late submissions of tenders, some project managers could find it difficult to reject a highly competitive tender, which has been submitted late, if it could save a substantial amount of money to the client. It is therefore necessary for the project manager to prepare such rules, articulate them to bidding contactors and, most of all, comply with them.

A FORM OF TENDER

Since a contractor's tender represents an offer, it is necessary for the project manager to specify, in conditions of tendering, the manner in which a contractor should make such an offer so that it forms a basis of a future contract. Many examples of a form of tender are readily available. They are commonly included in general conditions of contract. One specific example of a form of tender used in conjunction with an AS2124:1986 contract is given in Figure 7.1 (AS 1986: 47). Another example of a form of tender can be found in Horgan (1984: 194–5).

A PERIOD DURING WHICH A TENDER REMAINS OPEN

The usual period set aside for the evaluation of tenders is 30 days. The client expects that bidders will not withdraw their tenders during this period. However, unless bound by specific legal provisions (see Uher & Davenport 2002: 174), bidders are free to withdraw tenders that have not yet been accepted. What this means is that a period during which a tender remains open has been set aside not for bidders but for the client to select a winning tender. If this period is exceeded, offers lapse.

Bidding contractors should be given information about when the decision on a winning tender will be made.

Figure 7.1 A form of tender (adapted from SA 1986: 47)

Name of person, firm or company tendering	....
Address	of ....
	hereby tender(s) to perform the work for ....
Description of works	....
	............................(Contract No............................) in accordance with the following documents:
	....
	....
	....
List documents	....
	....
No general description	....
suffices	....
When the tender documents provide that the tender is to a lump sum only, (2) does not apply.	1. For the lump sum of .... ($............................); and
When the tender documents provide that the tender is to be a Schedule of Rates only, (1) does not apply.	2. At the rates in the attached Schedule of Rates.
If the tenderer is a firm, the full names of the individual members of the firm must be stated here.	....
Insert date.	Dated this ....day of ....
	....
	Signature of Tenderer

HOW WILL A WINNING TENDER BE SELECTED?

Although contractors may expect the project manager to recommend to the client to award a contract to the lowest tenderer, the client may decide to award a contract to any or no tenderer. To avoid accusations of favouritism and of using tenderers to merely establish a market price of a proposed project without any intention of developing it, the project manager must clearly specify the grounds on which a winning tender will be selected. This may also prevent the possibility of a legal claim made by an unsuccessful tenderer against the client. Such a claim may arise in situations where the lowest tenderer complies fully with the conditions of tendering but his tender is rejected. This may be because the client has adopted as a basis for acceptance of a tender some criterion that was not made known to the tenderer. An unsuccessful tenderer may be able to sue the client for breach of contract because conditions of tendering are regarded as forming a part of a contract. It is therefore important for the pro-ject manager to inform tenderers of all the criteria on which tenders will be judged. This is mandatory in Local Government contracting in New South Wales (Local Government (Tendering) Regulations 8(3)(b)). Clients who have published their own codes of tendering must ensure that their tendering processes comply strictly with such codes. The failure to comply may lead to litigation instigated by unsuccessful tenderers.

THE START DATE OF A CONTRACT

A contract between the client and the contractor is formed when the client has accepted the contractor's offer. A start date of the contract is commonly referred to as a date for site possession. It is commonly the date when the contractor gains access to the site and is required to start construction. Bidding contractors must be given this date so that they know how much time they have available for detailed planning of the project and mobilising of resources.

WHETHER OR NOT NON-CONFORMING TENDERS WILL BE CONSIDERED

When a number of contractors tender on full documentation, each is expected to price the same design. Assessment of such tenders is relatively simple since it is largely based on price.

Some contractors, however, may recognise that substantial cost and/or time savings could be achieved by altering an existing design. They may decide to submit to the client two tenders: a conforming tender and a non-conforming tender, based on an altered design, which offers cheaper or faster construction, or both. Although showing substantial cost and time savings, it must be remembered that a non-conforming tender is based on a different design. Consequently, comparing conforming to non-conforming tenders on price only may be extremely difficult. For example, a non-conforming tender may offer a cheaper construction cost but its life-cycle cost may be substantially higher.

From the moral point of view, is it fair and reasonable, in the absence of clearly defined rules on acceptability or otherwise of non-conforming tenders, to pass over the lowest conforming tenderer in favour of a cheaper, non-conforming tenderer? And what will be the likely consequence of such a decision? One potentially damaging consequence is the likelihood of legal action taken by the lowest unsuccessful tenderer against the client to recover the cost of tendering and the foregone profit. A prudent project manager must minimise the possibility of such legal action by clearly specifying a policy on non-conforming tenders.

WHETHER OR NOT A BID BOND IS REQUIRED

In some countries, for example the United States (but not Australia), contractors and subcontractors are commonly required to provide a bid bond at the time of tendering. The purpose of a bid bond is to prevent contractors and subcontractors from withdrawing their tenders during a period when tenders are being evaluated by the project manager. A bid bond also ensures that contractors and subcontractors would enter into a contract if one was offered. If a contractor withdraws a tender, which causes some loss to the client, the surety company will be liable to pay the amount of damages stipulated in the bond itself to the client. Under bid bonding, contractors and subcontractors are more reluctant to withdraw their tenders for fear of either paying a higher premium for future bid bonds or, in extreme cases when regarded as a bad risk by a surety company, being denied a bid bond and therefore being unable to tender for future contracts.

Conditions of contract

In order to prepare a competitive tender that reflects the true nature and value of a project to be constructed, a bidding contractor must know what rights and obligations will exist under the contract and the extent of risk a contractor would be required to carry.

It is the project manager's responsibility to ensure that conditions of contract are made available to bidding contractors. They are normally prepared in the design stage and fine-tuned at the start of the tendering stage. Although the legal consultant plays the leading role in drafting conditions of contract, the project manager's task is to ensure that while they reduce the client's exposure to risk, they are fair and equitable to the contractor. General conditions of contract were discussed briefly in Chapter 6. More information can be found in Uher and Davenport (2002).

Working drawings

Working drawings were prepared at the design stage. They show graphically and in detail the work to be undertaken. They are produced in the form of plans, elevations and sections to show the structure of a project, its architectural features, and details of services.

The project manager must ensure that a full set of working drawings is available for each tenderer. This requires printing enough sets of drawings to satisfy all bidding contractors. Printing of multiple sets of drawings is time-consuming and requires sufficient time to be set aside for it. It also involves cost that must be budgeted for. Alternatively, working drawings could be distributed to bidding contractors electronically.

Specification(s) and a bill of quantities

Like working drawings, specifications and a bill of quantities are prepared in the design stage. They must now be duplicated and made available to tenderers. Electronic distribution has become common practice in recent years.

Other documents

There may be other documents in the form of letters, sketches or technical reports such as a soil investigation, which may form a part of the contract. It is the responsibility of the project manager to ensure that any relevant information is made available to tenderers.

CLIENT'S COST ESTIMATE

At the end of the design stage the quantity surveyor calculated a final estimate of the cost of construction. This cost estimate represents a cost benchmark against which submitted tenders are assessed. The project manager in liaison with the client defines the limit of acceptable variability of submitted tenders from the client's estimate. The upper limit is set strictly within the bounds of a project cost budget.

In setting the lower limit, the project manager and the client must safeguard against erosion of a project's value when accepting a tender that is lower than the client's estimate. For example, a limit of +10 per cent of the client's estimate may be deemed acceptable because it is within the total cost budget. However, a tender that is, say, more than 7 per cent below the client's estimate is likely to be rejected on the ground that it returns too low a profit to the contractor, which in turn increases the probability of low quality standards and potential cost overruns. Clearly, in such a case the value of the project could be seriously diminished.

TENDER PROCESS

The tender process, which ends in the award of a main contract to the contractor, is complex and generally follows four distinct steps (Uher & Davenport 2002: 184–6):

Step 1:

Subcontractors and suppliers are invited by bidding contractors to prepare tender prices for specific project activities and materials. The structure of a bill of quantities allows a relatively easy break-up of a bill into specific trade sections for use by bidding subcontractors. A bill of quantities is also highly suitable for use by suppliers of materials.

Subcontractors and suppliers of materials prepare tender prices and quotations and submit them to bidding contractors.

Step 2:

A contractor compiles a tender either from prices submitted by subcontractors or from quantities in a bill. A contractor then submits a tender to the client.

Step 3:

The project manager evaluates submitted tenders and recommends to the client a tenderer with the best tender price. The client notifies the best bidder of acceptance and in doing so awards a contract to the winning contractor.

The term 'best tender' has different meanings. Some clients relate it to the lowest tender, some interpret it as the most competitive tender, while to others it probably means something else. Selection of a best tender will be discussed briefly in the following section.

Step 4:

After the award of the main contract, the winning contractor selects subcontractors and suppliers from those who have supplied bid prices to the contractor during tendering. The contractor is now in a strong position to negotiate competitive prices with subcontractors and suppliers. If the client forced bidding contractors to lower the bid price through bid shopping, it is quite likely that the winning contractor will apply the same bid-shopping practice to subcontractors and suppliers to recover the foregone profit. While it is not illegal, bid shopping is unethical and increases the client's risk. While the project manager may be unable to prevent the contractor from applying bid shopping to subcontractors and suppliers, he should discourage the client from initiating bid shopping in the first place.

TENDERING METHODS

The project manager may administer tendering using either an open or a select process. Let's examine these two methods.

Open tendering

Open tendering refers to bidding without restriction. Any suitably qualified bidder may tender for the work. Open tendering is traditionally applied to public projects to ensure fair accountability of public funds and to avoid accusations of favouritism. But in recent years the reliance on open tendering by government clients, particularly in Australia, has diminished, while select tendering has gained prominence.

Open tendering generates a large number of tenders. This is seen as an advantage to the client since submitted bids at the lower end tend to be highly competitive. However, variability of tender prices is often very large, which makes the selection process of a best bid a rather difficult and arduous task. Processing a large number of tenders is also time-consuming and costly. The overall cost of open tendering is obviously high.

Select tendering

As the name implies, select tendering is bidding by a limited number of bidders who are invited to tender. The Government Architect's Branch of the NSW Department of Public Works and Services has been operating a select bidding system since 1967. According to Thomson (1980) the select system has succeeded on the following grounds:

- The risk of failure of contractors and subcontractors has been reduced.
- Contracts by public authorities can be better regulated across the industry.
- The continuity of work for contractors and subcontractors has been improved.
- Costs of bidding have been reduced.
- The quality of work has been improved.

The project manager would normally invite between three or four tenderers to bid under a select tendering system. Or he may invite a large number of selected contractors to register interest to bid for a particular project. Based on the quality of submissions, he would then invite three to four contractors to prepare tenders. For more information on open and select tendering, see Uher and Davenport (2002: 183–4).

SELECTION OF A BEST TENDER

Competitive tendering encourages competition among bidders, and in so doing presents a range of prices from which the best tender is selected. The project manager's responsibility is to state clearly in the conditions of tendering the basis for selecting a winning tender; this may be as simple as 'any or no tender will be accepted'.

What 'best tender' represents is subject to different interpretations. It may be interpreted as the most economical tender, which is defined as a price at which it is economically sensible for the contractor to execute the work and for the client to invest in the development. In searching for the most economical tender, the project manager would need to consider a wide range of issues, broadly grouped as technical and commercial. Each such issue is then rated for each tender to produce an objective ranking measure of tenders.

'Best tender' could also mean the lowest tender. However, the lowest tender does not always result in the cheapest cost, particularly when a project is exposed to a high level of risk. A contractor on a low profit margin may have insufficient resources to control risk effectively. Delays and cost overruns then become inevitable (Bromilow 1970; Flanagan 1980).

Some project managers and clients tend to reject the lowest tender on the ground that it may represent too high a risk to the client by virtue of being the lowest. Under the open tendering method the project manager may be justified in rejecting the lowest tender particularly when:

- many bids have been submitted
- the variability between the lowest and the highest tenders is large
- the lowest bid is well below the client's estimate.

However, rejecting the lowest tender just because it is lowest under select tendering is difficult to justify, particularly when bidding contractors have been pre-selected. Rejecting the lowest tender under select tendering may cause a legal dilemma for the client. Cullen (1997) describes cases where courts have awarded damages to tenderers whose tenders have been passed over.

When the project manager rejects the lowest bid because it is lowest under select tendering, the question arises whether an unsuccessful lowest bidder should be compensated for the cost of tendering. There are valid arguments in favour of compensation payments (Collier 1969; Uher 1988), but the construction industry continues to be largely resistant to this rather radical idea.

'Best tender' may also mean the most innovative tender. For example, the competitions to select a design proposal for the Sydney Opera House and for the Australian Parliament House in Canberra were won by the most innovative design ideas, with little regard to their cost.

To avoid confusion as to the meaning of 'best tender', the project manager should qualify the terms in conditions of tendering so that bidding contractors know in advance the basis of tender selection.

The rejection of all tenders is a serious move on the part of the client. It is justified only if submitted tenders are well in excess of the client's estimate. However, calling for tenders to either ascertain a market price for the proposed development without any intention to proceed with it, or using the best submitted tender as a basis for 'negotiation' with a preferred contractor are unethical and unacceptable practices.

The manner in which acceptance of an offer is communicated to the contractor is important. In a legal sense, when acceptance is communicated to the contractor, a contract is formed and these two parties are in a legally binding contract. Acceptance must be unconditional. It means that it must not change any aspect of the offer. If it does, it becomes a counter-offer and it would be left for the tenderer to accept it. Only then would a contract be formed.

There is one other important aspect related to the manner in which acceptance is communicated. Since acceptance constitutes the formation of a contract and the start of a long-term relationship between the client and the contractor, it is vitally important that it is communicated by the client in person, preferably through a face-to-face informal meeting. This is a powerful gesture, rich in trust and openness and capable of creating an effective contractual relationship.

CONCLUSION

This chapter examined the role of a project manager in tendering. It reviewed important tender documents and stressed the importance of the client's estimate in tendering. It then briefly reviewed a tendering process and methods of tendering. In the final section, it discussed a process of selecting a best tender. The outcome of the tendering stage is the formation of a contract between the client and the contractor. This event signifies the end of the tendering stage and the start of the pre-construction stage.

EXERCISES

1 Why is it necessary for the client to disclose all the criteria on which tenders will be judged?
2 What is the purpose of bid bonds?
3 What are the phases of the tendering process that lead to the award of a main contract?
4 Under what conditions should the client select the lowest bid?
5 Develop your own model of effective tender administration. What will be its components?
6 What advantages could the client derive from paying bidders for the cost of tendering?

CHAPTER 8

MANAGING PRE-CONSTRUCTION

INTRODUCTION

The date on which the contract is awarded to the contractor signifies the start of the pre-construction stage. Its main purpose is to allow the contractor to plan and organise construction work and to mobilise necessary resources.

The project manager's task at this stage is to ensure that the client has secured funds, that the site is available for construction and that the contractor is ready to start work on the date for site possession. The project manager will also formulate a plan for monitoring and controlling the contractor's progress, and for administering the main contract.

The length of the pre-construction stage may be weeks or months, depending on the size of a project and its complexity. The actual period is specified in the contract as a length of time between the contract award date and the date for site possession.

Tender documents that the contractor used in formulating a winning tender have now assumed the status of contract documents. They specify and show the extent of work that the contractor is legally bound to execute. Any departure from the contract documentation would constitute a breach of contract on the part of the contractor.

This book assumes that the project manager is responsible for the management of all stages of a project lifecycle, including the pre-construction stage. However, the reader should note that in practice most traditionally procured projects do not employ a project manager for the management of the pre-construction stage. This role is usually performed by a design consultant or a superintendent.

This chapter initially examines the [...]
manager as the overall manager of the p[...]
reviews the specific tasks for which the[...]

RESPONSIBILITIES OF A PROJECT M[...]

A significant part of the project manag[...]
construction stage is to prepare an ov[...]
within which progress would be mo[...]
construction stage in terms of time, cost, quality or some other specific objectives. The remaining task is to liaise with and monitor the contractor's activities to ensure that he will be ready to start construction on the date set in the contract for site possession. Let's examine the specific tasks that the project manager will undertake.

Meeting with project team members

At the start of the pre-construction stage the project manager conducts a meeting with members of the project team, who include the client, design leader, quantity surveyor and the contractor. The purpose of the meeting is threefold. First, it will attempt to stimulate teamwork by inducting the contractor into the team and establishing open communication. Second, it will clarify to the contractor the project objectives and explain the client's expectations, the design philosophy and a system of information management. Third, it will attempt to identify specific activities that would need to be undertaken at the pre-construction stage and assign responsibilities for action. Potential problems will also be identified and appropriate solutions sought.

Availability of funding

It is now time for the project manager to verify that funds that have been pledged by financiers for the project are available. The total amount of funds borrowed must cover the cost of the pre-construction, construction and commissioning stages, including an allowance for contingency. It is also important for the project manager to ensure that the rate of borrowings matches the expected rate of project expenditure.

Availability of site

A site on which the project will be constructed may have already been purchased or leased by the client. If the client only holds an option to either buy or lease such a site, the project manager must instruct the client to take up the option immediately.

The site may presently be occupied by existing buildings, which will be either demolished or refurbished. The project manager must arrange for tenants of such buildings to move out and may even be required to find temporary accommodation for them.

Management of information

One of many initiatives of strategic planning undertaken at the conceptual stage was the development and implementation of an appropriate information management system. The purpose of such a system is to provide a central database of information for speedy transmission to project team members. Such information includes a production schedule of working drawings, the client's instructions and variations, a schedule of meetings, copies of the minutes of meetings, and so on.

It may well be that the project manager has already installed an appropriate information management system and that the team members are on line. In collaboration with an IT consultant, the project manager then assists the contractor to connect to it as soon as possible. As subcontractors become engaged, they too will be assisted by the project manager to connect to the system. To be able to connect, the contractor and subcontractors must obtain appropriate software. They too will need to undertake training in the use of the system, which the project manager will arrange. Among other things, they will learn about procedures and rules that they would be required to follow when using the system.

If there is no information system in place at this point the project manager must act swiftly in installing such a system and must ensure that the project team members are connected to it before the start of the construction stage.

Time management

The project manager will update a project schedule developed at the conceptual stage, which he will use to monitor the contractor's progress and performance in the construction stage. Of particular concern to the project manager is to ensure that the contractor will take possession of the site as given in the contract.

In liaison with the contractor, the project manager will examine ways of reducing the period of construction. This is particularly important in large projects with a high capital investment. A small percentage reduction in the project period can make a significant reduction in the total cost. For example, for a $100 million project built over five years with an annual interest rate of 10 per cent, the weekly interest charge will be approximately $200 000. A project time reduction of 1 per cent equals 2.5 weeks over five years with the corresponding cost saving (interest only) of $500 000 (or 0.5 per cent of the total cost). Significant cost savings can be realised through effective scheduling.

Often the contractor is required by the terms of the contract to submit a construction schedule to the project manager by the date given in the contract. Such a schedule is then regarded as a contract document. The schedule is prepared in the format of a critical path program and provides information about how the contractor proposes to build the project. The implication is that the contractor must not deviate from the schedule without reasonable cause or unless instructed to do so by the project manager. The schedule also assists the project manager in making timely decisions on which the contractor's progress depends.

The project manager reviews and analyses the contractor's schedule, paying particular attention to the critical activities and the amount of float. Since it is now a contract document, any instruction by the client or the project manager that delays the start of such an activity may give the contractor an opportunity to claim for a time extension and/or an extra cost.

At the construction stage, the schedule will assist the project manager to monitor progress and assess the contractor's claims. The project manager will require the contractor to update the schedule regularly in line with instructions and variation orders authorised by the project manager.

When the project manager requests the contractor to supply a construction schedule or asks him to supply some other information for approval, such as a quality assurance plan, the project manager's approval may relieve the contractor of some liability or give him a right against the project manager for negligence. When requesting specific information or documents from the contractor, the project manager must refrain from telling the contractor that such information or documents are approved or satisfactory, or otherwise give any indication that might lead the contractor to rely on such information (Uher & Davenport 2002: 235–6).

Contractors are generally highly proficient in understanding and interpreting the intricacies of critical path schedules. When a schedule is to become a contract document, the contractor may be tempted to artificially reduce its float and thus create more critical activities. This action would most likely enhance the frequency of the contractor's time extension claims. The project manager must have sufficient expertise in scheduling or seek expert advice to be able to safeguard against unethical use of the schedule.

Cost management

The most substantial part of the project cost budget, established at the conceptual stage and refined at the end of the design stage, is the cost of construction. It may seem that when a contract is awarded to the contractor for a specified contract price, the cost of construction is known for certain. But the contract price is rarely the same as the final contract cost. This is because it is commonly subjected to adjustments arising from variation orders, cost escalation, latent conditions, consumer price index increases, and any other cost adjustments. These adjustments could substantially increase the project cost (Bromilow 1969, 1970, 1971; Levido et al. 1981; Robinson 1987).

The crucial aspect of cost control in the pre-construction and construction stages is the effective management of design changes, which usually represent the main source of cost overruns. When a design change is authorised by the project manager in the form of a variation order, the contractor is able to make a claim for the cost of accomplishing such a variation order. Mistakes, omissions or ambiguities in the contract documentation for which the designer is responsible may also lead to design changes and variation orders.

Clearly, by minimising the number of design changes and eliminating mistakes from the contract documentation, the incidence of cost overruns would be significantly reduced. Since the client and the design consultant are the main culprits here, the effectiveness of the project manager in managing the conceptual and design stages will go a long way towards alleviating this problem.

At the conceptual stage, a preliminary analysis of the project's rate of expenditure was carried out and presented in the form of a financial graph knows as an S-curve (Figure 8.1). Its purpose is to align the expected cumulative rate of project expenditure with borrowings. The project manager will now prepare a final S-curve for the project based on the latest cost information and will use it as one of the tools of financial control.

At the end of the pre-construction stage, the project manager has established a firm budget for the cost of construction together with an appropriate contingency. He has also developed a process for collecting and reporting cost information in the form of the project's information management system.

Figure 8.1 An S-curve graph of cumulative expenditure

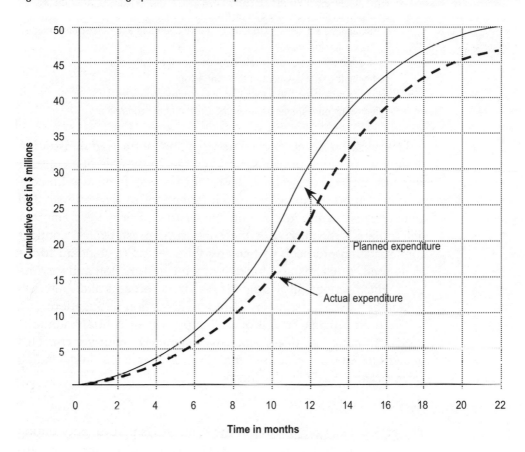

Quality management

The contractor is responsible for the development of a quality assurance plan for the project. The project manager's role is to ensure its adequacy and timely completion. In liaising with the contractor, the project manager determines how quality spot checks will be carried out during the construction stage and who will be responsible for making them.

Contract administration

One of the most important tasks of the project manager during the construction stage is to administer the contract on behalf of the client. This involves:

- issuing instructions to the contractor
- responding to a 'request for information' from the contractor

- assessing the contractor's claims
- certifying payments to the contractor
- certifying time extension, variation cost escalation and any other claims made by the contractor
- issuing a certificate of practical completion to the contractor
- certifying the final payment to the contractor
- releasing retentions and bank guarantees to the contractor
- adjudicating on disputes between the client and the contractor.

The work on developing an appropriate contract administration process starts in the pre-construction stage. In the role of a contract administrator, the project manager must become familiar with contract conditions, paying particular attention to operational and administrative requirements, and 'time bars'. The project manager will initiate a meeting with the contractor to establish a contract administration protocol, particularly with regard to standard forms to be used for claims and for requests for information and instructions. The meeting will also agree on an initial process of dispute resolution before disputes are escalated to arbitration or litigation.

The project manager will operate a process of contract administration through the project's information management system. This will ensure effective record-keeping, which is essential in defending against claims made by the contractor.

Performance measurement

The project manager will assess the contractor's performance during the construction stage in terms of whether or not it meets the project objectives of cost, time and quality. The task of measuring cost and time performance is relatively simple. The actual performance is compared with the budget and the variance, both positive and negative, is expressed in percentage terms. Assessment of quality, on the other hand, is more subjective and therefore more controversial. The project manager will rely not only on the contractor's quality assurance plan but also on regular spot checks.

The contractor's reward is the contract price or contract sum. Projects procured using the traditional method of delivery generally offer no bonus to the contractor for performance. In the absence of a reward for performance, there is no real incentive for the contractor to do more than what the contract calls for. It would be useful for the client and the project manager to consider offering a bonus or a reward fee for performance. Fehlig (1995) described the operation of award fee on a large engineering contract worth $US160 million as a financial incentive and reward to the contractor for good performance. The client established an award fee sum of approximately 3.1 per cent of the contract sum. The fee was divided into 13 fee periods of approximately four months each. At the end of each period, the client assessed the contractor's performance in four criteria: quality (50 per cent), management (20 per cent), timely performance (20 per cent), and public relations/community impact (10 per cent). The aim of both the contractor and the client was to earn and award 100 per cent of the fee, which would have established a win–win relationship for both contractor and client. On this particular contract, the contractor passed a portion of the award fee to the main subcontractors. There are other schemes designed to reward the contractor for performance. Some are discussed in Uher (1999).

Monitoring activities of the contractor and the designer leader

The project manager's main goal during the pre-construction stage is to ensure that the contractor will start construction on the date for site possession. The project manager will therefore closely monitor the contractor's progress, particularly with regard to:

- timely letting of subcontracts
- development of a construction schedule
- acquisition of resources, both human and physical, and particularly those that have long lead times for delivery
- development of a quality assurance plan
- timely organisation of the site
- arrangement of permits and approvals
- arrangement of insurance.

Although the design documentation was completed at the end of the design stage and used in tendering, it may not be complete and may require further changes. The project manager must ensure that work in finalising working drawings and specifications progresses on schedule. He must be satisfied that changes made to the documentation do not in any way alter the previously established conditions of contract.

TASKS PERFORMED BY A CONTRACTOR

Immediately after the award of a contract, the contractor selects a site manager and a site team who will be responsible for constructing the project in accordance with the contract. The contractor's site manager is commonly referred to as the 'project manager'. To avoid confusion, the title 'project manager' should only be given to the client's team leader, while the contractor's team leader on the site should be referred to as 'site manager'.

The first task of the site manager is to review the contractor's winning tender and the contract documentation. In preparing a tender at the tendering stage, the contractor would have developed a construction schedule showing a proposed construction strategy, completion dates of important project stages, and the final completion date. The site manager must become familiar with the construction strategy and the key decisions and assumptions that have been made. He must closely examine the viability and practicality of that strategy and decide if the project can be built for the agreed contract price and within the contract period.

A close scrutiny of the contract documentation will allow the site manager to get a better feel for the project's design, its main features, and the degree of complexity and risk. The site manager should also note errors, omissions and ambiguities in the documentation and report them promptly to the project manager.

The contractor's site manager will be invited by the project manager to participate in a meeting involving key team members such as the client, project manager, design leader and quantity surveyor. The meeting will provide the site manager with the wealth of information about the project, its processes and its important elements, including:

- the project's objectives
- the project's organisation structure, site personnel, and the lines of authority, responsibility and communication
- the design documentation, schedules and budgets
- crucial project tasks and strategies for carrying them out
- management systems and procedures that have been developed to administer the project and manage information
- a problem resolution process
- site conditions such as traffic flow, restrictions on access or offloading, overhead electrical wires and telecommunication cables, adjoining buildings, impact of noise on the local community.
- site access and security
- the requirement for site safety and amenities
- lead times for ordering of imported materials, plant and equipment
- the requirement for materials-handling equipment and personnel-handling
- the names of selected nominated subcontractors
- a process of production and checking of shop drawings

- permits for the supply of electricity, water and sewerage to the site
- statutory permits for a loading zone, hoardings, scaffolding, cranes, hoists, explosive tools, explosives
- insurance requirements in the form of public liability; contractor's all-risk and worker's compensation
- the type of reporting structure.

The site manager and his team are now ready to start planning and organising construction activities, some of which will now be discussed in detail.

Formation of subcontracts and supply contracts

Since most construction work is performed by subcontractors, one of the first and most arduous tasks for the contractor is to engage sub-contractors. Not all subcontracts are let in the pre-construction stage. Initially, the contractor lets subcontracts for work such as demolition and excavation that must be employed in the first phase of construction. The contractor then progressively enters into other subcon-tracts, ensuring that they are let well in advance of the scheduled start date of work. Finishing subcontractors such as painters and landscapers may not be engaged until much later in the life of the project.

Large projects require the engagement of a substantial number of subcontractors and suppliers, often well over a hundred. The contractor must effectively manage the letting of subcontracts and supply agreements strictly in accordance with a letting schedule that the contractor prepares at the pre-construction stage.

Although the project manager in liaison with the client selects nominated subcontractors, it is the contractor who enters into sub-contracts with them. Plumbing, air-conditioning, fire protection, lifts and electrical subcontractors are typical examples of nominated sub-contractors. They not only supply specialised labour but also materials, plant and equipment. Most plant and equipment, such as chillers, pumps, ductwork and lift cars, are custom-designed and manufactured by nominated subcontractors, and their delivery to the site often requires long lead times. Consequently, all nominated subcontracts should be let as early as possible, preferably at the pre-construction stage.

Apart from engaging subcontractors, the contractor will also need to enter into agreements for the supply of materials, plant and equipment. This is best achieved by first developing a schedule of suppliers of materials and one of suppliers of plant and equipment. The contractor will then be able to secure quotations and award contracts well in advance of the scheduled delivery to the site. Some materials, plant and equipment may need to be imported, which means that lead times for their delivery are likely to be long. The contractor must identify them as early as possible and award supply contracts forthwith.

Developing a construction schedule

The development of a construction schedule is probably the most important pre-construction task of the contractor. The aim is to formulate a construction strategy in the form of a logical sequence of tasks. A construction schedule is presented graphically as either a bar chart or a critical path schedule (see Uher 2003).

The contractor has already prepared a construction schedule at the tendering stage while formulating a tender price. However, because it was prepared for tendering purposes in mind and within a relatively short time, it should only be regarded as a preliminary schedule. The contractor will now develop it into a final construction schedule.

The first task is to examine the preliminary schedule carefully to ascertain whether or not its strategy is valid, and whether it is accurate and appropriate for the project. If it was prepared as a time schedule with no consideration given to the use of resources, its duration is likely to be overly optimistic. This is because time scheduling assumes that resources are unlimited and can be given to activities whenever needed and in whatever quantity. In reality, resources are always limited in their availability or capacity or in some other way. The fact that resources are limited must be reflected in the way in which construction activities are scheduled. Such an approach is referred to as resource scheduling and its outcome is a resource-based schedule. A resource-based construction schedule is developed by first, determining maximum limits of available resources and second, programming the work within such limits. While providing a realistic picture of how the work will actually be built, the major problem affecting a contractor in the pre-construction stage may be the realisation that a resource-based construction schedule is likely to be substantially longer than the preliminary time schedule. If the contractor's tender was indeed based on the time schedule, the contractor may be hard pressed to complete the project on time.

After reviewing the tender price and the tender schedule, the contractor is now ready to start developing a final construction schedule for the project. Important elements of scheduling will now be discussed.

OBJECTIVES OF A CONSTRUCTION SCHEDULE

The main objective of a construction schedule is to develop a detailed construction strategy that is properly sequenced and resourced, and to show the steps to be taken to achieve the project's goals. A schedule must be presented in such a way that those responsible for carrying out the work are able to implement it. It will be used during the construction stage to monitor progress and to help determine corrective actions.

A construction schedule is most commonly generated by a computer as a critical path program and printed in the form of a linked bar chart, something that most construction personnel are familiar with. It is essential that a construction schedule is highly legible and uncluttered, and contains only as much information as the site team needs.

Another important objective of a construction schedule is to assist the contractor in reducing the period of construction. Since the schedule is computer-generated, the contractor is able to speedily assess time performance of different construction methods and techniques, and examine the impact of overlaps among activities on time performance.

SCHEDULING WITHIN A FIXED CONSTRUCTION PERIOD

At the tendering stage, the contractor has an opportunity to find an ideal balance between the project cost and its duration in an effort to find the most competitive tender. Since cost and time are generally related, a shorter period usually results in a higher cost and, conversely, a cost may be lowered by extending a project's duration.

After a contract has been awarded to the contractor, however, the length of the construction period is fixed by the contract. It is defined as a period between the date of site possession and the date for practical completion. In such circumstances the contractor simply has to fit all the work under the contract into that fixed period, inclusive of a contingency for delays for which the contractor cannot claim time extensions. Under these circumstances, cost–time optimisation is not possible.

CONTROLLING ELEMENTS OF TIME

'Controlling elements' are major factors that control a construction period. In turn, they can be dominated by 'controlling trades'. Examples of controlling elements are:

- the structure: it controls the start of other tasks
- foundations: this is a time-consuming and risky activity with a high probability of time overruns

- delivery of specialised plant and equipment, and unique materials from overseas suppliers: long lead times are required for placement of purchase orders
- production of design and shop drawings: their late supply will cause project delays
- the number of workers to be moved horizontally or vertically on very large projects: the lack of personnel-handling capacity is likely to cause bottlenecks and delays
- the materials-handling equipment: its inadequate capacity may cause delays.

On the whole, experience will help the contractor to identify controlling trades that influence a period of construction. Common examples are:

- formwork in a reinforced concrete structure: its slow progress will delay subsequent trades
- the electrical work in any large project: failure to provide permanent power to specific activities such as the air-conditioning plant and the lifts will delay the project
- brickwork in a load-bearing structure: as with formwork, its slow progress will delay subsequent trades.

APPROPRIATE PRODUCTION METHODS

Production methods in construction fall into two broad categories:

1   assembly and installation methods
2   handling systems.

Assembly and installation methods are concerned with assembling or installing materials or components at their final position. They are capable of meeting specific time performance requirements, though they may be slowed down by outside factors such as the limited capacity of the personnel-handling equipment or the inability of the site team to co-ordinate several different but mutually dependent activities.

Handling systems refer to systems for moving materials and people on construction sites. A materials-handling system is expected to deliver materials and components from the point of their delivery or site storage to the assembly point. Since the volume of materials and components to be handled gradually increases as the project progresses, a handling system must be designed to safely handle the expected volume of work within the time-frame specified by the schedule. Designing an appropriate materials-handling system involves:

- the development of a materials-handling schedule that would meet the requirement of a construction schedule (it should show the volume of work to be handled each day)
- the selection of the most appropriate handling equipment based on a handling schedule (in terms of the type, capacity, speed, radius of operation)
- the location of the handling equipment on the site
- access to the handling equipment
- the frequency of repositioning the handling equipment.

Designing an efficient personnel-handling system is equally important on high-rise projects or projects that are large in area. Construction of a very tall building in excess of 60 or 70 levels may require a workforce of 2000 or more people. Designing a personnel-handling system that would move such a large number of people within a reasonably short time may not be economically feasible. It may even be impossible if the site is very confined. A personnel-handling system on a large project may well become a controlling element of time

When the contractor develops appropriate production methods, he records, for future retrieval, decisions made about activities' quantities, duration and resources that have been allocated to activities in method statements. For example, the decision is made to excavate the site using a backhoe. Knowing the volume of soil to be excavated and the rate of production of the backhoe, it is a simple task to calculate the duration of the activity called 'site excavation'. A method statement for this activity would then contain the following information:

- the activity name (for example, excavate site)
- the volume of soil to be excavated (in cubic metres)
- the type of plant (for example, backhoe) and its rate of production
- the activity duration (calculated from the volume of the excavated soil and the rate of production of the backhoe)
- the number of persons assisting
- any other relevant information.

Information on method statements can be found in Uher (2003: 33–7).

RESOURCES

Apart from showing a logical sequence of activities and their duration, a construction schedule must also be fully resourced. It means that required labour, plant and equipment must be allocated to each activity and their demand levels checked against the maximum limits set for each resource. When the demand level for a particular resource exceeds the available resource limit, a process of levelling will be applied to attempt to reschedule activities that compete for this resource by expanding their available float. If these activities have no float, and the contractor is unable to find other solutions to this problem, the schedule would be delayed.

It has already been noted that a construction schedule with its resources levelled is known as a resource-based construction schedule. When developing such a schedule, the contractor needs to carefully consider issues such as:

- availability of skilled labour and subcontractors
- availability of materials, and major items of plant and equipment
- impact on the project caused by delays in securing resources.

The contractor needs to carefully assess demand levels of the labour resource throughout the period of construction. This information is important for the provision of adequate site amenities, safety equipment, and equipment for distributing people throughout the site.

A SCHEDULE DEVELOPMENT

The contractor will develop a computer-based construction schedule in the form of a critical path method for the entire period of construction. This overall construction schedule must be realistic and must reflect the actual production process. It must be resource-based and show target dates specified in the contract (for example, a target date for the handover of a portion of the project to the client). It must include an allowance for public holidays and delays for which the contractor is responsible under the contract. Its time-scale will be in weeks and days.

Preparation of a schedule should be a team effort involving the site manager, an expert scheduler, and also 'doers', that is, people who will use it as a production and control tool. The input of such people is essential in ensuring the schedule's realism.

Apart from developing an overall construction schedule, the contractor also prepares a series of medium-range schedules that show construction activities in much greater detail. A medium-range schedule spans a period of three to five months. Its main purpose is to co-ordinate resources, such as plant and equipment and subcontractors. The time-scale is usually in weeks and days. The contractor usually prepares the first of these schedules in the pre-construction stage for the initial period of construction. Other schedules will be prepared later at the construction stage.

The contractor may be required by the contract to submit to the project manager a construction schedule within a specified period. Such a schedule then becomes a contract document. Its aim is to provide information to the client and the project manager that may be necessary for predicting cash flow, co-ordinating design information, monitoring progress, assessing claims for time extension under the contract, and ascertaining when funds are needed to meet progress payments. The benefits and pitfalls of using a construction schedule as a contract document have already been examined earlier in this chapter.

Schedules of resources

Apart from construction schedules, the contractor also develops schedules of various resources that will be employed in construction. These include:

- A schedule of materials that lists materials to be purchased, their quantities, source of supply, prices and dates when needed.
- A schedule of plant and equipment that identifies the type and quantity of plant and equipment to be purchased or hired, locates a source of supply, and finds out their prices and the dates when needed.
- A schedule of subcontractors that gives names of subcontractors, bid prices and dates by which subcontracts must be let. When a subcontract has been let, this schedule will also record the date of engagement, the contract price, the date when the subcontractor is required to start work on the site, and contact details for the subcontractor (address, phone numbers and e-mail address).

Cost management plan

The contractor's financial aim is to make profit on the project. To achieve this, he needs to:

- keep the project cost at or below the contract sum
- make legitimate claims under the contract for costs that he incurred but that are outside his control, for example the variation, cost escalation and the latent site condition cost
- maintain a positive project cash flow.

The contractor's cost management plan comprises a cost control system, a cash flow plan and a reporting system. These components of the cost management plan will now be discussed.

COST CONTROL SYSTEM

The purpose of a cost control system is to track costs during construction and report on cost performance.

The contract sum for which the contractor is required to build a project becomes the contractor's cost budget. The contractor's financial performance on a project will ultimately be assessed against this budget.

If the contract sum is not the same as his tender, the contractor's cost budget must be adjusted accordingly. The cost budget takes the form of a detailed cost estimate that the contractor has prepared at the tendering stage. The adjusted contractor's cost budget then forms the basis of a cost control system. Each item in the cost budget is uniquely coded, which enables costs to be tracked for each item or group of items. As actual costs are incurred during construction, they are entered into the cost control system under their respective codes. The system is then able to group cost items according to their codes and generate regular reports. The report shows the budget and actual costs for each cost item in the system, a percentage of the cost to complete and a variance.

The effectiveness of the contractor's cost control system depends on the quality of input data and the timely and accurate cost coding of such data. Thus effective collection of cost data is essential. It involves extracting cost information on:

- direct labour
- materials
- plant and equipment
- subcontractors
- preliminary items.

Direct labour costs are compiled from labour time sheets for which the contractor's site supervisors are responsible. Supervisors assign appropriate cost codes to activities performed by direct employees on an hourly or daily basis.

Material costs are compiled from coded delivery dockets for which site supervisors are responsible.

Plant and equipment costs are calculated from the weekly or monthly invoices of the hired plant. Costs of the contractor's own plant and equipment are compiled from time sheets kept for each such piece of plant and equipment on the site.

Subcontract costs are calculated from the weekly or monthly progress payments made to subcontractors.

The cost of preliminary items such as safety, amenities and insurance are compiled directly from the monthly invoices.

The important aspect of cost coding is accuracy, that is, assigning the right cost codes. This is commonly a problem when coding the labour time sheets and the delivery dockets of materials. Strict control over coding is therefore essential.

Another important element of the cost control system is a register of the project manager's instructions, claims for progress payments that have been made and received, and claims for variations, time extensions, cost escalations and other claims that have been made and received.

A CASH FLOW PLAN

Negative cash flow is often the cause of business bankruptcies. Contractors, in particular, must be good managers of cash since they don't receive the first progress payment for the work accomplished from the client until two to three months from the start of construction. However, since subcontractors perform most construction activities, some of the burden of funding the first period of construction rests with subcontractors.

A prudent contractor carefully plans cash inflows and outflows over the period of construction. Inflows are progress payments from the client and outflows are payments that the contractor makes for the cost of labour, subcontractors, materials, plant, equipment, preliminary items and overheads. A typical cash flow statement shows, at weekly intervals, cash balances of the contractor's expected inflows and outflows. This information assists the contractor to plan short-term borrowings for the negative cash flow periods, and short-term investments for the positive cash flow periods.

COST REPORTING

Reporting of costs incurred and comparing them to the budgeted costs is an essential part of the cost management plan. The reporting emphasis is placed on regular and timely assessment of costs in progress, on forecasting costs to complete, and on the analysis of deviations from the budget.

Quality control plan

The purpose of the quality control plan is to ensure that a contractor builds a project in accordance with quality requirements specified in the contract. In the past, the client's representative controlled the quality of construction by regular inspections. The present trend requires a contractor to self-assess quality, with only spot checks carried out by a project manager.

Most Australian construction clients require contractors to be accredited under the ISO9000/AS3000 standards. The accreditation implies that contractors have developed and implemented an effective quality assurance plan. Apart from inspections, a quality assurance plan involves a whole range of issues such as:

- creating a new organisation culture geared towards team environment and customer satisfaction
- changing attitudes to 'do it right the first time'
- training of employees to develop appropriate knowledge and skills
- employing suitable, safe and effective equipment
- creating suitable and safe working conditions
- improving people's motivation
- developing a quality system
- empowering people.

The contractor's quality assurance plan is the main vehicle of quality control during the construction stage. It will be discussed in detail with the project manager and with subcontractors, who will be expected to uphold it. The reader may find more information on quality assurance in Oakland and Sohal (1995).

Site organisation

How best to organise the site is a crucial question for which the contractor must find an appropriate answer during the pre-construction stage. In doing so, the contractor formulates a plan of action for the management of tasks such as:

- site layout
- site establishment
- site amenities
- temporary services
- safety and security provisions
- permits and insurance policies.

Let's examine those tasks.

SITE LAYOUT

Most urban construction sites are surrounded by existing buildings, roads, public spaces and a wide range of in-ground or above-ground services. Contractors are required to confine construction activities strictly within boundaries of sites, without unduly affecting, disrupting or interfering with the outside environment. Local councils may impose additional requirements on contractors by limiting hours of work per day, preventing work on weekends, and restricting hours for delivery of materials to sites.

Organising construction work within a confined site and without affecting the surrounding environment is a difficult task. One of the crucial decisions that the contractor must make that will affect production efficiency is to design an appropriate layout of the site. In doing so, the contractor would need to examine:

- access to the site
- movement of people and materials
- storage of materials
- location of amenities and site offices
- location of the materials-handling equipment.

The main aim is to provide an efficient, functional and safe work environment. Of particular importance in designing the layout of a site is an unloading space. The contractor must dedicate a sufficiently large space for unloading materials from trucks and for their storage, and have suitable materials-handling equipment on standby to avoid bottlenecks and excessive double handling of materials.

SITE ESTABLISHMENT

To ensure that the work is carried out strictly within the confines of the site, the contractor hires a land surveyor to verify the position of the site boundaries. The surveyor also helps to ascertain the location of existing services on the site. Although rare, there are examples of projects, particularly housing projects, that have been built on the wrong site.

SITE AMENITIES

The construction site is a place where all construction activities take place. It is a place of work of a large number of people for a considerable length of time. Apart from establishing a reasonably comfortable and functional accommodation for its own site management staff, the contractor is required by the law to provide:

- site accommodation for his workers
- sanitary conveniences
- a fully equipped first aid room.

The contractor may also provide site accommodation for subcontractors or he may reserve space for such accommodation facilities, which subcontractors would need to provide themselves.

TEMPORARY SERVICES

Unless already available on the site, the contractor will need to enter into supply contracts with various utilities for the provision of:

- electricity
- telephone and IT services
- water
- sewerage
- drainage.

The contractor should establish with the local post office a postal address for the site.

SAFETY

By law, the contractor is responsible for the safety of workers on the site and for the safety of members of the public in close proximity to the site. To meet this obligation, the contractor develops and implements a safety plan, which defines policies and specific safety requirements for:

- design and installation of general scaffolding and access scaffolds
- design and installation of fences and hoardings
- safe access to the site
- design and installation of handrails and kickboards
- overhead protection
- safety helmets, clothing and accessories.

An important component of a safety plan is a procedure for monitoring and controlling safety on the site. This will be the responsibility of the contractor's safety officer.

SECURITY

The purpose of site security is first, to prevent unauthorised access to the site that may result in personal injuries, and also theft and vandalism of property, and second, to avoid theft and pilferage by the site workforce. The contractor will achieve a reasonable standard of site security by:

- erecting secure fencing and hoardings around the site
- securing site amenities

- providing secure storage facilities for plant and equipment, building components and materials
- installing security lighting throughout the site
- installing alarms
- providing physical surveillance of the site
- installing electronic surveillance.

PERMITS AND INSURANCE

Laws and regulations closely control construction activities. In order to carry out construction work, the contractor needs to obtain a number of permits from statutory and utility organisations for tasks such as:

- the erection of hoarding, scaffolding, cranes and hoists
- the demolition of buildings
- the erection of temporary structures
- the use of explosives
- the connection to water, sewerage and drainage services
- the establishment of a loading zone for delivery of materials.

By law, the contractor is required to take up worker's compensation insurance, which protects workers against both medical and hospital costs and the associated loss of income arising from job-related injuries. The contractor must also check that subcontracting firms engaged on the site hold valid worker's compensation insurance for the protection of their workers.

By contract, the contractor is required to take up the public liability and the contractor's all-risk insurance for the amount specified in the contract. The public liability insurance covers members of the general public against personal injuries and damage to personal property. The contractor's all-risk policy insures the works and existing structures against damage.

Site supervision

The size of the project and the number of workers employed influence the way in which the contractor manages and supervises construction. A typical contractor's site organisation structure for a relatively large project is given in Figure 8.2. It shows several hierarchical levels of project supervision.

Figure 8.2 The contractor's site organisation structure

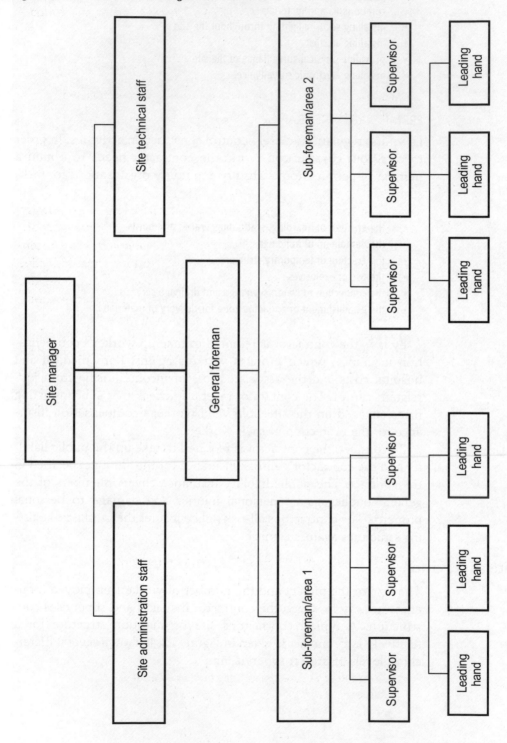

It is the contractor's responsibility to ensure that the project is effectively supervised during its construction. In doing so, the contractor needs to:

- develop a job specification, which gives the nature and scope of each supervisory position, and defines authority
- train supervisors in developing skills in team-building, leadership and motivation
- ensure that subordinates report to one designated supervisor only
- define an appropriate ratio of workers to a supervisor, which in most cases should not exceed 10 to 1.

Management of information

In order to construct a project, the contractor must interpret design information, understand contract conditions, and comply with various laws and regulations. In other words, the contractor is a receptor and processor of a very large volume of information. Effective management of information is thus vital to the success of a project.

The contractor may be required to connect to an electronic project information system through which the project manager manages information across the entire project lifecycle. The benefit of such a system lies in providing a central database of information for speedy transmission to project team members. Among other things, it tracks information such as a production schedule of working drawings, the client's instructions and variations, a schedule of meetings, copies of minutes of meeting, and so on.

In the absence of a project-based information system, the contractor develops a similar system for managing information. Such a system is designed to perform two vital roles in:

- tracking information
- keeping and filing records.

Examples of information requiring periodic tracking include:

- a production schedule of working drawings
- the client's instructions and variation orders
- a schedule of letting of subcontracts
- schedules of labour, plant and equipment, and materials resources
- the contract target dates
- a schedule of insurance policies and their expiry dates
- a schedule of permits
- a schedule of meetings.

An important task in managing information is to keep and file records for future retrieval. The most commonly used recording tool is a site diary that the site manager uses each day. Examples of information requiring recording, listing and filing are:

- emergency numbers of the police, ambulance, fire brigade
- the names of applicable regulations and codes
- technical reference handbooks
- photographic records of progress
- minutes of all meetings
- correspondence
- construction schedules and progress reports
- cost accounting and cash flow projection records
- contract administration information such as claims for variations, time extensions, payments
- insurance certificates
- approvals and permits
- a register of accidents and injuries.

Contract administration

The scope of the construction work, the duties and responsibilities of the contractor, and the reward that the contractor derives from carrying out the work are all specified in the contract. The contractor must become familiar with the contract conditions and must be able to interpret them and act on them. Clearly, his ability to administer the contract effectively will have a direct bearing on project outcomes. Equally important is the contractor's ability to administer subcontracts. The failure of only one of many subcontracts may cause irreversible damage to the project and to the contractor's financial profitability.

ADMINISTRATION OF THE MAIN CONTRACT

To receive a progress payment, or secure a variation order, time extension, rise and fall adjustment or any other cost adjustment, the contractor must first make a claim within the time specified in the contract. It is therefore particularly important to the contractor to have the knowledge of 'notices' and 'time bars' specified in the contract. For example, a substantial claim by the contractor for, say, a latent condition is likely to be disallowed if submitted outside a time bar specified in the contract.

Effective record-keeping of contract information is particularly important to the contractor since availability of such information will improve his chance of securing claims or winning disputes.

ADMINISTRATION OF SUBCONTRACTS

Administration of subcontracts is the contractor's responsibility with little or no interference from the project manager. While subcontracting is said to improve productivity through specialisation, it also increases the risk of contractual conflict, which may cause cost and time overruns. Considering that a large commercial project may employ over a hundred separate subcontractors and suppliers of materials, the contractor's task of administering so many subcontracts is extremely difficult. Problems with only one of those subcontracts may cause the project to overrun on cost and time. Although the client may argue that problems subcontractors may experience are the contractor's responsibility, the impact of such problems may nevertheless flow back to the client in the form of extra cost. This is because the contractor will attempt to pass some of the incurred costs to the client by exploiting contract conditions. Judging from past experience, it is highly likely that the contractor will succeed. The reason is that he usually keeps an accurate record of contract information that supports his claim.

The contractor is able to minimise the likelihood of problems with subcontracts by developing a strategy for effective administration of subcontracts. It involves:

- creating fair and equitable subcontract conditions
- meeting regularly with subcontractors
- ensuring that subcontractors' resources are adequate to maintain progress
- processing subcontractors' progress claims on time
- making progress payments to subcontractors on time
- responding in good time to claims for variations, time extensions and other claims
- responding in good time to 'requests for information'
- resolving disputes speedily.

Summary The contractor has used the pre-construction stage to develop a blueprint for constructing a project in the most effective and efficient manner in the form of a construction schedule, other schedules and a cost management plan, and by developing a strategy for site organisation and contract administration. The contractor has secured permits and insurance policies and recruited the necessary site management staff. The contractor is now ready to start building.

CONCLUSION

This chapter examined the management of the pre-construction stage from both the project manager's and the contractor's points of view. Issues that have been examined in detail included the development of a construction schedule, cost and quality plans, and the establishment of processes for effective contract administration and information management. Important aspects of site organisation, such as site layout, amenities, safety, security and supervision have also been examined.

The next chapter will discuss in detail the management of the construction and commissioning stages.

EXERCISES

1   Why is it important for the client and the main contractor to sign a formal contract as soon as possible after a contract has been formed?

2   What are important activities performed by the main contractor in the pre-construction stage?

3   What is the role of the project manager in the pre-construction stage?

4   Clients commonly do not interfere in the formation of subcontracts. What benefits, if any, would clients derive from preventing main contractors to impose onerous subcontract conditions on subcontractors?

5   What are the benefits and potential pitfalls to the client from requiring the main contractor to submit a construction program (assume that this construction program is a contract document)?

CHAPTER 9

MANAGING CONSTRUCTION AND COMMISSIONING

INTRODUCTION

The construction stage starts on the date of site possession, that is, the date when the contractor has been given access to the site, and ends on the date given in the contract as the date for practical completion. When the date of practical completion has been reached, the client takes possession of the project in both the physical and legal sense, although the project may not be fully completed. The extent of completion is measured in terms of whether or not the client is able to use the project for its intended purpose. The term 'practical completion' is unique to the construction industry, where clients are prepared to accept projects that are incomplete, contain defects and do not have fully functioning services. In comparison, consumers of manufacturing products would be unlikely to accept the same philosophy of 'practical completion' when buying, for example, motor-cars, boats or TV sets.

When the practical completion date is reached, the commissioning or defects liability period stage starts. During this period the contractor must complete outstanding parts of the project, must repair defects to the satisfaction of the project manager, and must commission the project's services and ensure their satisfactory performance.

This chapter is in two parts. A review of the management of the construction stage is followed by an examination of the main activities associated with the commissioning stage.

CONSTRUCTION STAGE

The date for site possession normally but not always signifies the start of construction. In a legal sense, it triggers the start of the contract period. While constructing a project, the contractor must safeguard against damage and/or injury to the existing structures on the site and in the near vicinity of the site, the project under construction, workers and members of the general public. The contractor is protected against such damage and/or injury through the contractor's all-risk, public liability and the worker's compensation insurances which the contractor secures in the pre-construction stage.

The management of the construction stage will now be examined from the perspective of both the project manager and the contractor.

Activities performed by the project manager

The project manager's main function during the construction stage is to monitor and control the activities of the contractor to ensure that the project is completed within the cost budget, on time and to the required quality standards. The project manager's plan of action was formulated in the pre-construction stage and was described in detail in Chapter 8. The project manager will now use this plan to control progress and measure the project's performance. Let's examine the tasks that require the project manager's close attention in monitoring and controlling the construction stage.

MANAGING INFORMATION

As part of monitoring and controlling a project's progress and performance, the project manager will concentrate on managing information flow in a timely manner across the entire project team. An information management system, installed at the pre-construction stage, will assist the project manager in this task. Regular meetings of project team members will play an important role in analysing feedback information and acting on it. Particular attention will be given to the timely supply of working drawings to the contractor and the speedy resolution of problems and disputes.

ADMINISTERING THE CONTRACT

The project manager's task in administering the contract between the client and the contractor has a significant bearing on the overall project cost and schedule. The project manager's approach must be efficient and timely in issuing instructions to the contractor, responding to requests for information, and processing claims for time extensions, variation orders, cost escalation and progress payments. It must follow the contract administration protocol established at the pre-construction stage. The project manager must act fairly and equitably in processing the contractor's claim in order to avoid the possibility of disputes. If contractual problems arise, the project manager must resolve them immediately before they develop into major disputes.

MEASURING PERFORMANCE

The project manager has the overall responsibility for achieving successful project outcomes in terms of cost, time and quality. At regular intervals the project information management system will generate progress information from which the project manager will establish the project's performance. The performance will then be reviewed at regular meetings involving project team members. Since most construction projects tend to overrun on cost and time, the ability to minimise the extent of overruns and contain them within the cost and time contingencies is a challenge for any project manager.

RESOLVING PROBLEMS

Construction projects are characterised by having a large number of individual contracts formed between different project participants. With so many contracts and so many organisations involved, problems are bound to occur. The project manager's role is to address such problems as soon as they surface and resolve them amicably and in good time. Open communication and a healthy respect and trust among the project team members is a condition for resolving such problems and reducing the possibility of costly litigation and arbitration.

At the end of the construction stage, when the project manager is satisfied that the project has been completed in accordance with the contract documentation and is ready for occupancy by the client or the client's tenants, the project manager will issue a certificate of practical completion to the contractor. This signifies the end of the construction and the start of the commissioning stage. In a legal sense, the project is ready to be occupied and the client is now legally responsible for it.

Activities performed by the contractor

At the pre-construction stage the contractor developed plans and strategies for building the project, organised the necessary resources, connected to the project information system and installed other management systems for monitoring and controlling the project's performance in terms of time, cost and quality. The contractor will rely on these plans, strategies and systems to build the project during the construction stage.

It does not, however, mean that the contractor's planning and organising tasks have ended. As construction of the project gets under way, he may need to change the construction strategy in response to changed project circumstances caused by the presence of unforeseen risk, design changes, latent site conditions and some other events. It means that the contractor's planning and organising efforts are ongoing through the construction stage. Apart from planning and organising, he will closely monitor and control the project's progress and performance.

The contractor's main tasks in constructing a project will now be briefly reviewed.

MANAGEMENT AND CONTROL OF TIME

The contractor has prepared an overall construction schedule in the pre-construction stage and will rely on it to build the project. If the schedule is a contract document, the project manager will use it to monitor and control the project's progress, and raise claims for time extension. The contractor must therefore regularly update the schedule to reflect changing circumstances.

For day-to-day production tasks the contractor will rely on medium and short-range schedules. The contractor has prepared the first medium-range schedule, which covers a period of about the first three months of the pre-construction stage. Other medium-range schedules will be developed progressively throughout the construction stage. The contractor will also prepare a series of short-range schedules for one to two week-long construction periods. Their scale is in days or even in hours. They show in detail what specific work will be carried out, when, and by what resources. Short-range schedules may be produced in the form of a bar chart, a multiple activity chart or a simple list of activities. Information on short-range scheduling can be found in Uher (2003).

The contractor meets with subcontractors regularly at weekly intervals to assess progress achieved to date and discuss a production plan for the next period. Specific problem areas are identified and appropriate solutions formulated. These are then incorporated into a short-range schedule for the next period. The regular meetings also assist the contractor in reviewing the medium-range schedule and the volume of its float, assessing the likelihood of meeting the target dates, and determining the appropriateness of committed resources.

MANAGEMENT AND CONTROL OF RESOURCES

The ability of the contractor to make profit very much depends on the contractor's ability to effectively manage committed resources. Of crucial importance is the contractor's ability to manage subcontractors. This requires close co-ordination of their activities, paying particular attention to:

- letting subcontracts according to a schedule prepared in the pre-construction stage, and discharging those that have come to an end
- meeting with subcontractors each week and agreeing on a schedule of work and the volumes of required resources
- resolving co-ordination problems
- monitoring performance of subcontractors
- responding immediately to requests for information from subcontractors
- processing payments to subcontractors on time.

The contractor also progressively enters into contracts for the supply of plant, equipment and materials. Plant and equipment no longer needed on the site must be promptly returned.

The engine room of any construction site is its materials-handling system. Its efficiency and effectiveness has a significant impact on the project's performance. At the pre-construction stage, the contractor developed a strategy for materials handling on the site, inclusive of the selection of appropriate plant and equipment. At the construction stage, this strategy is implemented and its performance monitored regularly.

Each week, the contractor prepares a materials-handling schedule, which shows in detail specific tasks that the materials-handling plant will be required to perform. The schedule's scale is in days and hours. Its development is based on feedback from subcontractors, who also comment on its appropriateness at regular weekly meetings.

Most construction sites have limited capacity to store materials. Consequently, their delivery, storage and handling must be carefully planned. The contractor has already prepared a schedule of materials needed for the project at the pre-construction stage. It will now be extended to show a delivery schedule of materials by closely observing the requirements of the construction schedule and the materials-handling schedule. The contractor must also monitor the effectiveness of the materials storage facility and assess how well or poorly it protects materials against theft and degradation.

At the pre-construction stage the contractor has prepared the labour demand schedule and demand schedules of individual labour resources. As construction gets under way, each day the contractor needs to closely monitor the volume of workers employed on the site to ensure adequacy of site amenities and safety provisions. Equally important is to monitor performance of specific labour resources. Trend graphs are commonly employed for this purpose. For example, the demand for the contractor's direct labour is graphically illustrated in Figure 9.1 in both the planned and actual demand levels. The visual nature of trend graphs assists in easy detection of undesirable trends.

Figure 9.1 A trend graph of the direct labour demand

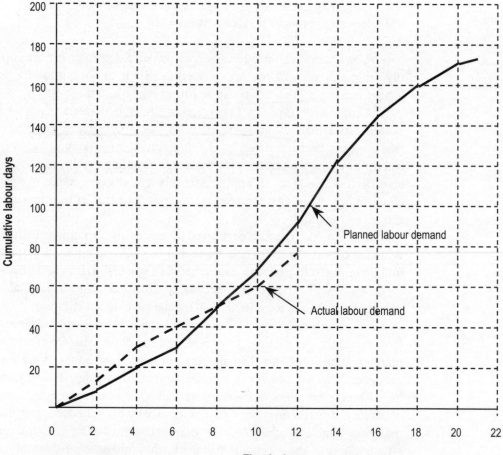

MANAGEMENT AND CONTROL OF COST

A cost control system was devised by the contractor during the pre-construction stage. Its effectiveness will very much depend on the quality of data collection and the accuracy of cost coding. These are important tasks that the contractor needs to address during the construction stage. Once cost information has been collected and analysed, a detailed cost report is generated at regular intervals.

MANAGEMENT AND CONTROL OF QUALITY

The contractor's quality assurance plan developed at the pre-construction stage should in theory ensure high-quality work and no rework. However, the mere existence of the contractor's quality assurance plan is no guarantee of quality. The contractor's most difficult task is to involve subcontractors in implementing such a plan. While it may be appropriate to rely on self-assessment of quality by subcontractors on some well-managed partnering or alliance projects, in other cases self-assessment of quality may be unrealistic. For as long as the construction industry and clients in particular continue to tolerate 'the defects liability period', it is unlikely that a new paradigm of 'do it right the first time' will be adopted. It is also difficult to promote total quality management (TQM) in the construction industry while the adversarial nature of the contractor–subcontractor relationship, fuelled by onerous subcontract conditions and the existence of bid shopping, continues to flourish. For as long as subcontractors are required to assume a disproportionately high level of risk, they are likely to continue to compromise on quality when their profitability is low.

It may therefore be prudent for the contractor to adopt a more traditional quality control approach through regular inspections of the work. At regular weekly meetings with subcontractors, a concerted effort should be made to identify faults in quality and devise appropriate solutions.

MANAGEMENT AND CONTROL OF SITE ACTIVITIES

When construction starts, the contractor is required to comply with various statutory and industrial laws and regulations governing safety, amenities, industrial relations and insurances. The contractor has already arranged for the necessary permits and insurances at the pre-construction stage. During the construction stage, the contractor will ensure that they are maintained and kept up to date for the period of the project.

The contractor's safety officer implements safety procedures on the site and regularly monitors adherence to safety. He takes whatever steps are necessary to maintain a safe working environment.

Managing Construction and Commissioning

The safety officer also organises training programs for site management and workers on a range of safety issues.

The contractor is by law responsible for the health and safety of people working on the site. On larger Australian sites, the contractor must employ a first aid officer who provides essential medical aid to injured workers and keeps a register of accidents and injuries. The contractor must also have first aid facilities on the site.

Industrial conflict is frequent in the Australian construction industry and lowers its productivity. The contractor must provide leadership on industrial matters to his own workers and subcontractors by strictly adhering to industrial awards and other industrial laws. The contractor's industrial relations officer is responsible for the development and maintenance of a strong employer–employee relationship. In liaison with the contractor's employees and subcontractors he is also responsible for the resolution of industrial relations problems.

MANAGEMENT AND CONTROL OF INFORMATION FLOW

The speed and the quality of work of the contractor and subcontractors very much depend on effective management of information throughout the construction stage. A project management information system implemented by the project manager is an ideal vehicle for managing information. The contractor is expected to actively contribute to the effectiveness of the system by providing timely information to the project manager and subcontractors.

MANAGEMENT AND CONTROL OF CONTRACT ADMINISTRATION

The objective of the contractor's contract administration process is to provide a framework for the most efficient execution of the main contract and subcontracts by complying strictly with the conditions of such contracts.

The contractor will implement a process that he developed at the pre-construction stage. It involves administering the main contract with the client and also subcontracts.

With regard to the main contract with the client, the contractor is particularly concerned with securing claims for:

- progress payments
- variation orders
- time extensions
- latent site conditions
- cost escalations and other types of cost claims.

To ensure that such claims are successful and are paid on time, the contractor must comply with time bars as specified in the contract. The contractor must also prepare such claims within the defined contract administration protocol.

When administering subcontracts, the contractor must treat similar claims made by subcontractor fairly and equitably, and in a timely manner as prescribed in subcontracts.

If or when contract disputes with either the client or subcontractors occur, they should be resolved as soon as possible through an agreed dispute resolution process. If, for whatever reason, a dispute escalates and is referred to arbitration or litigation for settlement, effective record-keeping can considerably improve the contractor's change of winning the dispute.

Summary At the end of the construction stage, the contractor will apply to the project manager for a certificate of practical completion, which will signify the end of the construction stage and the start of the next, commissioning stage. In a legal sense, once a certificate of practical completion has been issued to the contractor, he is no longer responsible for the insurance of the project. Furthermore, the client is required to release half of the retention sum to him. The retention sum is a form of guarantee of performance provided by the contractor to the client. It is the accumulation of deductions that the client makes from each progress payment to the contractor and is commonly capped at around 5 per cent of the contract sum.

COMMISSIONING STAGE

The commissioning stage or the defects liability period starts when the practical completion date has been reached and runs for a fixed amount of time specified in the contract, commonly between three and nine months. The purpose of the commissioning stage is to complete outstanding parts of the project, commission various services, such as air-conditioning, lifts, fire protection and security, repair defects to the satisfaction of the project manager, and bring the contract between the client and the contractor to an end.

Activities performed by the project manager

The project manager's role during the commissioning stage is to ensure that all construction activities have been finalised so that the client's contractual links with the contractor and the design leader can be discharged. The project manager may also assist the client in either selling the completed project or leasing it out. Let's explore the tasks performed by the project manager in more detail.

ENDING CONSTRUCTION ACTIVITIES

While practically complete, there may be some parts of the project still under construction. The project manager must ensure that the contractor has completed all the remaining parts of the project by the end of the commissioning stage.

At the start of the commissioning stage the project manager will issue to the contractor a written account of defects that the contractor must fix to the satisfaction of the project manager. The project manager will monitor the contractor's progress and compliance and may further direct him to repair additional defects.

In large buildings, services such as air-conditioning, ventilation, lifts, fire protection, security and IT account for over 50 per cent of the construction cost. The project manager's responsibility is to ensure that the contractor brings these services on line and that their performance matches the design requirements. The project manager must also get guarantee certificates from the contractor to cover the operation of the project's services, and arrange for contracts for the maintenance of such services.

SECURING A CERTIFICATE OF COMPLIANCE

The local council approves the construction of a project at the design stage on condition that it strictly complies with the design documentation, relevant building codes and regulations, environmental laws and any other conditions imposed by the council. When the project manager is satisfied that the contractor has completed the project, he will apply to the council for a certificate of compliance. By issuing such a certificate, the local council certifies that the completed project complies with building and environmental laws and regulations.

FINALISING EXISTING CONTRACTS

After receiving a certificate of compliance from the local council, and when he is satisfied that the contractor has fully completed the project, repaired the identified defects and successfully commissioned the services, the project manager must process any outstanding claims from the contractor and must settle the contractor's final account. It may well be that the final account shows that the contractor actually owes money to the client.

Upon settlement of the final account, the project manager releases to the contractor the remaining 50 per cent of the retention money held by the client, inclusive of an agreed amount of interest, and any bank guarantees provided by the contractor. From the legal viewpoint, the contract between the client and contractor has come to an end.

A retention fund and a bank guarantee serve as a form of contract security. Should the contractor fail to settle the final account, which may be in favour of the client, or should the contractor fail to rectify defects or be in breach of the contract, the client may use the retention fund and/or the bank guarantee to recover any incurred costs from the contractor.

The project manager also settles a contract between the client and the design leader. Upon receiving a set of 'as built drawings' for future reference, the project manager prepares the final account, which details the amount of the final payment to the design leader. Once the account has been settled, the contract between the client and the design leader is also at an end.

DEVELOPING A PLAN FOR MANAGEMENT OF THE COMPLETED PROJECT

The client may call on the project manager to assist in selling or leasing the completed project. In the former case, the project manager will find a suitable real estate agent. In the latter case, the project manager's involvement would be more extensive. He may be required to develop and implement a strategy for managing the completed project. Such a strategy may require:

- assisting in the selection of tenants and the preparation of tenancy contracts
- co-ordinating fitouts for the project
- selecting a facility manager for the ongoing management of the project.

Summary The end of the commissioning stage brings the project manager's contract with the client to an end. The project manager issues the final account to the client and on receiving the payment the contract is discharged.

Activities performed by the contractor

The contractor's main activities at the commissioning stage have already been described in relation to the work performed by the project manager. In summary, such activities involve:

- bringing all the construction activities specified in the contract to an end
- repairing any defective work as directed by the project manager
- commissioning services such as air-conditioning, ventilation, lifts, fire protection, IT and security, and ensuring their satisfactory performance
- securing warranties from services suppliers and installers, and handing them over to the project manager
- supplying an 'as built schedule' to the project manager (when required under the contract)
- submitting a final account under the contract to the project manager.

At the end of the commissioning stage, when the contractor has received the final payment from the client and the payment of the remaining portion of the retention sum, and after the release of the bank guarantee, the contractor's contract with the client is discharged.

CONCLUSION

This chapter examined the roles of a project manager and a contractor in managing the construction and commissioning stages. Important issues that contribute to successful completion of construction were highlighted. They include effective information flow across all project participants, regular short-range scheduling of construction activities and effective management of resources, particularly subcontractors. The need for effective and timely administration of the main contract and subcontracts was emphasised, as was the importance of a periodic monitoring and control of progress and performance.

At the commissioning stage, all construction activities are concluded, any identified defects repaired and a project's services commissioned. A strategy is developed for the future management of the project. Finally, the client's contracts with the contractor, the design leader and the project manager are discharged.

EXERCISES

1. Is it possible to eliminate the defects liability period from a project lifecycle? Give reasons for your answer.
2. What steps should a project manager take at the pre-construction and construction stages to minimise the volume of defects identified at the commissioning stage?

PART 3

PEOPLE IN PROJECT MANAGEMENT

CHAPTER 10

MANAGING GROUPS AND TEAMS

INTRODUCTION

As organisations grow they naturally develop internal group structures, which are differentiated on the basis of membership, function, purpose and/or geographical location. This is complicated in construction organisations because of the industry's orientation around projects that demand the formation of temporary teams from various distinct internal functional departments and external organisations. From an internal staffing perspective, construction companies structure project teams largely on the basis of who becomes available from other projects at the required time. In contrast, external organisations are normally employed on the basis of price, in a competitive tender. Economics also determines that there are overlaps between the membership of different projects in that most people are allocated to a number of different project teams, depending on the size of the projects and their capacity for work. In recent times, due to widespread downsizing and rationalisation within the construction industry, the number of projects that an individual or organisation may be involved in has grown considerably. This has caused its own problems of work-related stress and overload, which ironically have shown up in increased accident rates and lower performance.

The purpose of this chapter is to discuss a more intelligent approach to team composition within the multi-organisational setting of the construction industry. More specifically, the aim is to reinterpret the models of team-building developed for use within permanent organisations, where individuals are the unit of analysis, for use in a multi-organisation setting where firms are the unit of analysis. This first involves understanding the nature of groups as the basic building block of teams and in particular, how they form and how they vary.

DISTINGUISHING BETWEEN GROUPS AND TEAMS

There has been increasing emphasis in organisational research on teams (Robbins et al. 1994). Top management teams, work teams, task forces, project teams, quality action teams and so on have become an increasingly important part of organisational operations.

A group is commonly defined as two or more persons who work together to achieve certain objectives, while a team is defined as group of committed people with specific skills, abilities and interdependent roles who work together in an environment of trust, openness and co-operation towards achieving common goals. It should be noted that only a formal group could become a work team.

The difference between a group and a team is subtle in that a team contains the qualities of a group (a common purpose, interdependence, two or more people, assumption of different responsibilities) but in addition contains a greater degree of voluntary integration and individual commitment towards the shared goal (Miller 1971). A team is also differentiated by its members placing the welfare of the team above the welfare of themselves and wanting to belong and identify with it. A team has an identity or special purpose which members place above their own interests, which results in the combined efforts of the group having a synergistic effect. In effect, a team is a group but a group is not necessarily a team.

Teamwork is a popular buzzword in today's business world and organisations try to use teams extensively in one form or another. They may be permanent or temporary and are formed to perform a range of activities. Some popular examples of teams will now be briefly discussed.

Functional team

A functional team is commonly formed within the functional area of the organisation, for example in the sales, production or any other functional department. Its main task is to solve problems within such a department.

Self-directed or self-managed team

Traditionally, teams relied on managers to provide direction for tasks to be accomplished. An alternative solution is to empower teams to become self-directed or self-managed. In this case, rather than directing the team step by step towards attaining a goal, it is empowered to organise and perform the work by relying on its own management efforts. This approach gives team members a sense of ownership and tends to improve their morale and enthusiasm, which is reflected in better performance.

Project team

A project team is a cross-functional team that brings together specialists from different organisational departments whose task is to work on specific assignments. For example, the managing director of a property development organisation may appoint a project team comprising specialists from different departments to develop new investment strategies.

In the construction industry, as we have seen, project teams comprise representatives of different organisations who join the team at different stages of the project lifecycle. Their involvement may be for the entire duration of the project but is commonly limited to a specific period related to the extent of the work that they perform. Managing a project team with such a diverse composition of team members is an extremely difficult and challenging task for any project manager. An effective manager leads a project team by drawing on the four management functions of planning, organising, leading and control. He works closely with the team in setting goals, planning team activities, organising the work and resources, and periodically evaluating the team performance.

TEAMWORK IN THE CONSTRUCTION INDUSTRY

There has been increasing proof that the use of well-structured teams can contribute significantly to organisational success (Belbin 1984; Margerison et al. 1986; Barbara 1997). The potential benefits of team-working are particularly great when tasks are highly complex and interdependent, as they are in the construction industry (Hackman & Morris 1975; Galbraith 1977). But despite the nature of construction activity lending itself to a teamwork approach, little attention has been given to this issue in the construction industry (Rwelamila 1994; Luck & Newcombe 1996). The method of team selection in that industry has changed little and is primarily driven by accountability, which ensures that the main method of selection is competitive tendering and the ultimate criterion for bid selection is price (Latham 1994). This obsession with price is in part a reflection of the transitory nature of the industry, which perpetuates short-term managerial attitudes, unfair contractual practices and under-resourced, confrontational and crisis-prone projects that contain little sense of collective responsibility (Loosemore 2000). For those operating in such an environment, survival becomes a priority and the goodwill and tolerance necessary for effective teamwork is an unaffordable luxury. Unfortunately, the companies that thrive in such organisations are the unscrupulous ones and their dominance encourages mediocrity and at worst corruption and reduces the behaviour of the industry to the lowest common denominator. It is not surprising therefore that Latham (1994) cited a lack of teamwork as the main problem facing the construction industry.

More intelligence

A growing recognition of these problems have led to calls for more intelligent selection systems for contractors and consultants and a movement towards negotiated contracts, often within continuous partnering arrangements (Hatush & Skitmore 1997). However, while such systems are very important in improving the quality of project participants by screening them through a number of criteria, they do not address the issue of integration — that is, how project participants gel into an effective team. While various alternative models of team performance have been produced, it is widely accepted that such performance depends on composition and compatibility, so that the abilities and personal characteristics of team members are given scope and the team is structured in terms of the roles they play (Shaw 1976; Belbin 1984; Margerison et al. 1986; Szilagyi & Wallace 1987). Other factors that have been identified as important to team performance are group size, internal and external patterns of communication, leadership, norms, status, internal procedures for decision-making, communication, conflict management and feedback, resourcing constraints and external conditions relating to organisational culture, strategy, structure, regulations and technology (McGregor 1960; Argyris 1970; Hall 1971; Dunphy 1981; Likert 1984; Johnson & Johnson 1994; Archer 1996).

Despite the apparent value of research on team-building in the behavioural sciences, one should be cautious about transferring it to the context of a construction project. One reason is the construction industry's unique professional roles, employment practices, expectations, norms and traditions. While it is important not to treat construction as a special case, it is also important to recognise the possible influence of these unique characteristics on the team-building process. For example, Loosemore and Tan (1999) discovered the basis of strong stereotyped perceptions between different occupational groups within the construction industry that would have to be considered. For example, most occupations (particularly architects) saw contractors as aggressive and blunt, whereas engineers were widely seen as being systematic, proficient and reliable. Perceptions like these, whether true or false, play a significant role in team relationships, particularly during the early forming and storming periods (pages 225–6). However, it pays to be cautious because all team-building research has taken place within what Antony (1988) describes as permanent business organisations. From a team-building perspective, there are important differences between such organisations and the temporary project organisations used to produce buildings (Cherns & Bryant 1984; Bresnen 1990). In particular, Antony (1988) argues that temporary project organisations are typically more transitory and dynamic than permanent business organisations and are characterised by much more differentiation and conflict. Furthermore, the project organisations used in construction are multi-organisational in that they comprise a range of smaller component organisations which have their own norms and practices, reflected in their organisational culture. Collectively, these characteristics make team-building more complex, in that managers are dealing with a constantly changing array of people to integrate into teams whose suitability for different roles is determined not only by their individual personality but also by the culture or corporate personality of their employing organisation. This means that team-building techniques in a multi-organisation setting will have to operate on two levels: individual and organisational. Unfortunately, while considerable research has been done on individuals, there has been very little done on organisations. That is, we have very little understanding of how to combine different organisations into teams.

Achieving teamwork

Achieving teamwork in the construction industry is a challenging task because team members come from different organisations and are contracted to the project. Collectively they would want to work towards meeting the overall project goals, but individually they also need to protect their own interests. Since their participation is induced by contracts, under unfavourable or onerous conditions of contract some team members may carry a larger and inappropriate proportion of risk. In such circumstances they tend to be overly cautious and suspicious, and not trusting of other team members.

Developing trust in project teams working in the construction industry is vital for good results. Developing teamwork requires time and commitment. A project leader needs first to devise a strategy for achieving teamwork and then implement it by encouraging the full participation of team members.

A good starting point is to select a competent project manager with the appropriate skills who is experienced in managing project teams; the ability to develop teamwork is particularly important. The project manager is responsible for defining a team structure and the specific roles of team members, and will liaise closely with the client in selecting members. In reality, however, the project manager may have little control over the selection of the main contractor and subcontractors since they will most likely be selected on the basis of a low bid. Nevertheless, the project manager has an opportunity to exercise control over the selection of specific personnel from organisations contracted to the project. It is important that people experienced in teamwork represent such organisations. It is also important that they are empowered by their respective organisations to participate in a team decision-making process. A team has now been assembled and its work is about to commence.

A process of team development or team-building must start as early as possible. Ideally, it would be appropriate to address team-building in isolation from specific work-related activities of the team. This is best done through structured and facilitated training programs where participants learn to develop team-building skills by participating in selected games, case studies and role-plays. They would also learn to appreciate the importance of developing trust and openness in a team environment as the condition for achieving group synergy. Fryer (1990: 261–2) and Kezsbom and associates (1989: 276–81) provide a brief review of some of the better-known team-building approaches.

However, it is often difficult to run separate team-building training programs because of time constraints and the fact that team members come from different organisations. The most common solution is to integrate team-building into specific team activities. Partnering offers a workable model for combining team-building and specific work-related activities of teams (see Uher 1994). Its main components are start-up and follow-up workshops.

START-UP WORKSHOP

The purpose of a start-up workshop is to establish good communication and team-building skills between the key project team members.

The project manager together with the facilitator will develop a strategy for the workshop. This may involve:

- establishing a detailed agenda
- selecting project team staff
- nominating other important organisations such as subcontractors, suppliers, end users, local authorities, and so on
- establishing any training requirements for project staff.

The workshop should be held in a relaxed environment (preferably away from the project site to allow participants to get away from their daily duties and concentrate fully on the workshop activities) conducive to open discussion. Appropriate facilities are required to accommodate the whole group as well as subgroup discussions.

The number of participants should be kept to a size appropriate to promoting efficiency and teamwork. The length of the workshop should be at least two days.

A start-up workshop may include the following activities:

- Learning to know one another. Individual participants will be introduced, job relationships identified, and the notion of effective communication, negotiation, conflict resolution, team-building and openness introduced. The aim is to focus on changing the attitude of the participants from the traditional 'us' and 'them' to the team-spirited 'we'. This requires the development of trust.
- Developing mission statement and objectives. A mission statement will define what the team wants to be and to do, and the values or principles on which its work will be based. The team will commit to mutually developed project objectives that will be specific and measurable.
- Identifying problems, issues or opportunities and devising an action plan. In small groups, team members will identify and prioritise potential key issues or problems that may prevent attainment of project objectives and develop strategies for overcoming them. They will then develop a specific action plan to address such issues adequately.
- Understanding other team members' risks and concerns, and seeing where one's role in the contract fits in relation to others'; helping to build the essential team attitude. In the process, individuals grow to know and understand the personalities with whom they will be working before problems arise. This investment in the human dimension can have great benefits for the life of the project and potentially beyond.
- Developing a problem resolution/escalation process. The team will devise processes to ensure a rapid and satisfying solution to any conflicts that will arise during the course of the project. This will mean identifying lateral and vertical authority levels, and agreeing to resolve problems at the lowest level and in the shortest possible time.

- Developing an evaluation process. The team will devise a method of evaluating the team effectiveness against project objectives.
- Making and signing a project charter. On partnered projects, the team will draft a document, the Charter, which will include a statement of mission, a list of project goals and objectives, and the details of the issues resolution and evaluation processes. The Charter is not a legal contract but a moral or symbolic agreement of the team members' commitment to the project.

FOLLOW-UP WORKSHOPS

Follow-up workshops are useful for reviewing the performance of the project and the project team. They also serve the purpose of introducing new team members to the project. Follow-up workshops should ideally be held at three monthly intervals, but their frequency will need to be aligned with the overall project duration and the level of actual performance achieved.

THE NATURE OF GROUPS

A team-based approach to work is desirable in most organisations. It not only creates a better work environment for people but it also improves productivity. There is little to match the camaraderie, joy and fun of working together on something worthwhile – assuming that people can work co-operatively rather than competitively. Some notable characteristics of a team-based approach are (Tjosvold & Tjosvold 1991; Katzenbach & Smith 1993; Robbins et al. 2003):

- Promotion of co-operation and teamwork, which improves motivation, morale and job satisfaction of the team members and leads to improved productivity.
- The ability of self-directed teams to adapt to different situations and take advantage of workforce diversity in developing more innovative ideas. Management is free to focus on strategic thinking and long-term plans.
- Formulation of clear goals that teams understand and are committed to. Teams are able to work together to develop an appropriate strategy for achieving such goals.
- Pooling of relevant skills from competent specialists in a highly co-operative team environment, which are necessary for attaining the desired goals.
- Establishing mutual trust among the team members through the development of strong interpersonal relationships based on openness and mutual support. Organisations that promote openness, empowerment, worker participation and collaborative processes create an environment of trust.
- Developing unified commitment, dedication and loyalty.
- Developing strong communicating and negotiating skills. Since team members work together nearly every day, they understand and have respect for each other, and are able to resolve occasional conflict in a more relaxed manner.

- Developing appropriate leadership. Effective leaders facilitate rather than direct and control teams. They encourage, support and reward team members and generally focus their attention on achieving objectives.
- Providing support to teams in the form of resources, appropriate authority and responsibility, training, and reward and performance evaluation.
- The ability of teams to provide moral support to those who want to challenge existing practices and cultures.

We can begin to understand the behaviour of a work-group when we see that it is a subset of a larger organisational system. The explanation of the group's behaviour can be seen as based on the organisational conditions imposed on the group. See the explanation of a work-group at pages 219–20 below.

Strategy

Any organisation, including a construction project, has a strategy that defines what business it is in or wants to be in, and the kind of organisation it wants to be. It is set by top management, preferably in collaboration with lower-level managers. Strategy outlines the organisation's goals and the means of attaining these goals. It might, for instance, direct the organisation towards reducing costs, improving quality or delivering on time. The strategy that an organisation is pursuing, at any given time, influences the power of various work-groups, which in turn determines the resources that the organisation's top management is willing to allocate to it for performing its tasks. Typically, those groups that can contribute most to those goals become more powerful.

Culture

Every organisation has an unwritten culture that defines for employees what is acceptable and unacceptable behaviour. After a few months, most employees understand their organisation's culture. They know things such as what is appropriate dress for work, which rules are rigidly enforced, what kinds of unusual behaviours are likely to get them into trouble and which are likely to be overlooked, the importance of honesty and integrity, and which management goals really do count and which don't. While most groups have their own subculture with modified or unique standards, there is usually a dominant organisational culture that conveys to all employees those values the organisation considers most important. Members of work-groups have to conform to the standards implied in the organisation's dominant culture if they are to remain in good standing.

Managing Groups and Teams

Resources

Some organisations are large and profitable, with an abundance of resources. Their employees, for instance, will have modern, high-quality tools and equipment to do their jobs. Other organisations aren't as fortunate. When organisations have limited resources, so do their work-groups. What a group actually accomplishes is, to a large degree, determined by the money, time, raw materials and equipment that is allocated to the group by the organisation. An absence of resources will have a large influence on the group's behaviour.

Technology

The technology a group possesses is a strong influence on a group and its perception of how it operates. The use of facsimile equipment, electronic mail, mobile phones, personal computers and the technical nature of the work they do can affect the communication style, the number of people in the group and the frequency of contact between group members. For example, in a highly computerised project, employees are often lulled into a false sense of security in the false belief that more IT will improve communications. It is important to appreciate that IT may increase the amount of information flowing in a project but may not improve communications. Information overload is becoming a serious problem in many organisations and people also have a tendency to see technology as an end in itself and not a means to an end. This is not the fault of the technology but of the users, since most technologies are not designed to make decisions but to help decision-makers make better decisions.

Authority structures

All organisations have authority structures that define who reports to whom, who makes decisions, and what decisions individuals or groups are empowered to make. This structure typically determines where a given work-group is placed in the organisation's hierarchy, the formal leader of the group, and the formal relationships between groups. While a group might have an informal leader, the formally designated manager appointed by the organisation has an authority that others in the group don't have.

Formal regulations

Organisations create rules, procedures, policies and other forms of regulations to standardise employee behaviour. Those that have strict standard operating procedures reduce the discretion of work-group members to do things differently; the more formal regulations that the organisation imposes on its employees, the more the behaviour of work-group members will be consistent and predictable.

Personnel selection

Members of any work-group are, first, members of the organisation of which the group is a part. The criteria that an organisation uses in its selection determine the kinds of people that are to be in its work-groups. The selection factor is crucial in the construction industry, for instance when unions are a powerful force among the workers. In many construction industries around the world, unions play an active part in constraining an organisation's activities and work practices. Often the imposition of work conditions that a group considers unfair is dealt with by the unions, and a manager might have to moderate what he says to a group because of concern for the union's reaction.

Performance evaluation

Since work-groups are part of a larger organisation which provides employees with performance objectives and has mechanisms for evaluating and rewarding people's performance in relation to them, groups are inevitably influenced by this system.

FORMAL AND INFORMAL GROUPS

Groups can vary in many ways. Some are highly structured in terms of prescribed roles and responsibilities whereas others are random, impulsive, chaotic and flexible. Some are loosely coupled whereas others are tightly knit. Some groups have a much stronger sense of identity than others. The simplest way of combining all of these differences into one distinguishing attribute is by referring to their degree of formality.

Formal groups

Formal groups exist because someone with formal authority has created them. They can exist for a huge variety of reasons and do so at all levels of organisations so that information can be filtered upwards through a hierarchy, which considers ever more restricted and strategic issues.

The most common type of formal group is the work-group, which is composed of a supervisor and those who report directly to him. This type of group is used extensively within consulting practices to accomplish specific tasks associated with particular projects. For example, in a quantity surveyor's office work-groups would be created to produce a bill of quantities for a particular job and in an architect's office one would be established to produce a set of working drawings for a project. In a construction company, a work-group might be formed to complete a tender for a specific project. In organisations within the construction industry, these are typically temporary entities since it is difficult to predict the size and complexity of projects that will need attending to in the future. However, in organisations that operate within more predictable environments, a work-group may be more permanent; this has the advantage of strengthening the ties between its members and therefore the cohesiveness of the group as a working unit.

Another form of temporary formal group is the task force or think-tank, normally created specifically for the development of new products, the creation of new ideas, or the solution of specific problems. In contrast to work-groups, which tend to have little functional differentiation, task forces and think-tanks are normally multidisciplinary, reflecting the more complex nature of the task they are given. They are also likely to have a longer life span than work-groups and are the type of groups a construction project manager might construct to resolve specific problems on a project which demand multidisciplinary skills.

The longest life span of any formal group is seen in the committee, which also has the most formal and complex structure. All construction projects have committees, mostly at managerial level, to review and make policy decisions on the overall direction of the project. Poorly constructed or managed committees can become notorious talking shops, and to avoid this pitfall a committee's goals should be clearly specified in writing, as should its authority and scope of activities. The ideal size of a committee is five to ten people; a chairperson should be selected on his ability to run an effective meeting. A permanent secretary should be appointed to handle communications, to set an agenda and distribute all supporting material.

Informal groups

Unlike formal groups, which are a deliberate creation of management for a specific purpose, informal groups occur voluntarily, haphazardly, spontaneously and naturally as people interact day by day. For these reasons, they are often efficient and effective but notoriously difficult to manage. Carnall (1982: 21) has distinguished four main types of informal groups: interest groups, coalitions, collectivities and reference groups. All exist in construction projects, as they do in any organisation.

An interest group is a group of people located in a given structure, capable of articulating similar ideas, attitudes and beliefs. They have similar interests in respect of particular issues and are capable of acting together in an organised manner with agreed goals. In a traditional construction project, the two main interest groups are the design and construction teams, their different interests being determined by the nature of the procurement system, which separates them and their reward structures.

Coalitions are groupings of interest groups who are committed to achieving a common goal. They are more temporary than interest groups and often come together as a result of some event that causes them to have common interests. Indeed, they may be drawn from different interest groups. For example, on one construction project, which had a problem with excavation, the engineer took sides with the contractor in defending his argument that permanent earthwork support was required, whereas only temporary work had been allowed for in the bill of quantities by the quantity surveyor. Normally the quantity surveyor and engineer would be in the same interest group.

Collectivities are interest groups with external membership such as trade unions, professional institutions and so on.

A reference group is the group with which people compare themselves when assessing their own position with regard to financial rewards, job security, working conditions and status. For example, in a construction project, subcontractors are always comparing themselves with the main contractors on the basis of contractual risk distribution and rewards, which is often a source of demotivation for the former.

In addition to the above, Sprott (1958) pointed to the formation of social groups based around patterns of friendship and which can often overpower the influence of differing interests, although he acknowledged that patterns of friendship often develop because of similarities in interests. While groups normally only form for social reasons, they are an immensely powerful integrating tool for managers. For example, on one construction project a project manager effectively used social groups to break down cultural barriers by asking each nationality (which tended to socialise with themselves) to cook lunch for other cultural groups on a two-weekly roster.

The value and danger of informal groups

It is inevitable that informal groups evolve within organisations because they perform useful functions, particularly at times of crisis and change where formal structures may be inflexible, inappropriate and counterproductive. They are particularly prevalent during a construction dispute when conflicting interests are magnified. Therefore, in some types of construction projects or during certain periods of a project's life, they can have a greater influence on organisational outcomes than the formal system. For this reason it is essential that they are acknowledged and managed for constructive ends. Unfortunately, this is easier said than done because of their secretive and transitory nature, which makes them less identifiable within an organisation than formal groups. Indeed, the only characteristic that would differentiate an informal group from its surrounding organisation is that its members would interact with each other more than they would with anyone else. Thus the effective management of informal groups requires a project manager to be trusted, popular and continuously connected to the informal communication networks that exist at any time within the project's organisational structure. The problem in traditionally procured projects is that the project manager is often the architect, who is closely linked to a project interest group, and this can undermine any trust that members of opposing interest groups might have in him. There is therefore considerable justification for an independent project management role, particularly in high-risk projects where the probability of chaos, crises and conflict is higher. It is only an independent person who would be able to engender the sense of trust required for the successful management of informal groups.

The danger of groups

Whether informal or formal, groups clearly perform many important functions within organisations, not least enabling people to communicate more closely and solve tasks of much greater magnitude than they would be able to tackle alone. Indeed, if properly constructed, groups should have a synergistic property in that their collective capability should be far greater than the sum of the capabilities of the individuals who contribute to it. Groups also provide for people's social, identity and esteem needs and can provide a source of power for individuals in negotiations. The power-giving aspect of groups is very important in the confrontational environment of construction projects and is a major reason why they form. The reason for this is related to the high incidence of conflicting interests and conflict within construction projects. Finally, groups also provide an important source of security for individuals, enabling them to be more courageous and to take greater risks than they otherwise would. This is a valuable attribute, particularly in providing the creativity and innovations essential to the prosperity of organisations in an increasingly turbulent and competitive environment.

While groups can perform many useful functions for organisations, they can also be damaging. Poor management that fails to recognise and acknowledge them can result in them disjointing the organisation by ignoring important interdependencies, which people feel they need to satisfy their organisational and personal needs. Furthermore, some informal groups may exist for subversive ends and these need to be identified, investigated and disbanded if need be. Alternatively, by being aware of such groups, managers can better help them refocus their energy on the accomplishment of constructive organisational tasks.

Hornstein (1986: 103) has expressed alarm at the 'in-roads being made by the social ethic which espouses groups, not individuals, as the prime source of creativity and proclaims group membership, an experience akin to being in a family, as the ultimate need of an individual'. In Hornstein's (1986: 104) view, groups have a 'capacity for producing a special, perniciously subtle tyranny' which suppresses creativity and innovation. Marsh and colleagues (1978) have also noted the powerful influence that a group's norms can exert over its members. Their research into football hooliganism concluded that group norms (expected standards of behaviour) cause people to become less extreme in their own views and to exhibit blindly almost tribal violent behaviour that their personality and wider society would normally suppress. Furthermore, it has been found that the need to establish some form of consensus among group members can slow down decision-making and result in compromise decisions. These findings are of great importance since open-mindedness, flexibility, creativity and rapidity of response are becoming of greater value in today's increasingly uncertain and competitive business environment.

Janis (1971) referred to the tendency of some groups to over-emphasise the importance of consensus and agreement as 'Groupthink' and has argued that the more attractive the group is to its members the more potent is this effect, the stronger is the rejection of non-conforming individuals, and the less likely the group is to solve unique problems creatively. The main danger associated with this trend is that group members appear to be completely brainwashed by it and unaware that it is occurring. Janis produced a list of causes and cures for groupthink, which is given in Table 10.1.

TABLE 10.1 CAUSES AND CURES FOR GROUPTHINK

SYMPTOMS	CURES
Illusions of group invulnerability: members of group feel they are beyond criticism, attack or failure.	Expose the group to the risks they face and their vulnerability to them.
Rationalising unpleasant and disconfirming data: refusal to accept contradictory data or to consider alternatives thoroughly.	Provoke and encourage the sharing of views and opinions. Make people feel at ease and not threatened by outsiders.
Belief in inherent group morality: members of group feel it is right and beyond reproach by outsiders.	Expose the group to different views and ideas. Act as a manager or group leader and never favour or disfavour one particular course of action.
Stereotyping competitors as weak, evil and stupid: refusal to look realistically at other groups.	Encourage group members to think about and discuss the problems outside the group environment.
Applying direct pressure to deviants to conform to group norms: refusal to tolerate a member who suggests the group is wrong.	Divide the group up into subgroups to discuss the same issue and then bring them together to discuss the whole issue again.
Self-censorship: refusal by members to discuss personal concerns with the group as a whole.	Invite outsiders in to provide an objective view on discussions and act as a devil's advocate.
Illusions of unanimity: accepting consensus prematurely without testing it thoroughly.	Use second-chance sessions to allow people to contribute anything they may have previously left out; test consensus again before making a decision.
Mind-guarding: members of group protect others from hearing unpleasant ideas or viewpoints from outsiders.	Get rid of dominating and suppressive elements within the group that may be intimidating members and forcing them to conform.

The establishment of group norms depends on a number of factors, namely the extent to which they aid group survival, provide benefits for its members, simplify group processes, and add to the power of the group and express the central identity and values of the group.

Thus, paradoxically, it seems as if groups can be at their most valuable and dangerous during complex, innovative and non-routine projects. Effective management is therefore crucial and there is an argument that construction project managers should deliberately build divisions into project teams so that the consequences of groupthink are avoided. It is very important to appreciate that an effectively structured group is not necessarily a cohesive one.

THE FORMATION AND DEVELOPMENT OF GROUPS

An understanding of the group formation process provides the basis for their effective management. Turkman (1965) argued that although formal and informal groups and temporary and permanent groups develop differently, all groups go through four basic stages: forming, storming, norming and performing.

Forming

This is an exploratory period during which the initial ground rules and structure for the group's interpersonal relationships are developed. During this stage, members of a group become acquainted with each other and begin to establish relationships. Basic rules and procedures begin to emerge and people develop and communicate their expectations of each other.

Storming

As relationships grow clearer, the group starts to develop a sense of purpose and identity, which further cements the relationships between members. This enables personal relationships to develop in association with a clearer understanding of operating relationships. During this stage tensions are normally very high and periods of conflict may be common as group members jockey for position. Indeed, the tensions may temporarily divert attention from the group's task. This undesirable scenario is more likely in formal groups and can be avoided to some extent by considering the structure of group membership in advance of its formation. This is seldom done in construction projects or in construction companies when allocating staff and organisations to projects. In construction industry, price and availability are the primary selection criteria for team construction.

Norming

From the storming stage a group consensus and relationship structure should emerge within the group, as should consensus over its aims. The norming stage is the starting point for action, and although there may still be some instability in roles, responsibilities and relationships, the group strives to maintain harmony and work towards the accomplishment of its objectives. During this stage, a group culture develops through the establishment of acceptable standards of behaviour (norms), and people's individual roles within the group become firmly established.

Performing

This is the stage during which the group's focus moves wholly onto the achievement of objectives and where its structure, norms and behaviours are firmly established and accepted by all. The group is now mature, organised and effective and any problems which arise because of differences or misunderstanding can be handled with minimum conflict. Indeed, by this time the group will have identified and resolved most of its differences and the level of surprise should be at a minimum.

These stages of group development are illustrated in Figure 10.1 but there is also another stage in group formation known as adjourning which only applies to temporary groups and is associated with their eventual disbandment once their task has been completed. This is clearly of relevance to construction projects, which are temporary coalitions in their own right and which are made up of coalitions.

Figure 10.1 The stages of group formation

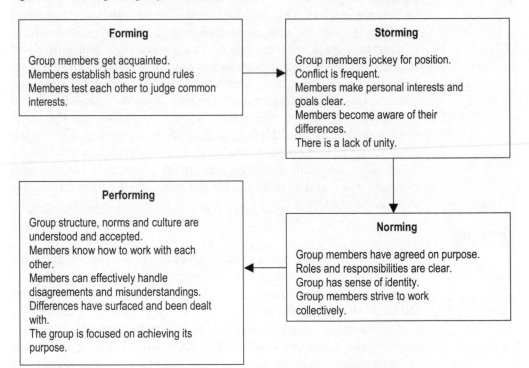

Forming

Group members get acquainted.
Members establish basic ground rules
Members test each other to judge common interests.

Storming

Group members jockey for position.
Conflict is frequent.
Members make personal interests and goals clear.
Members become aware of their differences.
There is a lack of unity.

Performing

Group structure, norms and culture are understood and accepted.
Members know how to work with each other.
Members can effectively handle disagreements and misunderstandings.
Differences have surfaced and been dealt with.
The group is focused on achieving its purpose.

Norming

Group members have agreed on purpose.
Roles and responsibilities are clear.
Group has sense of identity.
Group members strive to work collectively.

THE NEED FOR MANAGERIAL SUPPORT IN SUCCESSFUL GROUP FORMATION

It is important to point out that Figure 10.1 presents an idealistic model and that in reality groups progress through these stages at different rates. Indeed, some groups may never achieve full maturity. The task for managers is to monitor the formation process to ensure that the necessary support is provided to enable successful transitions to occur between each stage of the maturation process. This is particularly important in the development of formal groups, which have been constructed artificially by managers. For example, during the forming stage, group members will need help in understanding why they have been put together, what their mutual interests are, and how they can work together. Social events may be organised to establish and cement personal relationships and a sense of identity given to the group by the provision of a dedicated space. During the storming stage managers must be able to perform mediating roles between opposing parties and defuse any tensions that may develop. They must also help people to understand each other's goals, needs and expectations and to discover ways of increasing their compatibility. In the norming stage, managers must ensure that the group culture that develops is compatible with the overall organisation's culture and that conflicts do not interfere with task accomplishment. They must also act as an external intermediary in ensuring that a group has all the information or resources it needs to achieve its goals. A manager should also help to co-ordinate activities with other dependent groups which may be working in parallel. Finally, during the performing stage a manager should focus on maintaining information flow and help members address problems or constraints arising from outside the group boundaries. Within a construction project, the large number of groups that are constantly forming and disbanding may make this a time-consuming process and it may be necessary to take on a manager specifically for the purpose of managing group processes, though this rarely happens in practice.

GROUP DYNAMICS

Group dynamics are the patterns of interaction and behaviours that evolve in groups, which influence how well the members can accomplish their tasks and how satisfied and cohesive they are. Homans (1950) produced a model that is useful for understanding group dynamics and this is given in Table 10.2.

TABLE 10.2 GROUP DYNAMICS

	ACTIVITIES	INTERACTIONS	SENTIMENTS
Required behaviours	Develop work safety plans at the start of every project.	Require subcontractors to develop and submit work safety plans for their area of work.	Maintain a positive attitude towards safety.
		Report any problems to safety officer.	
Emergent behaviours	Collect any data or information necessary to construct effective safety plans.	Help subcontractors understand safety system.	Encourage others with enthusiasm.
	Work on weekends if need be to develop safety plans in time for start of job.	Work closely with subcontractors to solve problems as they emerge.	Work effectively with subcontractors and safety team.

Required behaviours are those actions which a group expects from its members as a condition of membership. In essence, they are the formally prescribed actions that are considered essential if the group is going to achieve its task. For example, certain members of a group are expected to meet at certain times and produce reports on progress. In construction projects, these behaviours are prescribed by the contracts of employment, which bind project participants together. In contrast, emergent behaviours are those actions that members do in addition to or in place of those required by an organisation. The balance between these types of action depends on:

- the receptivity of group members to prescription
- the appropriateness of the procedures prescribed
- the certainty, predictability and routineness of the task which determines the ability to prescribe procedures effectively.

Construction project managers should give careful consideration to the characteristics of their projects and to the people within them to enable them to understand the influence of contractual requirements formally imposed on their project team. In general, however, it would seem that the professional nature of construction team membership and the uncertainty of construction activity would ensure that emergent behaviours are predominant within construction project organisations. That is, project managers rely heavily on the flexibility, goodwill and ingenuity of its members and their ability to informally interact with minimum persuasion. This means that the construction process and the activities of construction firms are particularly vulnerable to the impact of a poorly constructed team. The maintenance of positive interpersonal relationships within a project team is clearly a critical aspect of a project manager's job.

In Homans' model, activities refer to the physical behaviours in which group members engage. For example, in a design team, architects and engineers would put pen to paper to produce the elevations, plans and details. Emergent activities relate to those performed beyond the call of duty such as shading or even colouring drawings to make them more attractive and legible. Interactions are the two-way communications between group members that the organisation requires. For example, construction contracts require that certain information has to be made available to the contractor at certain stages of the project. Emergent interactions relate to information offered voluntarily, such as taking time to explain drawings or notifying the contractor of potential problems or even possibly giving the contractor information before it is needed to assist him in the logistics of materials ordering and so on. Finally, sentiments refer to the feelings and attitudes people hold towards each other. For example all members of a group may be required to be polite and considerate towards each other but some may go further than this and go out of their way to be helpful. Sentiments within groups are very important to their performance, yet they are the most difficult aspect of a group to manage. This is because they are not easily specified and are largely a product of the nature of interpersonal relationships, which emerge over time.

The value of Homans' model is that it illustrates that managers should understand what people in construction projects expect of each other in terms of activities, interactions and sentiments. Furthermore, they should also understand that spontaneity and informality occur and that it is important not to rely too heavily on the required behaviours enforced by a formal construction contract.

GROUP ROLES

Benne and Sheats (1948) were the first researchers to empirically prove the importance of role structure to the performance of a group. A person's role, they said, is defined by the set of activities that a group member is expected to perform in his position in the group. Role structure is the total pattern of different roles identifiable in the group, and the relationships between them. Fifty years later, Tyson (1998) identified a number of potential problems that can develop in relation to role structure, and these are listed below.

- role ambiguity (differences between perceived and enacted roles and relationships with other roles)
- role conflict (two or more people performing the same or irreconcilable roles)
- role loopholes (an unallocated role)
- role underload/overload (degree to which the demands of a role fall short of or exceed the resources, time and expertise available to an individual)
- role stress (distress caused by giving an individual an inappropriate role or imposing too great expectations for a given role)
- role hunger (someone having a yearning to perform someone else's role).

A considerable contribution towards the reduction of such problems can be made by better matching of individual personalities (and in the case of construction projects corporate cultures) to specific roles. The first step in doing this is a consideration of the different types of roles that are needed within an organisation for its effective functioning.

Types of roles needed in organisations

Benne and Sheats' (1948) now classic study identified a number of roles that need to exist within an organisation. At the most basic level there are task and maintenance roles, the former being necessary for the accomplishment of the group's purpose and the latter for effective group functioning. Task roles perform a physical activity and maintenance roles manage interpersonal relations in an attempt to foster cohesion and good relationships.

However, if a group is to be effective then it must also develop a mix of formal and informal roles. One such mix has been proposed by Belbin (1983, 1994), based on extensive research in the United Kingdom. Belbin studied various groups and compared how they performed with different mixtures of roles. The findings suggest that effective teams are characterised by certain combinations of distinctive roles, which are listed below. Belbin's approach to effective teamwork suggests that all of these roles need to be covered by a team. In some cases a person may have to play two or more roles and different types may require more emphasis on certain roles.

- The chairperson/co-ordinator is a co-ordinator of the activities and skills of the team. This person encourages contributions from each team member and keeps the team directed to achieve the group's objectives. The chairperson can be so busy co-ordinating other people's inputs that he may lack creativity or intellectual ability.

- The company worker has a strong drive to achieve organisation or team goals. The company worker focuses more on the team goals than relationships within the team. He can be a great contributor to team performance but can sometimes lack concern for the welfare of the other team members.
- The shaper likes to contribute ideas and direction to the team but does not necessarily like to lead. He often challenges inertia or complacency in the team. Shapers are often intelligent, have a strong drive, and may be assertive and highly strung. They can become frustrated and impatient if their ideas are not accepted.
- The resource investigator maintains and develops contacts and liaison with people outside the group. He brings in information and new ideas and explores interests or resources that might be of use to the group. The resource investigator is usually extroverted, sociable and fairly stable. He can sometimes lose interest in ideas or new programs once they are started and want to move on to other things without completing the details.
- The monitor/evaluator role examines and tests the validity of the team's plans and thoughts. The monitor evaluator evaluates and criticises the methods and concepts of the group and forces the group to make the best decision possible. The monitor evaluator can sometimes be overly critical and can lack inspiration or innovativeness.
- The teamworker maintains harmony and team spirit between the team members. Sometimes the welfare of each team member and the sense of camaraderie are more important than the organisation's objectives. The team worker is often extroverted, sociable and gregarious. A team can have a number of team workers because they are the 'lubricators' of team spirit. Sometimes the team worker can be indecisive and unable to make a tough decision when needed.
- The plant/originator is often intelligent, independent and introverted. The role is to contribute unique ideas and innovative approaches to solving problems facing the team. If a plant's ideas are not practical and relevant to the team he can be considered 'up in the clouds' or vague.
- The finisher is a detail person and is ideal to check and ensure that the team has completed the task correctly or left out necessary aspects of the work. The finisher can sometimes pay too much attention to detail and can be too much a perfectionist. He can worry over small details and be unable to let things go.

These roles are based on the contribution an individual's behaviour makes to the work of the team, but Belbin warns that in practice, at the same time as making a contribution, each role has an 'allowable weakness' and an 'unacceptable weakness'. For example, in the case of shapers, an allowable weakness would be their tendency to become irritable or provoke other members, while an unacceptable weakness would be their inability or refusal to make amends or to apologise.

However, Belbin's most recent research provides greater insight into the way different roles interact in ways that are mutually productive or counterproductive. The great value of his work is that it provides well-tested tools (self-perception inventories) which can be used to select the members of a work team prior to formulation and establish the patterns of their relationships in the most appropriate fashion (Barbara 1997).

Margerison and colleagues (1986) and Margerison and McCann (1990) (see also Anon. 1992) have developed an alternative team framework to Belbin's, proposing teams roles such as explorer, adviser, organiser and controller. This approach, known as the 'Team Management Wheel', has been used by Hewlett Packard to improve teamwork between Australian pilots and flight engineers. It is based on two measures which index the extent to which managers and their teams cover the roles identified: the 'Types of Work Index' and the 'Team Management Index'.

CONCLUSION

This chapter has identified the crucial importance that project managers should attach to group formation within construction projects. Groups are an unavoidable and essential part of any construction project organisation and must be managed well if they are to be constructive. When managed poorly, groups can be a very destructive force. Well-managed groups can be an immensely powerful mechanism for managers to achieve organisational goals and this necessitates an understanding of different types of groups and their functions, the process of group formation, the behavioural dynamics of groups and the roles people play within them. While the evidence indicates that the construction industry has little understanding of these issues, this can be seen as an opportunity to achieve substantial improvements in productivity through more intelligent and thoughtful management practices.

EXERCISES

1   What is the difference between a team and a group?
2   What is the evidence that a well-structured team can contribute significantly to project success?
3   How are project teams normally formed in the construction industry?
4   How can better teamwork be achieved in the construction industry?

CHAPTER 11

MANAGEMENT LEADING

INTRODUCTION

Leading is one of the four main management functions. It is probably the most important function since the achievement of planning, organising and control is only possible through the actions of people. Management leading is about understanding how people behave in a particular manner in the work environment, what action to take to stimulate positive behaviour, and how employees are likely to respond to such an action.

People may behave entirely differently in a group environment than individually. For example, a normally reserved and quiet person may, within a group of friends at a football match, indulge in very noisy and rowdy behaviour.

Apart from the responsibility for the overall management of a project, the project manager also leads a project team. He must therefore understand how people behave individually and in groups, must know how to stimulate positive behaviour and how to motivate people, and must be familiar with fundamental principles of leadership.

The purpose of this chapter is to develop an understanding of the function of management leading and its relevance to project management. The chapter will examine individual and group behaviour of people, review relevant theories of leadership, and explain the concept of delegation. Since motivation is a substantial area of knowledge, it will be examined separately in Chapter 12.

INDIVIDUAL BEHAVIOUR

The project manager's role is to lead a team towards achieving stipulated project objectives. Ideally, he would want to work with people with positive attitudes and behaviour, who are highly motivated and are excellent team players. In reality, the attitudes and behaviour of project participants are usually diverse. It may well be that they are unable to work together as a team. It is the leader's job to spend the time to understand the individual behaviour of project participants in order to develop strategies for bringing about the positive change in behaviour necessary for establishing teamwork and group synergy.

Let's now examine the main components of individual behaviour.

Attitudes and behaviour

Robbins and colleagues (2003: 385) defined attitudes as 'evaluative statements concerning objects, people or events'. Using different words, attitudes could be described as good, bad, positive, negative, favourable, proactive, reactive, and so on. Attitudes basically describe how people feel about their work. Brooke and colleagues (1988) defined the most popular attitudes as:

- job satisfaction: how satisfied the worker is with the job and its relevant tasks
- job involvement: how involved the worker is in the job and in decision-making
- organisational commitment: how loyal the worker is to the organisation.

Although people don't often do what they say, they nevertheless don't want to be seen as contradicting themselves. Ajzen and Fishbein (1980) suggested that people attempt to reconcile the discrepancy between attitudes and behaviour by rationalising the inconsistency. For example, a person who is a frequent flier with one particular airline is known to be critical about the service provided by another airline. But he may be happy to switch alliance to that other airline if the employer agrees to pay a business-class fare for his travel.

A theory that attempts to explain the relationship between attitudes and behaviour is the theory of cognitive dissonance. It suggests that any inconsistency between attitudes and behaviour is uncomfortable and that people will try to reduce the dissonance (discomfort). Every person displays some degree of dissonance, both at home and at work. For example, a teacher who urges senior students to drive carefully and within the speed limit may in fact be a dangerous and speeding driver.

Although people may be aware of dissonance, they may not want to become more consistent. Robbins and associates (2003: 388) claim that the desire to reduce dissonance is determined by:

- the importance of the factors creating the dissonance
- the degree of influence the individual believes he or she has over those factors
- the rewards that may be involved in dissonance.

In general, a person is unlikely to reduce the dissonance if he believes that the factors that contribute to dissonance are unimportant or uncontrollable (for example, a directive by a superior to take a particular line of action may be an uncontrollable factor). Inconsistency also becomes less important to a person if he receives a significant reward for 'being inconsistent' (for example, a lawyer may suspect that the evidence given by his client is false, yet rather than declining to defend the client, he proceeds to win the case, which brings a substantial financial reward to his firm and to himself personally).

Managers often survey attitudes of workers to establish their morale and job satisfaction. This is because they believe that productivity is directly linked to employees' job satisfaction. Happy workers are commonly regarded as productive workers. But it may also be the other way around since the person's strong job performance may lead to feelings of accomplishment, greater involvement in decision-making, more pay, promotion, and greater job satisfaction.

Personality

'Personality is the unique combination of the psychological traits by which that person is described' (Robbins et al. 2003: 392). There are many psychological traits that are used to describe personality, for example shyness, aggressiveness, laziness, obnoxiousness, loudness, quietness.

A number of methods have been developed for linking traits to one's personality. Perhaps the best-known methods are the Myers-Briggs Type Indicator and the five-factor model of personality.

MYERS-BRIGGS TYPE INDICATOR (MBTI)

The MBTI method attempts to determine one's personality by quantitatively assessing responses to over a hundred carefully worded questions. Responses are placed at either end of one of four dimensions. These are:

Extrovert or introvert (E or I) Extroverts focus on the outside world. They are outgoing and like to talk. They like to take action since they want to change the world. They are often dominant and aggressive. Introverts prefer to focus more on their own inner world.

They tend to be withdrawn and prefer to operate within a well-organised and functioning work environment. They like to understand the world first before taking action.

Sensing or intuition (S or N) Sensing functions of eyes, ears and other senses help people to recognise what is actually happening. Sensing types like routines and because they accept what they see, they are realistic and practical. They are good at working with facts and details.

People who are intuitive rely on more than just their eyes and ears. They employ other senses in exploring meanings, relationships and possibilities. They like to look at the big picture and are generally impatient with routines and details. They have imagination and inspiration.

Feeling or thinking (F or T) Feeling types are aware of how people feel. They generally enjoy being with people. They tend to be sympathetic, caring and tactful.

Thinkers are less interested in how people feel. They think logically and make decisions on the basis of analysis of facts. They tend to be seen as unsympathetic and hard-hearted.

Judgment or perception (in terms of decision-making) (J or P) Judging people are planned and organised. They are decisive and focus on completing tasks ahead in an organised manner. They can be impatient with slow processes.

Perceiving people are adaptable, curious and spontaneous. They seek to understand life rather than control it. Since they are flexible, they tend to keep their options open.

By combining these indicators, a total of 16 combinations of personality types are established, with each combination describing a particular type of person. Table 11.1 briefly describes a sample of four personality types of the MBTI matrix.

TABLE 11.1 EXAMPLES OF MBTI PERSONALITY TYPES

PERSONALITY TYPE	DESCRIPTION
ISTJ	Well organised, practical, reserved, logical, realistic, dependable.
ESFP	Friendly, well organised, initiator, have common sense, practical, work well with people.
INTJ	Develop original ideas, methodical, highly independent, individual, critical and well organised.
ENFJ	Concern for people, responsible, social, popular.

SOURCE Adapted from Hirsh & Kummerow 1990.

PART 3 People in Project Management

MBTI is a popular method for understanding personality and predicting people's behaviour. It is used extensively by organisations for selecting employees for specific types of jobs.

THE BIG-FIVE MODEL OF PERSONALITY

John (1990) described different types of personality dimensions known as the big-five model of personality:

- Extroversion portrays a person who is friendly, sociable, likes to talk and is decisive.
- Agreeableness relates to a person who is good-natured, cooperative, friendly and likes working with people.
- Conscientiousness describes a person who is persistent, methodical, responsible, dependable and goal-oriented.
- Emotional stability relates to positive and negative aspects of emotional stability such as composure/depression, security/insecurity.
- Openness to experience describes a person who is imaginative.

OTHER PERSONALITY TYPES

Barrick and Mount established a relationship between the big-five model personality dimensions and job performance in specific job categories. See Barrick and Mount (1993) for details.

Robbins and colleagues (2003) reviewed the work of Organ and Greene (1974), Vleeming (1979) and others and highlighted a number of powerful personality traits that explain individual behaviour in organisations. They are:

Locus of control (focus of control) This relates to a person's ability or inability to control his own destiny. Some people display the internal locus of control (these are usually independent people who are in charge of their destiny) while other people (usually those who are dependent) believe that external forces influence their lives. They are said to display the external locus. People with the external locus of control are, in comparison with those with the internal locus of control, generally less satisfied and happy with their jobs, show less interest in their jobs and tend to shift the blame for their lack of performance on the others in their work environment.

Machiavellianism Robbins and colleagues (2003: 395) described 'mach' people as 'being pragmatic, maintain emotional distance and believe that ends can justify means'. People with high 'mach' and strong bargaining skills are generally productive as negotiators or salespersons.

Self-esteem The degree of like or dislike expressed by an individual for himself is known as self-esteem or ego and is related to expectations of success. People with high self-esteem are, in comparison with those with low self-esteem, confident in their own abilities, are prepared to take more risk in job selection, are prepared to explore unconventional solutions, are less susceptible to external influences, may not conform and are prepared to be unpopular if it means achieving a success.

Self-monitoring This describes the ability of people to adapt their behaviour to external influences. People who are high self-monitors adapt well to external situations and may often display contradictory behaviour in relation to their business as opposed to private activities. Those with low self-monitoring are unable to adapt to different situations.

Risk-taking Risk is perceived and responded to by individuals differently. Some individuals are willing to take chances (risk-takers) while others prefer to avoid risk (risk-avoiders). In comparison to risk-avoiders, risk-takers tend to make decisions quickly and with less information available without compromising the quality of such decisions.

Summary Knowledge of psychological traits is essential for managers to better understand the behaviour of individuals. If they can figure out why people behave in a particular manner, they may be able to redress the problem. Better knowledge of psychological traits can also help managers in job selection by matching personality types with jobs.

Character

The concept of a personality ethic is regarded as an important component of leading. It is popular because it offers a fairly simple recipe for achieving personal effectiveness and deep relationships with other people. However, Covey (1989: 19) sees the personality ethic as 'illusory and deceptive'. He believes that it is a 'get rich quick' scheme promising 'wealth without work', and is used by managers to create favourable impressions through charm and skill. Covey argues that although the personality ethic is beneficial and perhaps even essential in developing personality growth, communication skills, and education in the field of influence strategies and positive thinking, it is a secondary group of human traits.

Covey sees character ethic as the foundation of success in personal development. His concept of character ethic is based on the fundamental idea that there are principles governing human effectiveness or, to put it another way, natural laws in the human dimension. Covey (1989: 18) argues that 'there are basic principles of effective living, and that people can only experience true success and enduring happiness as they learn and integrate these principles into their basic character'. Covey adds that (1989: 35) 'These principles are deep, fundamental truths that have universal application. They apply to individuals, to marriages, to families, to private and public organisations of every kind'.

Covey describes principles as guidelines for human conduct that have permanent value and are self-evident:

- fairness
- integrity
- honesty
- humility
- fidelity
- courage
- justice
- patience
- modesty
- human dignity
- service
- quality or excellence.

Human character is constantly expressed through daily habits. What matters in the long term is what the person actually does rather than what he says he would do. In this context Covey (1989: 47) defined habits as 'the intersection of knowledge (what to do and why), skill (how to do) and desire (want to do)'. The manager may know the reason why it is important to maintain team synergy but may not have the skill or may simply have no desire to make it work. Creating a habit requires work in all three dimensions.

Covey extended the concept of character ethic based on habits to form a comprehensive model of personal and interpersonal development referred to as 'the seven habits'. They allow the development of personal and interpersonal effectiveness on a 'Maturity Continuum' from dependence to independence and ultimately to interdependence. These habits are:

1 be proactive
2 begin with the end in mind
3 put first things first
4 think win–win
5 seek first to understand, then to be understood
6 synergise
7 sharpen the saw.

A detailed description of these habits can be found in Covey's book *The Seven Habits of Highly Effective People* (1989).

Perception

Beardwell and Holden (1994: 47) defined perception as 'a complex process involving the selection of stimuli to which to respond and the organisation and interpretation of them according to patterns we already recognise. We develop a set of filters through which we come to make sense of our world'. People interpret what they see and call it reality but what they actually see may not be reality. It may only be perception. Individuals may perceive the same event differently. For example, a politician that offers tax cuts before the election may be seen by one section of the community as being caring, responsive to the needs of people and compassionate, while to other community groups the politician's action may be seen as nothing more than a blatant vote-buying exercise.

The reasons for people perceiving the same thing differently is related to the personal characteristics of the perceiver (such as attitudes, personality, motives, interests, past experiences and expectations), the features of the target that is being observed (for example, the fact that some people enjoy watching a particular advertisement on TV may have little to do with the product the ad attempts to 'sell'; rather it may be related to a catchy song or an interesting character that appears in the ad), and the situation or the context in which objects or events are being observed. A range of situational factors such as time of observing, location, weather and colour may influence attention (Robbins & Mukerji 1994).

It is important for managers to recognise that employees perceive the work environment and management actions through their own eyes and form views that are often substantially different from the reality. Low morale, disloyalty and poor performance may well be the outcome of negative perceptions of management actions. Open communication could overcome this problem.

Learning

Learning is a continuous process that leads to change in behaviour. It is important to understand how people learn in order to predict their behaviour.

People will generally adjust their behaviour in order to get what they want. For example, a child will keep its room tidy if it wants to earn some pocket money. The child's behaviour is likely to be repeated if rewarded (reinforced), but if unrewarded, it may revert to what it was before. Adult behaviour follows the same pattern. If people want something, they will adjust their behaviour to get it and if reinforced, the probability that it will be repeated increases. Desired behaviours can be encouraged by positive reinforcement and rewards; if not rewarded or if punished, desired behaviours are unlikely to be repeated. This learning process is referred to as operant conditioning. Its basic premise is that behaviour is a function of its consequences (Skinner 1971).

People also learn through observations and experiences. They learn from their parents, teachers, role models, media, colleagues, bosses and friends. This learning process is referred to as social learning theory. It assumes that behaviour is a function of its consequences but it also acknowledges that people respond to how they perceive and define consequences, not to the specific consequences themselves.

Managers need to work out how best to teach employees and project team members to develop behavioural patterns that would most benefit their organisations and the project. They can positively stimulate learning in the workplace through rewards and their own behaviour as role models.

GROUP BEHAVIOUR

In Chapter 10 important characteristics of groups and teams were discussed in detail. The aim was to explain how groups work and how they achieve project objectives. The previous discussion will now be expanded by focusing on important components of groups that contribute to group behaviour.

Important components of groups

Robbins and Mukerji (1994) defined roles, norms and conformity, status systems, group size and group cohesiveness as important group components. These will now be briefly reviewed.

ROLES

In organisations and project teams, people are expected to perform specific roles prescribed by their job specifications and defined by their superiors. But a range of job tasks that a person is required to perform may create role conflict. For example, a company executive might be expected to work long hours whereas his family wants him to spend more time at home. Clearly, this person faces conflict between his roles as executive and as husband/father.

NORMS AND CONFORMITY

Groups establish norms or standards that govern their activities, examples of which are effort, performance, dress and loyalty. These norms and standards then 'pressure' individual group members to conform. The most notable norm is performance. The group determines the level of performance and the group members are obliged to conform or face a prospect of being disassociated with the group. A similar pressure can be exerted on employees to maintain dress norm.

STATUS SYSTEM

Status is a standing that a person has within a group and it is recognised as a significant motivator. Status may arise informally or it may be conferred formally. Both types are equally important. Some individuals gain status within a group on the basis of some specific characteristics such as speaking skills, expertise or experience. When status is formally conferred, for example one member of a design group is appointed leader of the group, it is important to ensure that the group accepts that person's new status. If, for example, this particular leader had access to only economy-class travel while other group leaders in the organisation were able to travel business class, his authority could be downgraded or even rejected by the group.

GROUP SIZE

The group size may affect the group's behaviour. Large groups with over ten members are able to benefit from a greater diversity of views, which may contribute to developing new and unique strategies and policies. Communication and decision-making processes are expected to be longer than in smaller groups. When groups become very large, contributions of individuals may become difficult to monitor and measure. Smaller groups, on the other hand, are better suited to solving specific problems. They generally complete tasks faster but don't necessarily produce better output.

GROUP COHESIVENESS

Mullen and Cooper (1994) concluded that highly cohesive groups are generally more effective than those with less cohesiveness provided they are able to align the group's attitude with the organisational goals. When the goals are seen by individual team members as desirable, highly cohesive teams become more productive.

Factors that influence group behaviour

Some groups work better than others. Managers are naturally interested in learning the reasons for good or poor group performance.

While it is a complex issue, some definitive answers have emerged from extensive past research that has identified a number of components of group behaviour. Robbins and colleagues (2003) provide an excellent summary of major research contributions in this area, including the work of Friedlander, Shaw, Gibson, Hackman and Morris, and Waller. The key variables of group behaviour given by Robbins and associates are:

EXTERNAL CONDITIONS

These are various conditions imposed on project groups by external factors including:

- corporate strategies of group members' organisations
- laws, codes and regulations
- availability of resources
- the economic and social climate
- environmental policies, etc.

QUALITY OF GROUP MEMBERS

How well or poorly a group performs depends largely on the individual abilities, skills and personality traits of group members. However, groups comprising star performers do not always achieve strong outcomes because issues of egos, communication, morale, and willingness to work together may undermine the group performance.

GROUP INTERNAL STRUCTURE

The performance of a group may be influenced by its internal structure, which determines communication and decision-making processes, and controls members' behaviour.

GROUP PROCESSES

Groups implement various processes for communicating information, making decisions, resolving conflict, delegating authority and the like. Effectiveness of these processes is vital to the group's overall performance.

GROUP TASKS

When a group addresses simple tasks, it often applies standard procedures with little or no discussion taking place. When tasks become more complex, a group will devote a lot more time to discussion in order to find the best solution.

MOTIVATION

One of the most challenging tasks of managers is to figure out how to motivate people. From an employee's perspective, motivation is the willingness to work harder towards attaining organisational goals in satisfaction of some individual need. Motivation will be reviewed in detail in Chapter 12.

LEADERSHIP

The manager is expected to lead, to get other people to take action in helping to accomplish the organisational goals. The conventional point of view focuses on leadership of subordinates, but effective management leading requires more than this. The manager must get action not only from subordinates but also from people on his own level, from specific groups in the organisation, and even from agencies outside the organisation. Often the manager needs to get effective action from his own superior. It therefore becomes obvious that leadership is far more than the giving of orders. No matter whom the manager leads, he rarely finds it possible to get good results solely by exercising his authority as boss. He can force obedience from the subordinates but this 'followership' will be reluctant at best. The dictatorial approach is certain to fail.

Leadership is related to vision. Vision can be defined as 'interpreting life so that others may see it' (Vicere & Fulmer 1997: 2). Leaders are expected to have vision in order to define appropriate goals and plot a course of action to attain them. When a problem becomes fuzzy, they need to have the insight and the flexibility to alter the course of action and, if necessary, even the goals.

A leader's fundamental role is to lead a team of qualified and appropriately skilled people, to effectively communicate the goals and the course of action to the team, allocate appropriate resources, build the necessary reward structures to ensure adequate motivation, and develop teamwork. This is exactly what a project manager in the construction industry is expected to do.

Leadership can therefore be defined as 'the ability to persuade others to seek certain goals and the technique of taking them there' (Fulmar 1983: 302). It is also defined as the ability to influence others to attain goals. Effective leaders know how to influence others to perform beyond the actions dictated by formal authority (Kezsbom et al. 1989).

Leaders and managers

Before proceeding further, let's clarify the difference between managers and leaders. Managers can only be appointed, while leaders may either be appointed or emerge from within a group. Managers have power of authority and can direct or influence people within the bounds of the authority. Leaders, on the other hand, have the ability to direct or influence people beyond the bounds of the authority.

Covey (1989) defined management as a bottom-line focus which attempts to find the best approach for accomplishing a set objective and leadership as a top-line focus which attempts to find the most appropriate objectives that should be accomplished. Covey regarded leadership as the first (mental or right-brain) creation and management as the second (physical or left-brain) creation.

Peter Drucker and Warren Bennis distinguished between management and leadership in the following terms: 'Management is doing things right while leadership is doing the right things' (Covey 1989: 101). Covey's explanation of the difference between management and leadership is probably the most eloquent:

> You can quickly grasp the important difference between the two if you envision a group of producers (workers) cutting their way through the jungle with machetes. They're cutting through the undergrowth, clearing it out. The managers are behind them, sharpening their machetes, writing policy and procedure manuals, holding muscle development programs, bringing in improved technologies and setting up working schedules and compensation programs for machete wielders.
>
> The leader is the one who climbs the tallest tree, surveys the entire situation, and yells, Wrong jungle!!! But how do the busy, efficient producers and managers often respond? Shut up!!! We're making progress.

The question most commonly posed is: do leaders make good managers? The general consensus is that leaders make good managers as long as they are knowledgeable and skilful in the key management functions of planning, organising and controlling.

Do managers, then, make good leaders? There is no easy answer to this question. Most agree that a management title does not automatically imply leadership, though most leaders develop from the rank of managers (Walker 1996). According to Covey (1989), transition from manager to leader is not automatic and may require a major paradigm shift.

A review of theories of leadership

Research in the area of leadership has been extensive and has resulted in the development of numerous leadership theories and models. These continue to evolve in light of new experiences and new research information.

The early leadership theories such as trait and behavioural theories formed a basis for interpreting leadership issues through personal traits and behavioural patterns. Contemporary leadership theories, such as the Fiedler contingency model and the Hersey-Blanchard situational theory, have developed new perspectives on leadership and better understanding of leadership processes. Let's now look at leadership theories in more detail.

TRAIT THEORIES

The first research into leadership sought to identify certain traits by which effective leaders could be identified. These were of three types: physical, mental and personality traits.

Physical traits such as strength, height, appearance, physique, hair colour and the like were first thought to be important characteristics of leaders. This approach was soon discarded as unworkable. For example, no matter how physically strong politicians might be or how good their looks are, these physical traits are no substitute for good policies.

Leadership was also connected to mental traits such as intelligence, good communication skills and sound judgment. Mahoney and colleagues (1960) suggested that leaders have somewhat higher intelligence than the average of their followers. But history has shown that smart people are not necessarily good leaders.

Of all trait theories, personality traits offered a more reliable description of leadership. Personality traits cover a broad spectrum of factors such as charisma, self-confidence, enthusiasm, dominance, ambition and the like. Fulmar (1983) suggested that the leader's personality is the key to success, but no physical, mental or personality traits that all leaders possess have yet been identified. Kirkpatrick and Locke (1991) had more success in identifying six specific traits that generally distinguish leaders from followers:

- drive
- the desire to lead
- honesty and integrity
- self-confidence
- intelligence
- job-relevant knowledge.

Since trait theories alone were insufficient to explain leadership, subsequent research had focused on behaviour of leaders in an attempt to find answers.

BEHAVIOURAL THEORIES

Behavioural theories examine leaders' behaviour with the aim of finding some unique and contrasting factors for their characterisation. Some relevant theories will now be briefly reviewed.

Continuum theory This measures leader's attitudes towards employees along a continuum. An authoritarian leader is placed at one end of the system while a democratic leader occupies the other end, with a benevolent autocrat in the middle. At the extreme end of the continuum, beyond democratic leadership, is the laissez-faire style of leadership, which gives employees in a group complete freedom to determine their own policies, set objectives and make decisions. The continuum theory is shown graphically in Figure 11.1.

Figure 11.1 The continuum theory of leadership

Authoritarian	Benevolent autocratic	Democratic	Laissez-faire style
leader	leader	leader	leader

An *authoritarian leader* is job or production-centred and behaves according to Theory X (McGregor 1960) by dictating work methods and limiting employees' participation.

A *democratic leader* is people-centred and follows Theory Y (McGregor 1960) by empowering employees, encouraging their participation in decision-making and goal-setting, and allowing them flexibility in deciding how to reach the goals. McGregor's Theory X and Theory Y will be discussed in Chapter 12.

A *benevolent autocrat* is a leader who listens to opinions of employees and then makes a personal decision that may not represent a majority viewpoint. This type of leader communicates well with employees, shows genuine interest in their well-being and generally makes prompt decisions.

Some argue that benevolent autocracy is an undemocratic form of leadership. Others believe that the efficiency and effectiveness of a benevolent autocrat are more important in business than democratic principles. However, no consensus has been reached in the literature on whether democratic leadership or benevolent autocracy provides a more effective leadership model.

A *laissez-faire leader* gives employees total freedom to set their objectives and processes for achieving them. Many years ago Lewin and Lippitt (1939) found that a more democratic style of leadership was superior to other styles in terms of quantity and quality of production. This, however, has not been fully verified in later research except that in terms of satisfaction of employees, the democratic style of leadership is superior.

Tannenbaum and Schmidt (1958) defined specific styles of leadership on the continuum between job-centred (similar to autocratic) at one end and employee-centred (similar to democratic) at the other. The individual leadership styles on the continuum were:

- autocratic (at the extreme end of job-centred leadership)
- paternalistic autocrat
- bureaucrat
- diplomatic
- consultative
- participative team (at the extreme end of employee-centred leadership).

Kahn and Katz (1960) found that people-centred leaders achieved higher group productivity and higher job satisfaction than job or production-centred leaders.

The two-dimensional theory A research team at Ohio State University carried out experiments in 1948 to find out how a leader performs activities (Stogdill & Coons 1951). The team was able to isolate two dimensions of leadership behaviour: initiating structure (this dimension is task-oriented and is characterised by assigning to employees work tasks, objectives, standards of performance, and so on); and consideration (this dimension is people or relations-oriented and is characterised by the leader's concern for employees). In a general sense, the initiating structure dimension is similar to a job-centred style of leadership and the consideration dimension to a people-centred style in the continuum theory.

Each dimension of the two-dimensional theory was expressed in high and low values, and assembled in a four-quadrant matrix (Figure 11.2). Of the four possible combinations, only a high–high combination (high initiating structure and high consideration) was found to be a leadership style with consistently higher employee performance and satisfaction than any other combination. Although having questionable reliability in describing leadership behaviour, this theory nevertheless formed a solid base for further research, particularly in advancing the situational leadership theory.

The managerial grid theory Blake and Mouton's (1964) managerial grid was an extension of the continuum and two-dimensional theories. A managerial grid is formed by two leadership styles: concern for production, which is similar to the initiating structure and job-centred leadership; and concern for people, which is similar to the consideration and people-centred leadership. In the grid, each of the two leadership styles is divided into nine parts, which together form 81 grid positions, one for each particular leadership style. Blake and Mouton described the most dominant styles, four of which are located in outer corners of the grid and one in the middle. These styles are:

- minimal (1,1): this point on the grid identifies a style characterised by low concern for people and low concern for production. in this situation the leader relies on employees to do the work.
- country club (1,9): this style is characterised by high concern for people and low concern for production.
- team (9,9): this style is characterised by high concern for people and high concern for production. in this situation the leader works with employees towards achieving the goals.
- task (9,1): this is a leadership style that is concerned with production only; concern for people is low.
- middle of the road (5,5): this is a mid-point of the grid where both employees and production are moderately important to the leader.

Although not entirely conclusive, the findings of Blake and Mouton point to a 9,9 style as the best performing, which is similar to a high–high style of the two-dimensional theory.

Summary While trait theories tried to define leadership in terms of specific traits of leaders, behavioural theories focused on behaviour of leaders. Had these theories been successful, it would have then been

Figure 11.2 Four quadrants of the two-dimensional theory

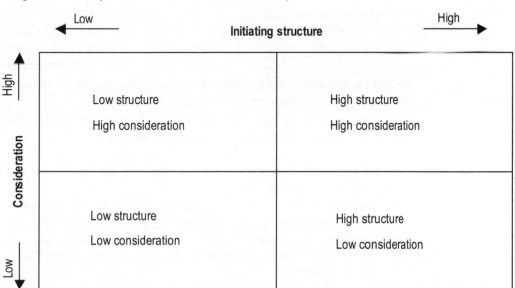

Management Leading

relatively simple to find and appoint a leader with the desired traits or simply train a suitably qualified person for a position of leader. However, neither trait nor behavioural theories could explain the relationship between leadership styles and performance. Behavioural theories advanced the knowledge of leadership and developed the groundwork for future studies that examined critical situational factors. Some of those will be discussed in the next section.

SITUATIONAL THEORIES

'The situational theory says, in effect, that leadership ability is dependent upon the individual's adaptive ability, i.e. the feeling the leader may have for sensing, interpreting, and treating the specific situation' (Fulmar 1983: 314). Numerous situational theories have been developed over the years of which the Fiedler contingency model and the Hersey-Blanchard Situational Leadership Theory are the most prominent. They will now be discussed.

The Fiedler contingency model Fiedler's (1967) approach assumes that leadership varies with the situation. His model is based first on determining the leader's personality and then on matching it with specific situations. Fiedler adopted task-oriented and relations-oriented styles of leadership from the behavioural group of theories. He devised a method for measuring a leadership style using 'the least-preferred co-worker questionnaire'. The questionnaire contained 16 pairs of contrasting questions and the respondent was required to relate his work experiences with an employee whom he identified as least pleasant to work with. Depending on whether the least preferred co-worker is viewed positively or negatively, the person surveyed is deemed either to be relationship-oriented (with a score over 63) or task-oriented (with a score below 58). A person with a score between those two extreme limits is viewed as socio-independent. Fiedler's approach assumes that depending on the situation, either of these types may be associated with effective leadership.

Fiedler then identified three situational factors for determining leadership effectiveness in terms of control and influence:

- leader–member relations: this factor shows the degree of confidence and trust employees have in the leader
- task structure: this factor gives the degree of formalisation of job tasks and procedures that the group has to perform (from poorly to well defined)
- position power: this factor determines the degree of the leader's formal authority (it is assumed that the leader with defined position power is able to secure followership).

Fiedler measured each of the above situational factors as either high or low. He was able to identify 'leader–member relations' or a supportive style of leadership as most important. He argued that a popular and well-respected leader does not need to use power or spend the time to provide details of the tasks to be accomplished to influence team behaviour.

When the three situational factors are combined, they form eight possible situations, which may either be very favourable, moderately favourable or unfavourable for the leader. These eight situational combinations are given in Table 11.2.

TABLE 11.2 FIEDLER'S EIGHT SITUATIONAL COMBINATIONS

SITUATIONAL FACTORS	1	2	3	4	5	6	7	8
Leader–member relations	High	High	High	High	Low	Low	Low	Low
Task structure	High	High	Low	Low	High	High	Low	Low
Position power	High	Low	High	Low	High	Low	High	Low

SOURCE Adapted from Robbins et al. 2003.

Through his research, Fiedler was able to determine that situations 1–3 were highly favourable, 4–6 moderately favourable, and 7 8 unfavourable for the leader. He then measured the performance of task-oriented and relations-oriented leadership styles in each of these eight situational groups. The results are illustrated in Figure 11.3.

Figure 11.3 The Fiedler model

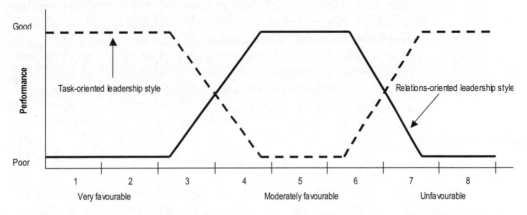

SOURCE Adapted from Robbins & Mukerji 1994: 372.

Management Leading

Fiedler found that when situations were either highly favourable or highly unfavourable for the leader, task-oriented leaders would perform better, whereas relations-oriented leaders would perform better in moderately favourable situations.

A full description of the Fiedler model can be found in Kezsbom et al. (1989: 189–91), Fryer (1990: 57–9) and Robbins and Mukerji (1994: 369–73).

The Hersey-Blanchard Situational Leadership Theory Most leadership theories focus on the leader and study the leader's attitudes and behaviour. The Hersey-Blanchard Situational Leadership Theory (2001) takes a different approach by focusing on followers since, in reality, actions of followers measure the leader's effectiveness.

The Hersey-Blanchard theory is a contingency theory that ties together three factors:

- direction given by the leader (task behaviour)
- socio-emotional support provided by the leader (relationship behaviour)
- readiness of followers.

The Situational Leadership Theory is an adaptation of the Fiedler contingency model and in particular the two leadership dimensions of task behaviour and relationship behaviour. The third factor adds an important element to the situational leadership theory by defining stages of followers' readiness. Together, these three factors describe a range of scenarios of leadership behaviour.

Readiness may be defined in many different ways. Covey (1989) defines readiness or maturity as an intersection of courage and consideration. However, for the purpose of explaining the Hersey-Blanchard theory, Fulmer's (1983: 315) definition will be adopted: 'Readiness [maturity] is the capacity to set high but attainable goals, plus the willingness and ability to take responsibility, and utilize education and/or experience'.

Hersey and Blanchard defined stages of follower readiness in four levels, given in Table 11.3.

TABLE 11.3 FOLLOWER READINESS

HIGH	MODERATE		LOW
R4	R3	R2	R1
Able and willing and confident	Able but unwilling or insecure	Unable but willing or confident	Unable and unwilling or insecure

Task and relationship behaviour of the leader are expressed in high and low values. When plotted on a grid, they form the four quadrants of the Hersey-Blanchard model (Figure 11.4). The level of readiness of followers is then projected across those four quadrants, resulting in four distinct types of leadership behaviour:

- telling: related to high task/low relationship behaviour (the leader tells employees what to do)
- selling: related to high task/high relationship behaviour (the leader directs and also supports employees)
- participating: related to low task/high relationship (the leader involves employees in decision-making)
- delegating: related to low task/low relationship (the leader has empowered highly matured employees to take responsibility for their actions).

Figure 11.4   The Hersey-Blanchard Situational Leadership® model

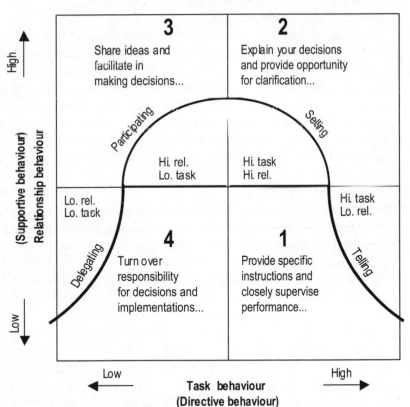

Management Leading

The Hersey-Blanchard Situational Leadership Theory is probably the most widely used leadership model and its full description can be found in Kezsbom et al. (1989: 191–4), Walker (1996: 146–7) and Robbins et al. (2003: 489–91).

Summary of leadership theories

The knowledge area of leadership is being continually expanded through research, resulting in the development of new theories and models; examples are the path-goal leadership theory, the leader-participation model and the action-centred leadership model. These and a number of other leadership theories are explained in Fryer (1990) and Robbins et al. (2003).

Despite the availability of many leadership theories that explain the behaviour of leaders, there is no general consensus on the composition of crucial leadership factors. Among most quoted leadership factors are the leader's power, credibility, character, and ability to create trust and empowerment.

The leader's task is to create the conditions that will help employees to find satisfaction in their work. The leader exercises the power to involve people in participation, goal-setting and goal attainment. The leader makes the work of employees important, challenging and rewarding. The leader motivates employees to attain organisational as well as personal goals. In return, under effective leadership, employees want to be accountable, assume responsibility, make decisions, show initiative and achieve high standards.

Leadership in project management

The foregoing discussion explained leadership in general terms. While the reviewed theories are relevant to both general and project management, their specific application to project management will now be briefly explored.

In the context of this book, a leader is a project manager who, by working together with other project participants, is responsible for attaining project goals. Achieving successful project outcomes is generally attributed to the ability of project participants to work effectively together as a team under the leadership of a project manager (Burke 1999). While technical and administrative competence is important, a project manager must have the essential ability to integrate and co-ordinate specialist skills of different project team members and to create effective teamwork and team synergy (Cleland 1995).

Attempts have been made by many to define skills that a project manager should possess. Posner (1988) established a list of such skills in a descending order of importance as:

- communicational
- organisational
- team-building
- leadership
- coping
- technological.

Donnelly and Kezsbom (1994) defined qualities of project leaders in terms of managerial, analytical, integrative, collaborative and organisational competencies.

The past view of project managers in charge of construction projects as 'head kickers' is no longer relevant. Accepting that the manager's fundamental role is to attain project objectives through effective teamwork, a greater value must be placed on developing the project manager's leadership, motivation, team-building and communication skills than on reinforcing the outdated approach of the authoritarian leadership style based on task orientation. If a project manager is to function effectively, in a sense being a 'conductor of an orchestra', he should not be burdened with routine administrative tasks. Support staff could more effectively perform these.

Leadership styles of project managers

A manager of a typical functional organisation leads employees of that particular organisation. In comparison, a project manager in charge of a construction project leads a team that comprises representatives of different stakeholders' organisations. These representatives may themselves be leaders of their respective organisations. It means that a project manager may in fact be a leader of leaders.

The nature of work for which the project manager is responsible is often different from that of a manager of a functional organisation. Unlike a typical manufacturing process, construction projects are finite. They progress through a number of lifecycle stages and involve many participants from different organisations, who do not often stay together for the entire project period. Projects are driven by specific objectives and the project manager is responsible for their attainment. This requires integration and co-ordination of activities of many participants throughout the entire project lifecycle. Yet the project manager is required to accomplish all of this with only a limited amount of power over the project team. This is because members of the team are neither direct employees of the project manager nor are contracted to him. The matter is further complicated by the fact that the project manager is usually external to the client's organisation. Although the client is an integral member of the project team, he may not surrender the full extent of the decision-making power to the project manager.

The project manager's task is made even more difficult by the frequently changing composition of the project team. As a typical construction project passes through a series of lifecycle stages, new participants join the project team when their work becomes due, while others depart when their work has ended.

Clearly, leading a construction project team is a difficult and a highly challenging task for any project manager. Although there is no leadership theory dedicated to construction project managers, existing theories are sufficient to provide guidance.

What style of leadership is best suited to project managers? Burke (1999) suggested that project managers could use all the leadership styles defined by the continuum theory as circumstances dictate. Walker (1996) believed that when leading a team of mature and experienced specialists, the project manager would use a more democratic leadership style with low task and higher relationship focus. However, if a project team comprises less mature and less experienced members, the project manager would apply a more autocratic leadership style with a higher task focus. Keszbom and colleagues (1989) and Walker (1996) recognised that different leadership styles were needed at different stages of the project lifecycle. The former concluded that since a project manager's leadership style is contingent on a range of factors, contingency leadership theories such as the Fiedler model and the Hersey-Blanchard Situational Leadership Theory are appropriate leadership styles for project managers.

Seymour and colleagues (1992) examined the leadership style of 46 project managers working in contracting, petrochemical and project management organisations in the United Kingdom using the Fiedler contingency model. They found that project managers in all organisations surveyed to be highly relations-oriented. Their findings were reinforced by a similar study in Hong Kong by Rowlinson and associates (1993). In contrast, another study in the United Kingdom by Bresnen and associates (1986) found contractors' project managers to generally exhibit task orientation.

Clearly, the nature of a project manager's work is such that it requires adaptation of different leadership styles for different situations. Let's now briefly explore situations that may occur throughout the project lifecycle and try to formulate appropriate leadership styles for such situations.

A wealth of information on leadership has been generated over the years, and has benefited mainly functional organisations and functional teams. Leadership of project teams, on the other hand, has attracted little attention. The first attempt to explain behaviour of a project leader over the project lifecycle is found in Gilbert (1983). Gilbert combined Blake and Mouton's managerial grid, formed by two leadership styles — a concern for production and a concern for people — with Hersey-Blanchard's readiness curve, which Gilbert believed closely followed a project leader's style of leadership over the project lifecycle. This is illustrated in Figure 11.5.

Figure 11.5 Leadership styles in a project lifecycle

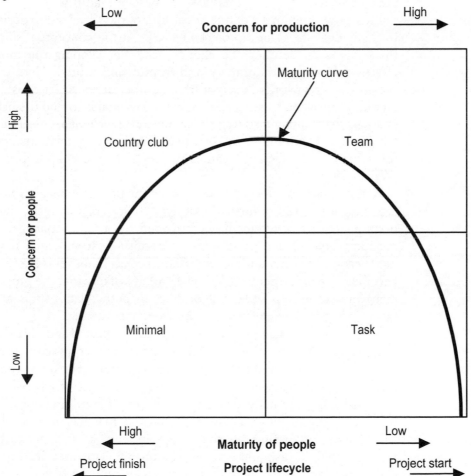

SOURCE Adapted from Gilbert 1983.

He suggested that in the early stages of a project, more emphasis is placed on tasks and team development. A project manager's leadership style would gradually move from a 'high concern for production-low concern for people' type to a 'high concern for production-high concern for people' style. Once teamwork develops, a project manager allows team members more freedom to work with limited supervision. His leadership style then gradually moves through a 'low concern for production-high concern for people' type to a 'low concern for production-low concern for people' type.

While conceptually sound, Gilbert's model is overly simple. Leadership of construction project teams is far more complex. Project lifecycle stages are more distinct from each other and the presence of contracts gives more emphasis to tasks than to relationships. Furthermore, a project team membership varies from stage to stage, which hinders the development of readiness or maturity.

Let's now attempt to define most appropriate leadership styles for individual stages of a project lifecycle. At the conceptual stage the focus is on defining the client's needs, developing the most effective development strategy and formulating a brief. Here the project manager works closely with a number of specialist consultants in an open and trusting environment. His goal is to find the best possible development strategy that meets the client's objectives. Clearly, the emphasis is on teamwork and strong relationships, for which a democratic leadership style is most appropriate (Walker 1996).

At the design and documentation stage the project manager's task is to integrate and co-ordinate the design activities of architects, engineers and others with feedback provided by a value manager, a quantity surveyor and other experts. Because cost and time have now become the constraining factors, the project manager is likely to become more task-oriented (Walker 1996). It is likely that the project manager's leadership style will be high in both relationships and tasks since it is essential to maintain strong teamwork.

At the tendering stage a general contractor is selected and awarded a contract. At the pre-construction stage the contractor then mobilises resources and prepares for construction. It is quite common in these two stages to find the first indications of adversarial relationships, particularly when the contractor's conditions of contract are being negotiated. Since the project manager acts as the client's agent and represents the client's interest, he is likely to be more focused on task and less on relationship. This may lead to a shift towards a less democratic style of leadership.

In the construction stage, the main focus is on constructing a project and meeting the key objectives of cost, time and quality. Since subcontractors perform most construction tasks, the contractor's role is to co-ordinate and integrate their work. Time and cost slippages and elements of contract disputation commonly emerge at this point, increasing pressure on the project manager. What leadership style will he adopt here? It is likely that in the above scenario the project manager will be more autocratic than democratic. His focus will be on administering the contract, which represents task orientation. Contractual disputes with the contractor may lead to a growing level of mistrust. With the erosion of the relationship aspect, the project manager's leadership style is likely to be more autocratic with a high task focus.

In conclusion, a project manager is faced with a complex range of situations for which a highly contingent approach to leadership is required. The fundamental characteristic of a project manager must be his ability to be flexible in adapting to different situations. Although the Hersey-Blanchard Situational Leadership model comes close to describing a project manager's leadership style, the continuum theory offers a range of leadership styles that a project manager is likely to adopt in the different stages of a project lifecycle.

DELEGATION

Delegation is entrusting responsibility and authority to others for specific activities, and establishing accountability for results. It empowers subordinates to make decisions.

An effective manager concentrates his efforts on those tasks that only he is able to perform, and delegates other tasks to subordinates. While, for example, a project manager is responsible for the management and completion of the entire project, it is neither possible nor desirable for him to perform every task for which he is responsible.

A project manager is principally responsible for goal-setting, the formulation of scope and an appropriate development strategy, managing each stage of a project lifecycle, and monitoring and controlling progress and performance. Those are the tasks that a project manager must retain, while delegating largely operational tasks down the line.

Roadblocks to delegation

The inherent benefits of delegation are the reduction of a project manager's workload and the improvement of subordinates' motivation. These benefits should make the use of delegation widespread. However, project and other managers often fail to delegate, mainly because of psychological and organisational barriers.

PSYCHOLOGICAL BARRIERS

Managers often decline to delegate because they believe that they could perform specific tasks faster and better themselves. Some managers are simply afraid to delegate. They may have no confidence in subordinates' ability to do the job properly and may prefer doing the work themselves. In the short term, the fear to delegate is justified if subordinates are poorly motivated or their skills are inadequate, but in the long term it is a manager's job to take positive action to overcome these deficiencies.

Managers may also fear that subordinates will do a better job, which could result in a loss of face. But this fear is unjustified. Managers must recognise that subordinates with specialised knowledge and skills in operational tasks are bound to produce better results. Effective managers are those who focus on developing people.

ORGANISATIONAL BARRIERS

Sometimes job tasks are not well defined and a manager is unclear what authority he has for making decisions. Under such circumstances, the manager would be unlikely to delegate parts of his responsibility and authority. In general, failure to define responsibility and authority prevents delegation. The obvious solution involves clear and precise definition of the limits of responsibility and authority within which project and other managers can safely operate.

Poorly defined organisational structure with unclear lines of communication also prevents delegation. If it is difficult for a manager to establish who, for example, authorised a particular task and to whom the manager needs to report the findings, the manager is again unlikely to delegate.

Types of delegation

Covey (1989) defined two types of delegation: (i) gofer and (ii) stewardship. Gofer delegation focuses on methods. The manager directs the subordinate step by step to perform a particular task in a specific way. This task-oriented approach is clearly an inefficient form of delegation.

Stewardship delegation is much more powerful. It focuses on results rather than methods. The subordinate chooses the most appropriate method to accomplish the task for which he is responsible.

Principles of delegation

The first element of delegation is responsibility. It is the work assigned to a position. To delegate, the manager assigns part or all the work for which he is responsible to others.

The second element of delegation is authority. It is the sum of the powers and rights assigned to a position. The manager must be given enough authority if he is to assume responsibility for the execution of the work. For example, a football coach must have authority for the players' selection and possibly even for their recruitment if he is to accept responsibility for the team's performance. To delegate, the manager must assign sufficient authority to a subordinate to perform the delegated work

Accountability is the third element. It is the obligation to perform responsibility and exercise authority in terms of established performance standards. While the manager can delegate responsibility and authority, he cannot delegate accountability. The manager is accountable whether or not he performs the task directly or delegates it to a subordinate.

Steps in effective delegation

Effective delegation involves a number of logical steps. Let's briefly examine them.

DEFINE RESPONSIBILITY, AUTHORITY AND PERFORMANCE STANDARDS

The manager first defines a task to be delegated to a subordinate and states what is to be accomplished. He then gives the subordinate a written statement of what is to be done, by when, the extent of responsibility and authority, the decisions to be made, the results to be accomplished and the performance standards to be achieved. Consequences of good or bad performance should also be clarified.

ORGANISATIONAL COMMITMENT TO DELEGATION

For delegation to work effectively throughout the organisation, it must start at the top. By delegating, top managers signal their trust and confidence in subordinates' abilities. Setting challenging tasks with appropriate levels of responsibility and authority to subordinates are a major motivator and should be encouraged by managers.

LIMITS TO DELEGATION

The manager needs to clearly define the parameters within which subordinates perform delegated tasks. The manager should avoid telling subordinates how to do the task, but he could provide guidance on where the potential problems might be and what not to do. In the end, subordinates are responsible for results.

The manager needs to be prepared to temporarily withdraw delegated authority and responsibility if the subordinate's performance diminishes. He should consider reinstating it when the problem has been resolved.

MONITOR PERFORMANCE AND ESTABLISH FEEDBACK

The manager needs to regularly monitor the progress subordinates make in performing the delegated task to make sure they understand what results are to be accomplished, by when and the performance standards by which their work will be assessed.

CONCLUSION

Management leading is a crucial management function, which ensures effective implementation of other management functions of planning, organising and control. The project manager needs to understand the behaviour of individual team members in order to influence positive behaviour towards effective teamwork. The project manager's most important task is to provide leadership. He must stimulate and motivate team members to take action to accomplish project goals by adopting an appropriate leadership style for any given situation.

Managing a project team in the construction industry with a diverse composition of team members is an extremely difficult and challenging task for any project manager. Since the construction project environment is highly situational, the project manager must be able to adapt his leadership style to different situations. The Hersey-Blanchard Situational Leadership Theory and the continuum theory generally describe leadership styles of project managers in the construction industry.

Delegation is often given little importance by managers, but if effectively implemented it not only frees managers from performing routine operational tasks but it also provides empowerment to subordinates, which improves their motivation.

EXERCISES

1   Explain the difference between attitude and behaviour using cognitive dissonance theory.

2   Distinguish between personality and character.

3   Explain how the MBTI model could help in matching people to jobs.

4   Why do people perceive the same thing differently?

5   What is the difference between operant conditioning and social learning theory? How can these two models assist a manager to understand and predict the behaviour of subordinates?

6   Does the size and cohesiveness of a group affect its performance?

7   Why do some groups work better than others?

8   What is the difference between groups and teams?

9   Distinguish between functional, self-directed and project teams.

10   How would you develop teamwork in project teams?

11   What are the characteristics of effective teams?

12   Define leadership. Distinguish between leadership and management.

13   Are trait and behavioural theories relevant today?

14   What advances have situational theories brought to leadership?

15   What leadership style should a project manager adopt?

16   What is delegation? What are the three important components of delegation?

17   Distinguish between gofer and stewardship delegation.

CHAPTER 12

MOTIVATING PEOPLE FOR PROJECT SUCCESS

INTRODUCTION

Organisational objectives are best accomplished when organisational members want to achieve them. Ability alone is not enough without a desire to achieve, and it is this desire and enthusiasm to act that is the essence of motivation. Construction project managers, like a manager in any industry, need to understand how to motivate their workforce, yet surprisingly, the subject has attracted little attention compared to other more fashionable subjects such as conflict management and risk management. While there is no reason to believe that management in the construction industry is any different from management in other industries, there are some complications introduced by the peculiarity of the construction industry's professional structure and organisational practices. For example, there is a long history of division between the various specialists who contribute to the process, which manifests itself in different languages, cultures and strong stereotyped images. This introduces an element of mistrust and suspicion into the workplace which is exacerbated by the industry's procurement and contractual practices. Furthermore, most work is undertaken by subcontractors, which separates operatives from those who seek to manage them. This makes it difficult to control people at an individual level without first having to work through the organisations that directly control them. The purpose of this chapter is to discuss these problems in more detail and to put forward some recommendations to overcome them. More specifically, it is to review the major theories of motivation and to interpret their value in the context of project management.

DEFINING MOTIVATION

Motivation is a psychological phenomenon which has received considerable attention in mainstream management literature. In essence, motivation is concerned with a set of cognitive processes that arouse, direct and maintain human behaviour towards a goal. In a managerial sense, it is about encouraging a willingness in people to exert effort towards an organisation's goals; the quality of effort is as important as the intensity.

MOTIVATING PEOPLE

It is generally accepted that everyone has the potential to be motivated and in this sense it is wrong to label people as lazy. Rather, managers need to appreciate that people differ in what triggers their motivational drive and it is important to try and identify what these triggers are. This is illustrated in Figure 12.1.

What makes the subject challenging is the variable nature of people's needs and aspirations, which underpin their motivation to work and which can also change over time. People work for all sorts of reasons such as for money, self-esteem, enjoyment, company and prestige, and in this sense a manager is best able to motivate people individually. But it is not always efficient to do this and it is impractical to try to understand every organisational member's drives, particularly in larger organisations. While the manager's job can be made easier by employing people of similar needs, this is also difficult and most managers turn to standard approaches that enable them to create an environment where people can align their personal objectives with the organisation's. These approaches are often based on general theories of motivation developed in the management sciences and it is to these that we now turn.

MOTIVATIONAL THEORIES

There are numerous motivational theories based on different assumptions about people's behaviour. They can be categorised under five headings:

- Need theory, based on the assumption that motivation derives from the satisfaction of needs and that people strive to satisfy frustrated needs.
- Expectancy theory, based on the assumption that motivation derives from what people expect to happen as a result of their actions.
- Equity theory, based on the assumption that motivation derives from the

equity people perceive to exist between their circumstances and comparable situations.

- Reinforcement theories, based on the assumption that motivation derives from the consequences of action.
- Job characteristics theory, based on the assumption that an employee's motivation, performance and job satisfaction is higher when he has personally performed well on a task that he cares about.

Need theories

If people strive to satisfy frustrated needs then managers can motivate them by seeking to understand what those needs are and by presenting opportunities to achieve them. Maslow (1954) made a significant contribution to knowledge when he argued that people have

Figure 12.1 Motivation

	Sex		Age				Income level				Job type				Organisation level		
	Men	Women	Under 30	31–40	41–50	Over 50	Under $12 000	$12 000–$18 000	$18 000–$25 000	Over $25 000	Blue-collar unskilled	Blue-collar skilled	White-collar unskilled	White-collar skilled	Lower non-supervisory	Middle non-supervisory	Higher non-supervisory
Interesting work	2	2	4	2	3	1	5	2	1	1	2	1	1	2	3	1	1
Full appreciation of work done	1	1	5	3	2	2	4	3	3	2	1	6	3	1	4	2	2
Feeling of being in on things	3	3	6	4	1	3	6	1	2	4	5	2	5	4	5	3	3
Job security	5	4	2	1	4	7	2	4	4	3	4	3	7	5	2	4	6
Good wage	4	5	1	5	5	8	1	5	6	8	3	4	6	6	1	6	8
Promotion and growth in organisations	6	6	3	6	8	9	3	6	5	7	6	5	4	3	6	5	5
Good working conditions	7	7	7	7	7	6	8	7	7	6	9	7	2	7	7	7	4
Personal loyalty to employees	8	8	9	9	6	5	7	8	8	5	8	9	9	8	8	8	7
Tactful discipline	9	9	8	10	9	10	10	9	9	10	7	10	10	9	9	9	10
Sympathetic help with personal problems	10	10	10	8	10	9	9	10	10	9	10	8	8	10	10	10	9

Please note: Ranked from 1 (highest) to 10 (lowest)

SOURCE Adapted from Hackman & Oldham 1976: 259.

a hierarchy of needs and that they attempt to satisfy their basic needs first before moving onto those at higher levels. Managers can motivate people by providing opportunities in work for the satisfaction of these needs; the various measures to achieve this are depicted in Figure 12.2. Physiological needs refer to basic bodily needs such as hunger, thirst and shelter. Safety needs refer to things such as security and protection from physical and emotional harm. Social needs refer to affection, belongingness, acceptance and friendship. Esteem refers to self-respect, autonomy, achievement, status and recognition. Finally, self-actualisation refers to achieving one's potential and destiny through self-fulfilment and personal growth.

Figure 12.2 Maslow's needs hierarchy

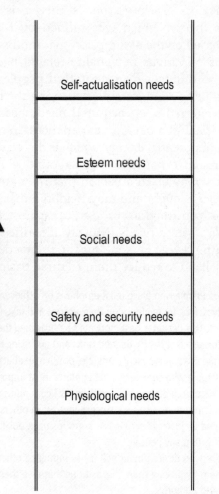

Motivating People for Project Success

In providing for these needs, however, managers must be careful because they cannot provide all needs at work, particularly at the higher levels. Furthermore, the complete satisfaction of needs results in a person who is beyond motivation, which means that a 'carrot and stick' approach is therefore justified. In theory, the ideal workforce is one that is slightly frustrated but living in hope of being able to achieve the aims they have yet to meet. Care must also be taken because people are different: it may be easier to satisfy certain needs in some than it is in others. This is particularly so across the many cultural boundaries that pervade modern multicultural organisations. In this sense, while valuable in understanding people's behaviour, Maslow's hierarchy is of limited value in a practical managerial context since its application would involve time-consuming individual assessments of satisfied and unsatisfied needs.

An alternative to Maslow's needs theory is Murray's (1938) acquired needs theory, which was refined by McClelland (1961). This theory is based on the assumption that motivation is a function of the strengths of various personality-related need factors. These may be active to different degrees within people and a manager's energies should be focused towards satisfying the more active ones. The needs are referred to as acquired needs because they change over time as a result of a person's experiences. In other words, people have a pool of potential energy which is directed by needs, which are learned through experience. For example, a recent experience with redundancy may teach a person that they are not as indispensable as they once thought and may heighten that person's need for security. On a construction project, a spate of accidents may heighten people's safety needs. Murray identified 20 needs which were innate within every person, but the impracticality of assessing them led McClelland to reduce them to three basic needs:

- *Achievement.* People who place most emphasis on achievement, set difficult goals for themselves, look for feedback on their actions and take responsibility for solving problems. This person is best motivated by providing them with responsibility for the performance of clearly defined tasks and by linking pay to their achievement since the reward represents tangible, positive feedback on performance.
- *Affiliation.* People who consider social contacts most important will seek personal assurance from others, build widespread communication networks, conform to group norms and show a sincere interest in others. These people are best motivated by placing them in work-groups or in positions where they come into contact with many people.
- *Power.* Those who desire power will try to influence others. Managers can motivate them by placing them in positions that give them a certain degree of autonomy and authority.

This theory is more useful in a managerial sense than Maslow's because a manager only has to look at a person's previous experience to understand what types of needs they will have learned. The argument is that over time, people naturally fall into working roles which reflect their personality needs and that those roles mould their needs to a certain degree. For example, a person who has experience in training will have dealt with many people and will probably have an affiliation bias. Those who have held positions of leadership will probably have a power need bias. This has important implications for managers because it implies that people can be taught through the provision of certain experiences to acquire the needs that will enable them to be successful in various types of jobs.

✗An extension of needs theory was provided by Hertzberg's (1974) two-factor theory, which takes all of the needs identified by Maslow and Murray and categorises them as either motivational or hygiene factors. Motivational factors are those that produce satisfaction such as responsibility, achievement, advancement and recognition. In contrast, hygiene factors are those that do not motivate people but prevent dissatisfaction, such as pay, working conditions, policies, job security and relationships with managers and peers. Hertzberg was interested in what people wanted from their jobs and he asked people to describe situations where they felt bad and good about their work. The results are presented in Table 12.1, which clearly indicates that the factors that satisfy are very different from those that dissatisfy. If one assumes, as Hertzberg did, that satisfaction and dissatisfaction are very different things, then it follows that removing dissatisfiers does not necessarily increase satisfaction levels and vice versa. Furthermore, satisfiers tend to be intrinsic factors whereas dissatisfiers tend to be extrinsic. Hertzberg argued that the problem with most managers is that they focus on the extrinsic hygiene factors and not the intrinsic motivational factors because the former are much easier to control. For managers this has profound implications because it indicates that without a more in-depth and thoughtful approach to management, they can only control levels of dissatisfaction.

While useful, Hertzberg's theory has been questioned by more recent research. For example, the representativeness of his sample of engineers and accountants has been questioned. Furthermore, it has been shown that hygiene factors have been associated with both satisfaction and dissatisfaction and that there is an important relationship between extrinsic and intrinsic factors. This interdependence means that the stimulation of one will affect the other. This

TABLE 12.1 HERTZBERG'S MOTIVATION-HYGIENE THEORY

MOTIVATORS (SOURCES OF SATISFACTION)	HYGIENE FACTORS (SOURCES OF DISSATISFACTION)
Achievement	Supervision
Recognition	Subordinates
Advancement	Company policy
Work itself	Interpersonal relations
Responsibility	Job security
Growth	Work conditions
	Salary
	Personal life

finding gave rise to cognitive evaluation theory, which has shown that when extrinsic rewards are given for certain tasks, the intrinsic rewards derived from those tasks are reduced, which means there is no increase in productivity. This clearly has major implications for managers by indicating that it is extremely easy to waste money in trying to motivate people, particularly on jobs with high intrinsic satisfaction. Managers should only focus on the use of extrinsic rewards with jobs that have little intrinsic satisfaction. Finally, research has indicated that some people cannot be motivated to perform above a certain level however much attention is given to motivational factors by management.

While it has attracted some recent criticism, the advantage of Hertzberg's theory was that it showed how people can be satisfied and dissatisfied simultaneously. It also pointed out that traditional focuses on monetary rewards would not motivate people and that managers should focus on motivational as well as hygiene factors in maximising the driving-force behind organisational efforts. The ideal organisation is where there are low levels of demotivation and high levels of motivation. The neutral organisation has high levels of demotivation and high levels of motivation and the ineffective organisation has the opposite.

NEED THEORIES APPLIED TO CONSTRUCTION PROJECTS

In managing construction projects there are complications in using need theories to motivate people. The problem is that people operate under the influence of a range of needs through a matrix organisational structure. These include the client's needs, their own personal needs, their employer's needs and any needs of interest groups or temporary coalitions they may belong to at a particular time.

In this sense, motivating an individual by attention to needs alone may be difficult and there may be a tendency to focus too much on the commercial needs of employing organisations rather than the needs of individuals. In other words, an individual's personal needs can become swamped in a sea of other more powerful commercial needs, thereby depersonalising the whole organisation. Indeed, a project manager can only indirectly satisfy many of an individual's needs, such as the desire for promotion, and in this sense his hands are to a great extent tied. A further problem a project manager has is a lack of continuity and time to satisfy longer-term needs. A project is by definition a temporary phenomenon and most contributors to it will be even more temporary. In this dynamic, short-term and commercial environment, matters of motivation are difficult to attend to and they are often reduced to a financial basis in the form of the provision of better working conditions and monetary rewards and punishments. The intrinsic motivation of individuals tends to be left to their employing organisations. That is, the current way in which projects are procured in the construction industry, and in particular the growth of subcontracting, restricts the focus of motivation by project managers to extrinsic factors. This means that project managers are constantly trying to reduce demotivation rather than motivate, which means that they are constantly fighting a losing battle and have a very negative emphasis.

Expectancy theories

These theories are based on what people expect and perceive to happen as a result of their efforts. They have important implications for the construction industry because of the high degree of uncertainty under which it operates. Expectancy theories were originally formulated by Vroom (1964) and later refined by Lawler (1973) and argue that motivation is equal to the product of expectancy and valence.

Motivation = Expectancy × Valence (M = E × V)
where
E = prediction of the probability that a specific action
    will have a certain result (expectancy), and
V = positive or negative value attributed to that result (valence).

In essence, what this theory is saying is that if a person considers a task to be achievable and worthy of carrying out in terms of potential rewards, then he will be motivated to do it. The value of this theory from a construction project perspective is that it introduces the whole concept of risk-taking into motivational theories. That is, it argues that people assimilate the possibility of an action not producing a desired result and base their motivation on this. For example, if a person believes that it is 100 per cent certain that a particular action will result in some positive outcome then he will be highly motivated to perform that action. A lower probability of a positive outcome will result in lower motivation unless the potential rewards are so enormous as to make it worthwhile attempting even at a low level of probability of success. In contrast, people are demotivated by events that have a potentially negative outcome. The concept of risk-taking complicates matters because different people (or organisations in the case of construction projects) have different attitudes towards risk-taking. For example, person A, as a result of working for a risk-taking company, may be willing to take a 60 per cent risk, but person B, working for a risk-averse company, may not. In this sense, motivation is also a product of a person's risk-aversion, a higher aversion (dislike of risk) effectively reducing levels of motivation for a particular event probability.

where

$$M = \frac{E \times V}{R}$$

R = risk aversity

The implications of this theory for managers are interesting since they suggest that managers would be best placed by employing organisations that like to take risks since they are more easily motivated. It also places considerable emphasis on managers focusing on realistic goals and making potential rewards clearly defined, worthwhile and linked to the performance of specific and measurable events. The clearer the relationship between specific events and rewards, the stronger motivational forces will be. If a person sees an event as both achievable and worthwhile in terms of rewards, then they will be motivated to perform.

These implications have important application to construction projects because of the low margins that organisations are increasingly being expected to operate on. This has become so bad that some organisations have left construction and diversified into other more profitable industries. In this sense, project outcomes are rarely worthwhile in profit terms and therefore provide little motivation.

Furthermore, this demotivation is exacerbated by the fact that outcomes are often unclear because project goals are often ambiguous due to problems with the briefing process or poorly drafted contracts. They are also often difficult to define in measurable terms, particularly in the case of function and quality. Furthermore, much of the emphasis is on punishments for non-achievement, and there are precious few contracts that provide for potential rewards for those who seek out and take advantage of opportunities to increase performance levels. Of course, while a threat of punishment may cause someone to carry out actions to minimise the chances of punishment, this must not be confused with the concept of motivation, which it seems the construction industry has done. Motivation is the desire to achieve and maximise potential benefits, whereas the action resulting from threats of punishment is better described as action to avoid or minimise potential losses.

Equity theory

Equity theory has important implications for the construction industry because of the perceived inequities built into the way we manage and procure construction projects. As Loosemore (1999b) found, in construction projects it is normally the case that the parties with the least power and the lowest rewards have the highest levels of responsibility and risk. Indeed, the whole issue of equity has become of heightened importance within the construction industry with the publishing of reports such as that by Latham (1994), which stress the importance of trust and fairness within construction project relationships.

Equity theories were first developed by Adams (1963) and are based on the assumption that motivation derives from the equity people perceive to exist between their circumstances and comparable situations. In other words, if people feel under-rewarded or even over-rewarded for their efforts in comparison to others who make similar efforts, then they will be dissatisfied or satisfied respectively.

The theory is based on the argument that people intuitively seek to balance their inputs and outcomes with those of others. Inputs refers to factors such as intelligence, training, education, money investments, effort, experience, seniority and so on, and outcomes include pay, promotions, praise and perks. The relevance of equity theory to construction project organisations is limited by the variable nature of each project and of each person's or organisation's contribution to the process. This makes comparisons difficult. However, since making comparisons is natural for most people, it is highly likely that points of comparison and benchmarks will be found, such as previous projects; if people feel that factors on either side of their equation are changing without an appropriate change on the other side, then they may make efforts to redress it by reducing effort, cutting corners or being more claims-conscious. Indeed, Adams (1963) argued that input/outcome comparisons are particularly likely to be made around money because it is easily measurable. For example, if people think they are being paid too little for their skills (compared to other people with similar skills) then they will seek to rebalance the input/outcome equation by either requesting an increase in income (outcome) or by reducing their level of commitment or work effort (input).

The implications for construction managers are that they must recognise that people will make comparisons and that they should be able to justify differences in rewards over time and between different tasks. At present there is a perception in the construction industry that it is the most powerful rather than the most responsible who are the most highly rewarded. Most the problems in construction are related to equity and revolve around the issue of risk distribution within construction contracts. This is particularly so between subcontractors and main contractors, the latter having been accused of passing ever greater amounts of risk onto the former without an accompanying increase in rewards. Equity theory makes it plain that the consequences of such actions will be attempts by subcontractors to redress the imbalance by mechanisms such as increased claims (seek to increase outcomes) or reduced commitment to the project (reduce inputs). The fallacy of this approach is also that contractors are asking for inputs which subcontractors are not able to supply because many do not have the resources to underwrite these increasing risks. No party wins from following such a strategy since bankruptcy of a subcontractor will merely transfer the risk back to the contractor who sought to offload it. It would seem that greater attention needs to be given to the issue of equity in construction projects.

Reinforcement theories

The theories reviewed above are cognitive theories. That is, they focus on people's feelings, attitudes, perceptions and expectations. In contrast, reinforcement theories are behavioural in that they focus on actions, behaviour and tangible responses. Essentially, reinforcement theories are based on the assumption that people react to what they get. That is, rather than assuming that underlying attitudes and values determine behaviour, behaviour is seen as determined by what a person tangibly receives for doing something (the consequences of their actions). Skinner (1953) was one of the greatest advocates of this school of thought and popularised the term 'operant conditioning' to refer to the process of controlling behaviour by manipulating its consequences. This is a controversial theory which is based on the assumption that managers should focus on the use of rewards and punishments to control people's behaviour. That is, they should positively reinforce desirable behaviour with rewards and negatively reinforce undesirable behaviour with punishments. Inaction is dangerous, particularly in the case of undesirable behaviour, because failure to respond has reinforcing consequences. Similarly, a failure to reward positive behaviour and a focus on punishment will at best result in merely satisfactory performance. For this approach to work, employees must clearly understand what will be punished and rewarded since without this information the correct behaviour will take time to discover and learn. Furthermore, managers must provide a supportive role in telling employees why they have been punished and how they can avoid being punished again. In this way, even punishments can be made to have a positive appearance and motivating effect.

It would seem that in a construction project setting, it is the reinforcement theory that has had the greatest influence because motivation is most commonly attempted through conditions of contract that try to prescribe clearly what parties will be rewarded and punished for — although the emphasis has been primarily on punishment. With the exception of target-based contracts, rewards are generally set in stone at the beginning of the contract and, surprisingly perhaps, can only be increased by what would be negative behaviour to the client — that is, increasing in some way the amount the project costs by being able to blame some other party. This can only result in mediocrity and fundamentally contradicts the aims of reinforcement theory and indeed, all motivational theories, which is to align the goals of those concerned with those of the organisation. Another problem in the construction industry is in the use of contracts themselves as communication documents. Construction contracts are often unintelligible to the layman and therefore widely misunderstood, which ensures that construction project participants are often unclear about which actions induce rewards and punishments.

Job characteristics theory

Hackman and Oldham (1976) have developed one of the major

frameworks for defining job characteristics, which are important to employee motivation. This is called the Job Characteristics Theory, which argues that an employee's motivation, performance and job satisfaction is higher when he or she has personally performed well on a task that he cares about. Based on this theory, a predictive index has been developed which is based on the following factors:

- Skill variety refers to the range of skills that a job requires from an individual.
- Task identity refers to the clarity of the task and its desired outcomes.
- Task significance refers to the importance of the task to the person (which is influenced by its importance to the organisation).
- Autonomy refers to the degree of flexibility and independence given to a person to carry out a task.
- Feedback refers to the provision of information about how well the task is being performed.

The formula used to derive this predictive index is shown below (Hackman & Oldham 1976: 266).

$$\text{Motivating Potential Score (MPS)} = \left[\frac{\text{Skill variety} + \text{Task identity} + \text{Task significance}}{3} \right] \times \text{Autonomy} \times \text{Feedback}$$

While there is some debate about the precise influence of each factor in different situations, this model has received widespread interest in recent years. From the perspective of a construction project manager, it is clearly important since being involved from inception to completion provides huge opportunities to manipulate job characteristics and to maximise motivational levels. What is worrying is that the current trend in the construction industry towards specialisation, automation and ever-greater fragmentation is reducing the project manager's ability to build skill variety, task identity and task significance, autonomy and feedback mechanisms into the work of project participants.

PARTICIPATORY MANAGEMENT

A practical approach to management, which combines need, expectancy and reinforcement theories, is participatory management. This is built on the assumption that people have a need to feel involved in decision-making processes that affect their future, and that by doing so they can more easily determine worthwhile and appropriate rewards and thereby make their expectations and perceptions more accurate. The foundations of participatory management were laid by McGregor (1960), who argued that management style is a function of the manager's attitude towards the nature of work behaviour. He proposed two contrasting models of managerial approach based on assumptions about work and people: Theory X and Theory Y. Managers with Theory X beliefs are job-centred and work on the assumption that the average human being is stuck in early adolescence, has an inherent dislike of work and responsibility, cannot be trusted, prefers to be directed, has little ambition, wants security above all, and will only work under external coercion and control. In contrast, managers who advocate Theory Y believe that the capacity to exercise imagination, ingenuity and creativity is widely distributed in the population and is typically only partly utilised. Theory Y managers also believe that the expenditure of mental and physical effort in work is as natural as play and rest, that the average human being was not necessarily averse to work but would see it as a source of reward or punishment depending on controllable conditions, and that control and threats of punishment are not the only way to bring about dedication to organisational goals. Indeed, people will exercise self-direction and control and under the proper conditions can learn not only to accept but to seek responsibility.

According to McGregor, a participatory style of management could only be successfully practised by managers who hold Theory Y beliefs about human behaviour. Subsequent work has also added the conditions that participatory management is best used in conditions where tasks are not easily measured and offer little intrinsic satisfaction. While few managers yielded to the Theory Y view during the 1960s, participatory management has recently grown in popularity in the belief that it can lead to a greater acceptance and commitment to organisational goals. Once again this is reflected in contemporary construction management trends in the growing emphasis on building trusting relationships between construction project participants through techniques such as partnering and total quality management. For example, one aspect of TQM is the use of quality circles, which are problem-solving groups in which workers and their supervisors meet to identify, analyse and solve productivity problems. The argument behind such schemes is that as people gain a better understanding of a situation, become more responsible for managing themselves and their jobs are expanded to include more responsibilities, they become more productive and better motivated.

While there is some limited evidence of the use of participatory techniques such as quality circles in construction companies, in a construction project context the growing popularity of participatory management styles has been reflected in the development and increasing popularity of alternative procurement systems. Systems such as design and build, construction management and management contracting were designed to break down the traditional barriers that existed between design and construction teams within traditional procurement systems. In reflecting on the care that must be taken in the application of participatory management styles, it appears that in the correct circumstances they are able to achieve higher levels of productivity and performance than the traditional system was capable of.

CONCLUSION

This chapter has reviewed a number of motivational theories and critically appraised their applicability to managing construction projects. It is clear that there is no best way to motivate and that managers should carefully assess the characteristics of their organisations before deciding on the most appropriate combination of approaches. While none of the theories above apply to every situation, collectively they do increase a manager's understanding of human behaviour and therefore their ability to motivate employees. A manager's method or strategy will vary depending on whether the problem seems to stem from individual needs, expectations, perceived inequities or rewards, or a combination of these factors. In addition to considering the nature of the problem, managers must consider what they can actually control in order to determine which theory to use. Rewards may be at the heart of the problem, but many managers do not control rewards such as pay rises or promotions; in these cases the managers will have to rely on other methods of motivating employees, such as offering more autonomy or more support, depending on the employee's needs. In conclusion, managers should use the theory or combination of theories that best suits the situation and that focuses on the aspects of the situation that they can control.

EXERCISES

1   What motivates people to work?
2   Describe the basis of five different motivational theories?
3   How does each theory apply in a construction project setting?

CHAPTER 13

CONFLICT MANAGEMENT

INTRODUCTION

There is much research that provides evidence of increasing conflict within the construction industry, some quoting fourfold increases in the amount of litigation since the 1980s (Fenn & Speck 1995). But care is needed in drawing conclusions. In particular, problems of measurement are posed by the diversity of the construction industry, the informal manner in which many disputes are resolved, and difficulties in defining conflict. Furthermore, increasing educational standards within the industry, the emergence of specialist construction lawyers and changes in the law which make it easier to litigate may have resulted merely in an increase in formally recorded disputes that would have previously gone unpublicised. Despite these measurement limitations, the issue of conflict has been elevated to one of the most important contemporary challenges facing the construction industry. Considerable momentum has been added by reports such as CSSC (1988), Gyles (1992) and particularly Latham (1994), which have portrayed conflict as a cancerous force, with huge opportunity costs which need to be reduced and ideally eliminated from the construction process. The result has been the diversion of considerable resources and energies towards the reduction of its incidence.

The purpose of this chapter is to explore the causes of conflict in the construction industry and the different approaches to managing it constructively. However, before considering this matter in detail, it is important to place the construction industry's experiences in historical perspective, because conflict has been a perennial feature of organisational thought since its conception. Such an exercise will also, at the outset, reveal the difficulties in attempting a definitive definition of conflict.

CONFLICT: A PERENNIAL PROBLEM FOR ORGANISATIONS

Arguably, the earliest manifestation of conflict research in organisations can be traced back to the work of Adam Smith, Robert Owen and Charles Babbage (Sheldrake 1996). They were concerned with the productivity benefits that could be achieved from specialisation and mechanisation but they also recognised the potential for conflict in the tedium of the factory production systems that had developed during the Industrial Revolution. Adam Smith chose to live with the problem, considering that the advantages of such a system to society as a whole outweighed the disadvantages to individual workers. However, during the depths of the Industrial Revolution, the Luddite campaigns of intimidation and machine-breaking brought about a more conscious effort to deal with the problem of industrial conflict. For example, Robert Owen, a self-made cotton manufacturer of the early 19th century, successfully experimented with a paternalistic style of management. He showed that the provision of better working conditions and welfare facilities for employees and their families such as housing, free education and a shorter working day could defuse tensions between managers and employees and be good for business. In contrast, Charles Babbage's panacea for healing the antagonism of factory production systems was the extension of profit-sharing schemes, underpinned by a belief in the mutuality of interests between managers and employees which, given proper attention, could transcend any superficial differences between them.

THE TRADITIONAL POSITION

Later, during the early 20th century, the rise of American industry and the scale of operations that developed in factories forced managers to explore new methods of controlling labour. Initially, FW Taylor, the Gilbreths and Henry Gantt sought to reproduce the successes of engineers in manipulating physical materials in a human setting. The principles of science and engineering lie at the heart of this approach, which was designed to replace a traditional craft and judgment-based system with the certainty and sterility of what became known as scientific management. Conflict was seen as wholly undesirable and as arising from indiscipline and poor supervision. Consequently, every effort was made to eliminate it, mainly by providing monetary rewards and penalties, the scientific selection and development of workers, greater standardisation, specialisation, mechanisation and close control. Although this was an attempt to produce harmonious organisations of co-operative structures and to achieve common objectives with no conflict of interests, the opposite occurred: reduced worker autonomy and deskilling, and increased antagonism between unions and employers. Furthermore, the increasing alienation of supervisors from workers and the replacement of the former's autonomy with prescriptive rules and procedures also caused supervisors to reject it. In the long term, this caused more conflict than it resolved.

THE CONTEMPORARY POSITION

These events highlighted the human costs of scientific management and prompted a change in management thinking that was founded on greater sensitivity to human needs and an emphasis that was more theoretical than empirical. This movement was led by scholars like Henri Fayol, Max Weber, Mary Follett, Elton Mayo and Abraham Maslow. Together they sought to redefine management by giving it a psychological foundation and moral base in an attempt to reunite workers and managers through more open, interactive and considerate management. The result was a greater understanding of the psychological and sociological aspects of organisations and an emergence of the view that conflict was an inevitable aspect of organisational life to be accommodated rather than suppressed. Organisations represented an arena within which, through formal and informal systems, competing interest groups battled for limited resources as a means of increasing their status and wealth. The most fundamental difference that pervaded the work of these researchers was the view that conflict was primarily a psychological rather than an economic phenomenon. Consequently, from a practical perspective, there was an emphasis on the role of groups, the importance of taking a holistic view of organisations, the view of an organisation as a dynamic network of social relationships, the advantages of human co-operation over conflict, and the benefits of integration over competition within organisations. In particular, Weber and Follett paid special attention to the issue of power, authority and legitimacy in organisations and laid the foundations for the political science tradition in organisational theory which first challenged the view that conflict was a wholly disruptive force in organisations (Simon 1948; Selznick 1957; Crozier 1964).

The contemporary view of conflict is a direct descendant of this position. It is now accepted that conflict is the norm rather than the exception in organisations and that absolute harmony is impossible. Conflict develops through a number of phases and is an essential and healthy part of organisational life in that it performs many important and positive functions for its participants (Robbins 1974; Likert & Likert 1976; Argyris 1990). For example, conflict is an effective stimulant to debate through which new ideas can be generated and differing perceptions recognised and resolved. Without conflict, many simmering underlying tensions would otherwise remain hidden from a manager's view, acting to undermine all efforts at control. In this way, conflict can be an effective mechanism for change, creativity and innovation and can perform an integrating function within an organisation. The managerial challenge is to harness these potential positives.

RADICAL VIEWS

The most radical contemporary views on organisational conflict appear to be coming from those who argue that it should be positively provoked (Pascale 1991). This is called the interactionist view and is founded on the belief that conflict is the doorway of opportunity to learning and to the fulfilment of organisational and individual potential. Harmonious, tranquil and peaceful groups are prone to become apathetic, static and unresponsive to customer needs and the need for change and innovation.

A DEFINITION OF CONFLICT

While it was tempting to start this chapter by providing a generic definition of conflict and its causes, the previous section shows that this would have grossly oversimplified a challenging and complex concept, which has developed over time with the evolution of management theory. Each school of thought that has contributed to the development of management theory provides a unique perspective on organisational conflict, which is often difficult to reconcile with some logical progression towards a universal definition. One of the main difficulties in arriving at a clear definition arises from the progressive nature of conflict. For example, Philips (1988) describes a self-perpetuating progression from simple disagreement to contention, dispute, limited warfare and, finally, all-out warfare. All-out warfare is characterised by one part trying to destroy another at all costs. When a conflict reaches this point, emotions take over and people can become irrational in the pursuit of self-interest. Philips points to an increasing momentum that accumulates over time and thereby indicates the importance of nipping a conflict in the bud.

While universal definitions are difficult, De Bono's (1991) description is valuable in that it is simple and appears to encompass the varying perspectives that exist. He sees conflict arising from a situation where there is an incompatibility of economic interests and/or psychological values between people which interferes with the attainment of their needs. He is at pains to point out that incompatibility creates the underlying potential for conflict but that interference is the condition that precipitates it. In metaphorical terms this description portrays the image of organisations having tranquil surfaces but existing with a constant internal tension which occasionally, as a result of some trigger event, bursts through the surface in an explosive release of tension. In traditional terms, it is the responsibility of managers to minimise the internal tension. However, in a contemporary sense, while there is agreement over the minimisation of internal tension, on some occasions managers may find it valuable to stoke up the internal tension in order to bring about some explosive release of creative energy or create a crisis that will induce some change. In this sense, contemporary management thought requires that modern managers incorporate a reactive element into their strategies.

PROJECT-BASED CONFLICT

Most research relating to organisational conflict has taken place within permanent business organisations and has adopted an inter-personal stance. While useful in the consideration of construction conflict, it is important to appreciate that the multi-organisational nature of construction projects means that construction conflict is primarily inter-organisational in nature (Cherns & Bryant 1984). While it is impossible for people to emotionally insulate themselves from a construction conflict, project participants are acting mainly as representatives of their employing organisations and are therefore constrained by their situation, culture and policies. In this sense, the forces that influence the way people behave during a construction conflict may be more complex than what is portrayed within standard management texts. As Antony (1988) points out, projects are typically characterised by a much wider range of interests than permanent business organisations, meaning that it is a much more unstable phenomenon. Furthermore, projects are typically more dynamic in that at any one time there may be a multitude of disputes in progress, each creating a different interest structure. In this sense, contrary to popular belief, static interest groups do not exist within construction projects but only a constantly shifting array of coalitions (temporary groups of people with the same interest). This is a characteristic of construction project organisations, the appreciation of which has a profound impact on the way they are managed. In depicting a static situation, current texts encourage managerial complacency, but the dynamic nature of interest groups demands a continuous, flexible and responsive approach to conflict management.

Thus conflict occurs at the interpersonal level between people, the overlap of their interest representing the potential for mutually satisfactory resolution. However, each individual also exists within the interest structure of their different employers and temporary coalitions. They both exist within the influence of the client's interests, whether they acknowledge them or not. In this sense, each person involved in a conflict operates under a multitude of conflicting forces from their own side and that of their opponents, some more covert than others, and it is the difficulty in resolving these that is the challenge in managing a construction conflict. Under such conditions, not only is it difficult to discover what is motivating people in their actions but it is very easy to miscalculate and precipitate an unintentional escalation.

CONFLICT MANAGEMENT IN THE CONSTRUCTION INDUSTRY

There are two approaches to the management of conflict: proactive and reactive. Proactive management is concerned with prediction, prevention, and thereby the reduction and eradication of conflict. In this sense, it is in tune with the traditional approaches to conflict management. In contrast, reactive management is concerned with damage limitation and opportunity maximisation after the event and is far more in tune with contemporary approaches to conflict management.

In the construction industry, the traditional approach has dominated and continues to do so, with momentum being added by reports such as that by Latham (1994), which called for a significant reduction in the level of conflict in the industry. This is indicated by the popularity of research into concepts such as business process re-engineering, partnering, benchmarking and value-engineering, by the preponderance of literature seeking to reduce construction conflict through the development of predictive models and the establishment of its incidence and causes, by the relative lack of behaviourally focused research, and by the general perception that conflict is fundamentally an economic rather than a psychological phenomenon (Clegg 1992; Gardiner & Simmons 1992). This approach has a number of motivators which include the industry's engineering values (Seymour & Rooke 1995) and the increasing resource and time constraints that now characterise construction projects. Within this context, conflict is seen as a damaging force that needlessly wastes scarce organisational resources and diverts managerial energies away from progressive tasks. Indeed, the theoretical basis of construction management research in general would also seem to reflect the state of management theory in the early 20th century in that, until now, it has substantially been based on empirical evidence and has weak theoretical foundations (Betts & Lansley 1993; Loosemore 1996; Runeson 1997).

Despite an apparent lack of theory, in recent years the construction management literature has begun to reflect the intermediate view that conflict is inevitable. Many researchers have given their attention to the efficacy of conflict resolution techniques such as litigation, arbitration, expert determination and alternative dispute resolution (Brooker & Lavers 1995; Fenn & Speck 1995; Davenport 2000; Uher & Davenport 2002). Reflecting the same position, there are those who have emphasised the contemporary view that conflict in not necessarily dysfunctional for an organisation and can be beneficial if properly managed (Loosemore & Djebarni 1994; Gardiner & Symonds 1995). Therefore the problem for the construction industry is not in the existence of conflict but in the way it is managed. Most of this work, however, is rhetorical and little research has been done in how to manage a conflict constructively.

The most radical view that has emerged recently in the construction management literature, and which reflects the most contemporary management thinking, is that conflict should be positively encouraged within construction project organisations (Hughes 1994). The contention is that conflict can be managed to positive effect and therefore its reduction incurs an opportunity cost for construction clients. Indeed, the creative and high-risk nature of construction projects ensures that conflict is a necessity rather than an evil and that the emphasis should be on facilitating more constructive conflict management as a foundation for encouraging conflict, a view that is diametrically opposed to Latham's (Hughes 1994). However, while these contemporary views are increasingly being propounded, Loosemore and colleagues (2000) have argued that a receptive attitudinal and sociological base does not exist within the construction industry and therefore such efforts may be counterproductive rather than beneficial. That is, construction managers do not have the desire, attitudes or skills to manage conflict constructively. In this sense, contemporary calls for the encouragement rather than reduction of conflict would seem to be premature.

REACTIVE CONFLICT MANAGEMENT

The above discussion would seem to justify the current emphasis on the reduction of conflict. However, in reality it may be a short-term, unthinking and misguided one. A more beneficial strategy in the longer term may be to act on the underlying attitudinal and socio-structural aspects of construction projects in order to provide a better foundation for constructive conflict management. As it stands, the preventive emphasis, if successful, would incur significant opportunity costs for the construction industry by reducing the possibility of bene-fits arising from the effective management of construction conflicts. Furthermore, an increasing emphasis on prevention would paradoxically increase the need for better reactive management. As Wildavsky (1988) argues, continual advances in proactive management techniques tend to produce an overreliance on strategies of prevention and by deflecting attention away from the need for organisational resilience, make organisations more vulnerable to conflict becoming costly when it does occur. Indeed, there is every indication that this 'vulnerability paradox' is already at play in construction, because once a conflict has slipped through the preventive net, there is little skill in managing it to prevent its costly escalation (Loosemore 1996).

Therefore, although prevention is better than cure, it is important not to see increased attention to reactive managerial strategies as a sign of managerial failure in prevention. Rather, it is realistic to recognise that managers cannot create a conflict-free environment and that there is a danger of the vulnerability paradox if there is an over-emphasis on proactive strategies. There is also an opportunity cost associated with the potential benefits conflict could bring.

There is therefore a strong justification for a more balanced approach to conflict management in construction which incorporates a reactive as well as a preventive element. This does not mean increased ingenuity in the development of ADR techniques because they deal with the consequences of failed managerial efforts. Rather, it means better management of people to avoid disputes escalating to the point where expensive and time-consuming third party intervention is required. As Robbins (1978: 143) pointed out, 'conflict management and conflict resolution are not synonymous terms'.

PROACTIVE CONFLICT MANAGEMENT

The key to prevention is to discover the causes and cut them out. There is no shortage of literature relating to the causes of construction conflict, although it is largely a fragmented one with a lot of repetition and little consensus over a definitive list of causes. Most of the models that have been produced also suffer from the problem that many of the causal factors are difficult to define and measure in a practical sense. Furthermore, few give consideration to the dynamics of conflict development, an exception being Gardiner and Simmons (1992, 1995).

The value of Gardiner and Simmons' work is that it dispels the traditional perception that conflict is a characteristic of the construction phase of projects. They suggest that it is far more common in the design stages. While the construction phase is characterised by the highest level of conflicting interests, particularly during traditionally procured projects, the high incidence of conflict during design is probably a product of the emotive and creative nature of the design process and the sheer volume of decisions that have to be made from first principles.

A further interesting point is that while the incidence of conflict is higher in design, the costs of dealing with it are highest during construction. There are several possible reasons for this. For example, it is likely that conflict during construction has its roots in design and that a lack of attention to it has enabled it to become more deepseated. Furthermore, it is also likely that by the time construction has commenced, most project decisions will have been made and built on by subsequent dependent decisions. Any subsequent changes in decisions will be likely to result in a widespread network of repercussions which will involve substantial abortive work and costs. As Gardiner and Simmons point out, the ideal pattern of conflict on construction projects should show most conflict being resolved in the earliest possible phases of a project. This is often difficult because of the time pressures, the desire to progress, and the euphoria and amicable relationships that typify the start of most projects, which tend to suppress any underlying tensions. The skill of the project manager is being constantly sensitive to underlying tensions by tapping into the informal organisational networks of communication, by being politically oriented in reading between the lines of people's communications and by drawing any underlying tensions to the surface, however uncomfortable this may be. In essence, all of their evidence represents a stark warning to project managers about complacency in managing conflict and about the need to 'nip a conflict in the bud'.

THE IDENTIFIED CAUSES OF CONFLICT

Although the literature relating to the causes of construction conflict is voluminous and repetitive and therefore impossible to cover in its entirety, it is worthwhile to spend some time discussing the common threads that run through it.

A lack of trust

There is widespread consensus that the reduction of conflict is dependent on a culture of mutual trust and collective responsibility in that this increases the effectiveness of communications within organisations and thereby reduces misunderstanding. This can be bought about by greater attention to the language and structure of contracts and to equitable and clear risk distribution within them (Barnes 1989, 1991; Uff 1995).

The client's role

There is also little doubt that the client's attitude towards paying for risks and their integration into the project is essential in reducing conflict (Kometa et al. 1994). Continuing with the theme of client involvement, the importance of the briefing process in defining the project and its boundaries has received considerable attention for some time and it is accepted that rushing this phase can lay the seeds of problems that may show up later in a project (Kelly et al. 1992).

Too much emphasis on price

Recently, attention to abuses of competitive tendering by reducing margins to restrictively low levels and over-emphasising price as a team selection criterion has led to calls for more intelligent tendering assessment systems for the selection of contractors and consultants and movement to negotiated contracts, often within continuous partnering arrangements (Hatush & Skitmore 1997a, 1997b).

A lack of participation

Newcombe (1994) has also pointed to the hierarchical, class-based structure of traditional construction projects and argued for flatter construction project organisations that facilitate more participative management styles. To facilitate greater participation, some have questioned the role of standard contracts, arguing that they discourage people talking about their roles, relationships and responsibilities and thereby build in misunderstanding from a very early point in the construction process (Murdoch & Hughes 1992). Indeed, many have attributed the success of construction management to the very fact that there is no standard form of contract, which means that people are forced to talk through and thereby take part in clarifying their contractual responsibilities. It is plainly wrong to assume that because a contract is standard it is understood clearly and similarly by different parties to it. Nevertheless, there are still many people who believe that there are significant risks associated with using non-standard forms of contract. Their argument is that since such contracts have not been tested in the courts, parties are more likely to interpret them differently. Of course, this is only true if people working on projects and interpreting contracts are aware of and up to date with the latest court decisions and are able to interpret them clearly, which is unlikely.

Time and cost pressures

Others have pointed to unrealistic time and cost pressures as a cause of conflict, particularly during design, which results in rushed decision-making and incomplete, ambiguous and inconsistent contract documentation). However, in response some have argued that it is the traditional procurement system's fault for not being able to satisfy demands for faster times that is the problem rather than the demands themselves. The problem with this system is that it is highly sequential.

Subcontracting

The growth of subcontracting has also been identified as a source of conflict in the complexity of the contractual arrangements and challenges to managers it produces (NEDO 1983, 1988). With the growth of subcontracting has come fragmentation, instability, short-termism, reduced customer orientation and the accompanying problems it produces in communication, motivation and quality.

Procurement systems

There has also been some inconclusive debate about the relationship between procurement systems and conflict (Naoum 1991). But this debate is more likely to have been motivated by the attractive possibility of a simple explanation for conflict than a thoughtful investigation into its true causes. Indeed, a focus on procurement systems could be dangerous and misleading in that it could grossly oversimplify the complexity and variability of construction project organisations and of the potential causes of conflict within them. The point that is missed in such discussions is that a procurement system is not an entity in itself but is merely a function of a particular combination of decisions relating to issues such as those discussed above. Indeed, one may be justified in saying that projects are so variable in the huge range of decisions which have to be made that it is difficult and simplistic to use procurement system categorisations. The concept of a procurement system is one that has been much maligned and misunderstood and although it performs a convenient categorisation function, overreliance on it could oversimplify the true variability of construction project organisations and thereby induce inflexibility and lack of thoughtfulness in their management. In this sense, the concept of procurement may be more damaging than helpful in the conflict debate.

The size of the market

At a more fundamental level, others have pointed to the simple explanation that the industry is too big for its market, particularly with the recent influx of foreign competition. In such an environment, where too many companies are chasing too few jobs, the market operates to force margins and contingency allowances down. This results in lean, poorly resourced projects which are susceptible to conflict because of their inflexibility in being able to deal with unexpected problems that impose extra resourcing demands (Richardson 1996). Within such organisations survival is the priority and good will an unaffordable luxury. This is particularly so in high-risk projects where the increased occurrence of problems will erode contingency allowances more rapidly and reduce the ability and/or willingness of companies to accept their resourcing responsibilities.

Unfortunately, the companies that thrive in such an environment are the unscrupulous ones who 'go in' low with the aim of 'coming out' high. In this sense, it is an environment which encourages mediocrity and at worst corruption and reduces the behaviour of the industry to the lowest common denominator. In many respects, this is where the role of clients is critical in that they determine the way in which the market operates. Unfortunately, there is considerable evidence to suggest that they have succumbed to the temptations of low price which such conditions bring, thereby perpetuating the problem. More recently, however, there are signs that some clients are leading the way in changing the emphasis of employment criteria to value rather than price (NSW DPWS 1993).

Conflict is institutionalised and inherent within construction activity

There are those who argue that conflict is inherent in the construction process by pointing to the nature of the process in comparison with other less conflict-ridden industries. For example, construction's temporary, project-based nature means that most project teams are newly formed and experience teething and learning curve problems early on. This increases the chances of misunderstanding during the most critical period of a project. The opportunity to maintain consistent project teams through a partnering arrangement is limited to major developers, particularly in the government sector, but the majority of construction clients (by number) are not of this kind.

Other reasons for the institutionalisation of conflict relate to the nature of the production process, which is concerned with the creation of one-off products, is small-batch in nature, and takes place within a relatively uncontrollable environment (Winch 1989). Furthermore, the industry does not lend itself to mechanisation and is essentially a labour-intensive industry which is subject to all the problems associated with human idiosyncrasies and needs, of which conflict is one.

Finally, others have pointed to the history of the construction industry and to the roles of the professions in maintaining the damaging stereotyped culture of division and mistrust, which induces attitudes that tend to conflict. Related to these origins are the industry's engineering values, which cause the industry to develop prescriptive and inflexible systems that seek to reduce and suppress conflict. These are counterproductive in an industry that is fundamentally human in nature and characterised by a considerable degree of uncertainty and change.

THE ROOT CAUSES OF CONFLICT

While this knowledge about the causes of conflict has the potential to reduce its incidence, De Bono's (1991) work suggests that the problems we have created in the construction industry cannot be wholly solved at the same level of thinking we were at when we created them. Since conflict is essentially an interpersonal phenomenon, De Bono argues that the most fundamental cause of conflict is that people see things differently. Covey (1994: 40) indicates the same when he asserts that 'the way we see the problem is the problem'. Covey goes on to argue that the cause of most relationship difficulties is rooted in conflicting or ambiguous expectations about roles and goals, which lead to misunderstanding, disappointment and the withdrawal of trust. In this sense, to successfully resolve conflict, a manager must look at the paradigms through which different people see the world as well as the world he sees. People's paradigms are the source of their attitudes and behaviours and represent the 'lens' through which people interpret everyday life. These are important in the construction industry because of the strongly differentiated professional cultures, which have developed and the stereotyped images which have accompanied them. De Bono went on to consider the forces that shape people's paradigms, thereby highlighting the importance of team formation as a low-cost, proactive mechanism for reducing the potential for conflict within construction projects. These forces are discussed below.

Personality and past experience

Past experiences, personal, practical and educational, represent a powerful shaper of opinions and perceptions about situations and other people. This is particularly so in the construction industry where the professional institutions have successfully nurtured strong stereotyped images of each other in order to defend their traditional roles with the procurement process. Seymour and Rooke (1995) argue that such images play a particularly important role in construction projects because at the start of a project, the newness of the team means that there is little else to base relationships on. All construction projects are therefore characterised by an early uncertain and cautious period of relationship-building, which is important in reinforcing or dispelling these stereotyped images which people initially hold. In this sense, to minimise conflict, project managers should pay particular attention to the perceptions of team members towards other disciplines and to the past experiences of team members with each other.

De Bono also points to people's differing personalities, which suggests that personality tests may also form a valuable part of the project manager's toolkit for constructing project teams. It is well known that in building successful teams there are certain combinations of personalities which are more productive than others (Belbin 1984). Merely selecting psychologically compatible people with positive attitudes towards each other and the problems the project poses could make a significant contribution to project success.

Logic bubbles

De Bono uses the term 'logic bubble' to refer to the concept of organisational culture, in the knowledge that people are embedded within employing organisations whose the culture will inevitably influence their perceptions and behaviour. For example, it is well known that some organisations are more claims-conscious than others and that their employees are constrained to reflect this policy in the way their performance is measured. It is also well known that contractors differ widely in the way they perceive subcontractors, the blacklist of contractors produced by CASEC in the UK (a subcontractor's confederation) being testimony to this. The multi-organisational nature of construction projects makes this an important consideration and indicates that the construction project manager's task, in assessing the experiences and personalities of a project team, must be widened to organisations as well as individuals. Unfortunately, the relative newness of organisational culture as a concept ensures that there are few techniques for measuring culture and less understanding of the cultural combinations that produce effective results.

Universe

De Bono uses the concept of a universe in a similar way to the logic bubble, although he widens it in the knowledge that people exist within a hierarchy of organisations whose culture influences their behaviour. In the construction industry, this brings us back to the values instilled by professional institutions, which differ markedly. For example, a visit to the institutional meetings of architects and builders will vividly illustrate the stark differences that exist within the construction industry's professional structure. In contrast to the builders, whose values would reflect realistic, practical and earthy themes, architects exude conceptual, creative, artistic values that may appear pretentious to the former. This is even reflected in distinct professional languages and in differing modes of dress and behaviour.

At an even higher level there is the influence of the construction industry as a whole. It is increasingly being recognised that different industries have their own unique cultures, and while little is known about the construction industry's, the most commonly cited characteristics are its macho and confrontational nature. Indeed, many have argued that conflict has become institutionalised within the construction industry to the extent that people expect and therefore may encourage it when they take part in a project. At a higher level still is national culture, which is becoming increasingly relevant because of the increasing internationalisation of the construction industry. Deresky (1997) gives an amusing example of how cultural differences have the potential to cause unintentional conflict through simple misunderstanding of differing expectations, gestures and words, which hold different meanings in different cultures. This is covered in more detail in Chapter 4. However, in terms of managerial control, these represent broader cultural influences which are largely beyond the construction project manager's influence in assembling a team, although there may be people who straddle the boundaries of these institutional and international cultures more easily than others. Furthermore, it may pay dividends to use more people from other industries in an attempt to water down the confrontational influence of the construction industry's institutionalised values. One of the authors was recently involved in a very successful project which was project-managed by a person who was formerly employed as a manager for a large car manufacturer.

Information, truth, falsity and corruption

De Bono recognises that everyone enters a dispute with different information, which will shape the way they perceive and define it. Recently, advances in the behavioural sciences have thrown light on the unlikelihood that people's perceptions will be the same. For example, Tsoukas (1996) points to the incredibly complex and unpredictable way in which information is dispersed within organisations which in turn makes it impossible for anyone, including the manager, to know the differing information and therefore perceptions that people hold at any one time. He gives the example of an organisation of seven people, which has the potential for 242 different structural combinations of relationships. In this sense, the challenge to managers of conflict is to conflate highly dispersed information in order to equalise information differences between people and thereby allow a common definition of the problem to emerge. In a construction project this task may be made difficult due to the numbers of people involved, but more specifically due to their dynamic membership. This ensures that people come and go, which creates the possibility that they take crucial information with them, thereby preventing any real convergence. Further problems may be caused by the habit of separately distributing construction risks, which produces conflicting objectives and a lack of collective responsibility. In this climate, people are more likely to hold onto information as a potentially important power source in negotiations and only release information, which helps to persuade others on the validity of their personal perspective on the problem. De Bono points out that in extreme circumstances, when people's existence is threatened, desperation may induce people to fabricate or falsify evidence in order to positively deceive and distort the truth in their favour. Therefore, it is evident that a lot can be done in the formulation of contracts to control people's behaviour and make the job of information equalisation easier when a dispute arises.

Human desire for contradiction and consistency

This refers to people's tendency to see things in black or white. De Bono argues that when we think, we deliberately search out contradictions and opposites and categorise them into universally applicable solutions to problems because this prevents uncomfortable ambiguity and confusion, which demands thought and time and therefore money to resolve. Indeed, this is likely to be a particular problem in construction projects because of the engineering values that underpin the field, which induce a particularly acute discomfort with uncertainty. A further encouragement to this tendency are the time pressures that have come to characterise construction projects, which reduce the time people have to think. Finally, the way in which construction contracts separate construction risks also induces a cut-and-dried perception that there must be a winner and loser in a dispute. Mentally, this concept is much easier to deal with than a win–win solution, which inevitably takes more time to construct. But paradoxically, in the long term, the reality is precisely the opposite in that the latent resentments of losers may undermine subsequent conflict management efforts.

De Bono argues that the problem with this approach is that it results in aggressive, non-compromising signals, which increase the chances of polarisation, stalemate and conflict. In this sense, it is the responsibility of managers to encourage people to think in tones of grey rather than black and white. Translated, this means encouraging people to look at problems with a sense of cultural responsibility. However, such a task is likely to be difficult because of the pressures which increasingly characterise construction projects and people's natural association of compromise with weakness, particularly in Western cultures.

REACTIVE MANAGEMENT STRATEGIES

Knowing the vulnerability paradox, the impossibility of providing a conflict-free environment and the opportunity costs of reducing conflict, all organisations should incorporate a reactive element into their conflict management strategies. This is particularly so on high-risk projects where there is an increased chance of problems that create the potential for conflict slipping through the preventive net.

The aim of reactive management

The aim of any reactive strategy should be to minimise the potential costs of a conflict to an organisation and maximise the potential benefits. These start to accumulate, in the form of opportunity costs of diverted energies and the administrative costs of management, as soon as a conflict emerges, even if it is in the 'just begun' phase. Ansoff (1984) has produced a useful representation, which can be adapted to show how these costs accumulate over time. Its value is that it highlights the exponential nature of loss accumulation and the importance of a swift reaction, which will nip a conflict in the bud. If left unresolved, the costs associated with a conflict can escalate to such an extent that they can threaten the very existence of an organisation. Ansoff's model is illustrated in Figure 13.1.

Figure 13.1 The accumulation of conflict costs

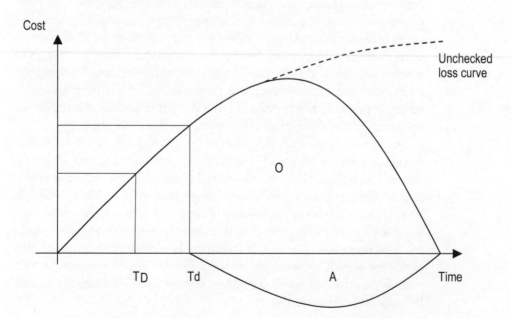

The origin of the graph represents the event that triggers the conflict. The tensions that represent the origins of a conflict often go unnoticed for a time, lost within the normal fluctuation of relationships within a project. Furthermore, the people involved in a dispute may decide not to report it to management in the hope that they can resolve it informally by negotiations between themselves. Certainly, managers would not want to get involved in all disputes that arise within a construction project.

Time (T_D) passes until the conflict is noticed by management or brought to their attention due to an inability to resolve it between the parties involved. This time will probably depend on the quality of relationships on a project, better relationships prolonging the period by reducing the rapidity of escalation and therefore the need for third party intervention by management. Then there may be another delay (T_d) while management investigates the history of the conflict and the issues at stake. By this time, losses will have accumulated in the manner shown by the unchecked loss curve, and the administrative costs of managers of dealing with the problem will also have begun to accumulate. This is indicated by the curve below the x-axis. It is likely that the efforts of managers will take time to have effect and therefore that losses will continue to accumulate before they reduce with the dissipation of the dispute. The conflict is resolved when there are no further energies being invested in it, and this is represented by the point of coincidence of the cost curve with the x-axis once again. The total cost associated with the resolution of the dispute is represented by the areas O and A.

The limitation of this model is that it only considers the tangible direct costs of conflict and ignores the substantial indirect and covert costs associated with damaged relationships, which may show up later in a project in the form of further disputes which would have otherwise not occurred or been so acrimonious. These are of course impossible to quantify but should not be ignored.

Problems in reactive management

There are many forces that can interfere with the efficiency of reaction in mitigating the costs of conflict. For example, many conflicts may remain undetected because their causes emerge slowly over time. However, sudden conflicts may also remain undetected as managers become flooded in a sea of information. Poor communication, which is often a characteristic of construction projects, can also cause conflicts to remain undetected. Indeed, in certain circumstances, people may be tempted to ignore any potential conflict in the hope that the problem will go away. In psychology this phenomenon is known as repression and it involves people burying unpleasant facts or memories in their subconscious. Another reason for conflicts going undetected is the tendency for all people to exhibit, to some extent, a natural resistance to change. No person likes a challenge to the status quo because of the discomfort and work it involves, and this is particularly true for people who may lose from the new order or be implicated in blame. This causes people to either deliberately hide problems from managers or, at the very least, play them down. There is also the intermediate behaviour where people craft messages in a way that allows them and their superiors to ignore the problem in comfort, so that they can blame each other if it escalates into a crisis. Argyris (1990) has referred to these behaviours as defensive routines. There is also the issue of managerial style in encouraging openness and deceitfulness. For example, some managers naturally adopt a laissez-faire attitude towards the resolution of conflict, which would not encourage people to report it. Indeed, neither would an autocratic manager, who would be likely to take complete control of a situation in imposing solutions on disputing parties.

Therefore, the task of the manager must be to encourage an open environment which facilitates communication and to do this they must establish an air of trust and independence. However, in the knowledge that they will never create a completely open environment, managers must also be sensitive to the hidden messages within people's communications and to any underlying tensions that exist or may develop within their organisation.

Before progressing, it is important to point out that while a swift response to a conflict is desirable, there is some logic in managers not encouraging the reporting of every minor problem that arises. While all may hold the potential for conflict, intervening in every case would consume so much energy that it would prevent any progress in managing the main project tasks. Indeed, psychoanalysts argue that some delay in dealing with a problem can be useful in convincing people that there really is a serious problem that needs resolving. Without the accumulation of this concern, the problem is less likely to be resolved successfully.

Conflict behaviour

Assuming that some kind of reactive intervention is needed from a manager, what types of behaviour could they expect to have to deal with? In the first instance there would be shock and panic as both sides frantically try to construct their cases based on their own interpretations of contractual clauses and the events that led to the conflict. This may well be driven by a genuine belief in their own version of events or by a strategy of deliberate deception on the part of potential losers. This behaviour may be designed to delay damages as long as possible or perhaps even bankrupt their opposition before they are successful. A manager must be sensitive to the latter strategy and not permit it since it is unethical and will severely damage all concerned with the project, including the client. The origins of such behaviour are often found in prevailing economic conditions since during lean times companies will find any way to survive. It can also be found in the way that risks are handled in construction contracts. Many smaller contractors and consultants, and many larger ones also, pay little attention to the risks they are being given, and so when a problem arises they are surprised by their responsibilities. They may either totally neglect to read their contracts or merely misunderstand what has been stated. In any event, the result will be that they will have not allowed for their risks in the original tender, meaning that the only way of avoiding serious loss is to resist the responsibilities imposed. Certainly, the legalistic manner in which contracts are written is not conducive to clarity and few companies will invest in the services of a lawyer at the tendering stage when they have yet to win the contract. This is the advantage in two-stage tendering since it provides a rough contract as the basis for an initial bid, which can then be negotiated in more detail with a successful company who will be prepared to invest in a lawyer to understand what is being stated. A further advantage of this approach is that people can take time to talk and thereby identify and resolve any misunderstandings that may exist between them. A further reason for this type of behaviour relates to the way risks are distributed and in particular to the habit of not paying parties for the risks they take on, giving them risks they cannot bear and risks they are unqualified to control. It is essential that managers give attention to the way risks are distributed during the early stages of a project in order to ensure that people are paid a fair premium for their risks. Furthermore, it is crucial that parties are given a level of risk they can afford and the type of risks they can control.

Whatever the causes of this behaviour, it is unfortunate that it faces the manager dealing with conflict with two paradoxes. The first is that communication breaks down at a time when open communication is of increased importance. The second is that at a time when teamwork and collective responsibility are increasingly important, it is less likely. This is because information becomes a source of power in negotiations and people are therefore more secretive with it. Communication also becomes more formal in that the writing of a letter says something in itself. Furthermore, during a conflict, face-to-face communication becomes uncomfortable and places weaker parties in a vulnerable situation. Indeed, communication breakdown might occur because the organisation becomes flooded with such information, which is all directed towards a few people who are involved in the dispute. The result is information overload, which causes tension, pressure and possibly stress, conditions that also add to the manager's problems.

CONCLUSION

This chapter has addressed the issue of construction conflict from a proactive and reactive perspective. In summary, the discussions lead us to a number of recommendations which can help construction project managers deal with conflict more effectively:

- Nip it in the bud – prevent it accumulating momentum (driving a fast car is more difficult than driving a slow car).
- Bring it to the surface rather than ignore or suppress it.
- Clarify it by developing a common definition between interested parties.
- Get to the bottom of it by investigating its underlying reasons.
- Pay attention to the maintenance of effective communication channels.
- Try to encourage mutual sensitivity between differing interest groups.
- Try to be impartial.
- Depersonalise the conflict to reduce its emotional base and increase its rationale base.
- Seek compromise solutions.
- Keep energies focused on the main project tasks.
- Try to draw out the positives in the conflict by looking for win–win collaborative solutions rather than a win–lose compromise solution. The discovering of a mutually satisfactory solution can rapidly defuse a very volatile situation.
- Be aware of people's bargaining tactics and the escalating impact these can have.

EXERCISES

1   What are the main causes of conflict in the construction industry?
2   What is the argument for encouraging more conflict in the construction industry?
3   Why is it important to have a proactive and reactive approach to the management of conflict?

CHAPTER 14

NEGOTIATION IN CONSTRUCTION PROJECTS

INTRODUCTION

Negotiation is the initial and informal means by which parties attempt to resolve their differences during a construction dispute. However, knowledge of these processes in construction projects is scant, with most attention being given to developing more formal, costly and reactive means of dispute resolution. This chapter questions this approach in favour of a more efficient one, based on a better understanding of negotiation behaviour and improved negotiation skills. To this end, the behavioural complexities of the negotiation process during construction disputes are discussed and it is shown that most construction disputes are unintentional and escalate as a result of misunderstandings and tactical miscalculations during the negotiation process. A series of recommendations are made to reduce the potential for unintentional escalation during a construction dispute.

AVERTING DISPUTES THROUGH NEGOTIATION

The previous chapter pointed out that the potential for disputes arises when the limited resources available to an organisation become exhausted by an unexpected problem. When risks are distributed unevenly, as they commonly are in construction contracts, the responsibility for supplying extra resources falls unevenly on project participants, leading to the emergence of distinct winners and losers and a redistribution of resources between them. Consequently, such an event is likely to be met by a range of positive and negative behaviours, with potential beneficiaries encouraging the change and potential losers resisting it. It is within the resultant tensions that conflict is born, a phenomenon that escalates through a number of phases from 'just begun' to 'dispute' to 'contention' to 'limited warfare' to 'open warfare'.

Most of the construction literature concerned with dispute resolution focuses on the techniques of conciliation, mediation, adjudication, expert determination, arbitration and litigation (Whitfield 1994; Sawczuk 1996; Davenport 2000). Surprisingly, few authors have given attention to the process of negotiation, which is likely to be the initial and informal mechanism by which people attempt to resolve their differences. As a result of this neglect, there is little understanding of the negotiation process in construction, an important omission in the light of contemporary demands for a reduction in the opportunity cost of managing construction conflict. A greater understanding of the negotiation process, and in particular of how conflicts escalate, would help to improve negotiation skills, reduce the probability of escalation, and thereby avoid the need for costly intervention from independent third parties.

NEGOTIATION: THE PROCESS

Negotiation can be thought of as a game of strategy analogous to poker or chess. The need to negotiate implies a difference in interests, objectives and expectations and is concerned with reaching accommodations between them. In essence, it is a process that involves a struggle between adversaries who attempt to move, stepwise, towards an agreement over resource redistributions which is in their own favour. In this struggle, people may make sacrifices in pursuit of some overriding goal and negotiation can often become more a matter of forestalling the consideration of certain unattractive solutions than a matter of extracting a change of position from an adversary. Often, as in the case of construction projects, negotiators belong to and represent the interests of distinct profit-making organisations. In this situation, people's attitude in negotiations is determined by restrictions imposed by those organisations, restrictions that reduce decision-making authority and therefore a negotiator's autonomy at the negotiation table. Thus, during a construction dispute, negotiators must monitor their constituents as well as their opponents and in this sense pacify tensions within a complex web of interpersonal and inter-organisational relationships.

NEGOTIATION TACTICS

The essence of the negotiation process are the 'tactics' or 'moves' that parties use to influence each other to move closer to their own position. Although a wide range of negotiation tactics, within and outside a construction context, have been identified by the likes of George (1991) and Whitfield (1994), it is possible to place them all at some point along a continuum from accommodative to coercive. Essentially, accommodative tactics involve one party moving their own position closer to an opponent's by making a concession. In contrast, coercive tactics involve the use of force and threats of punishment to pressure an opponent to change their position against their will to become closer to one's own. Rogers (1991) argues that the tactical approach that a party adopts in the negotiation process depends on their 'negotiation code', that is, their set of beliefs about an opponent which influences the way they interpret and respond to their messages. These cognitive beliefs are, in turn, dependent on the personalities, experiences and attitudes of people, the policies of their employing organisations, and the nature of their power relationships with adversaries. Rogers argues that in a negotiation setting, the cognitive beliefs of negotiators can differ in: the image of an adversary, including beliefs about their objectives, decision-making style and typical negotiation strategies; the image of dispute dynamics, including beliefs about the nature of dispute escalation and the manner in which war might erupt; and general beliefs about the optimal mixture of coercion, accommodation and persuasion in a negotiation strategy. On this basis, Rogers groups negotiation codes into four broad categories, types A, B, C and D, which have the characteristics depicted in Table 14.1.

TACTICAL MISCALCULATIONS AND UNINTENTIONAL ESCALATIONS

Rogers (1991) regards negotiation as a process whereby negotiators make tactical choices guided by expectations about an opponent's response and outcomes. The probability of escalation depends on the accuracy of their diagnosis and the skill with which they construct appropriate moves. Unfortunately, people's belief systems are vulnerable to misperceiving their opponent's motives, something that can lead to tactical miscalculations and an unintentional escalation of a dispute. Argyris (1962) suggests that this becomes an increasing danger as the dispute escalates because people's perceptions of other become less rational and more emotional. Furthermore, as a dispute escalates, people tend to become progressively embedded in their own negotiation position as increasing investments of resources reduce their willingness to compromise and make winning at all costs increasingly important.

TABLE 14.1 NEGOTIATION CODES

CODE	COGNITIVE CHARACTERISTICS
A	The adversary is seen as aggressive; it is believed that only intentional war is possible and that any escalation is easily controllable; a fait accompli or the strong use of force is seen as the best way to resolve a dispute; little consideration is given to an adversary's possible response and its potential for escalation; success is defined in military rather than diplomatic terms.
B	It is believed that the control of a dispute is possible by understanding the dynamics of escalation and thereby avoiding the point where control is lost; it is assumed that probabilities of escalation can be assigned to various tactics and strategies; incremental small-step escalations are seen as timid and a sign of weakness that are likely to lead to an escalation; it is believed that a failure to show resolve is the most common cause of war.
	TYPE B-I
	Coercive diplomacy (verbal threats of extreme actions and all-out war) and bluffing are considered the best means of dispute resolution; escalation is assumed to come from failure to communicate a determination to protect one's vital interests at any cost; it is considered dangerous not to brandish the ultimate weapon.
	TYPE B-II
	Bluffing and threats are considered dangerous since they may induce a counteractive aggressive response inadvertently; it is best to use limited force to avoid an all-out war rather than use threats of all-out war.
C	It is difficult to determine whether an adversary is offensive or defensive, meaning that tactics must be cautious and context-driven rather than automatic; it is believed that there are many unpredictable paths to escalation, that it is difficult to avoid slippery slopes, and that the brink cannot be recognised in advance; it is only partly possible to control a dispute and threats or use of power are dangerous; the best strategy is to tread carefully and compromise with an opponent; a carrot-and-stick approach is advocated as a means of manipulating an adversary.
D	It is assumed that an adversary operates in a defensive mode and that the control of a dispute is extremely problematic if not impossible if there is even a modest emphasis on coercion; the best approach to dispute resolution is a highly cautious one based on accommodation and compromise; the entire strategy is based on the fear of uncontrollable escalation and avoiding negotiation situations.

SOURCE Adapted from Rogers 1991.

George (1991) elaborated on the idea of unintentional escalation by arguing that an important constraint on the negotiation process is the fear of war, a self-perpetuating and extremely damaging form of conflict. In other words, a party's primary objective is to get their way but at the same time they want to do this without precipitating war — they are as much concerned about avoiding war as they are about getting their way. Like Rogers (1991), George (1991) argued that most conflicts escalate unintentionally as the result of miscalculated tactics by those involved in negotiations. He held that associated with each negotiation tactic was a risk of escalation and that some tactics were more risky than others. However, risky tactics, if successful, can bring about a swifter end to a conflict and this produces a temptation to employ them. For example, blackmail is a coercive strategy which involves demanding something on pain of punishment if it is refused. While it has the potential to achieve a quick decisive gain, it is a highly risky tactic because the blackmailer may misjudge his opponent's resolve. The opponent might make a surprise response by calling the blackmailer's bluff or may employ some counter-action and thereby escalate the situation rapidly. To reduce the risk of such strategies, some parties may initially employ a tactical probe, which involves testing an opponent's resolve or response to an impending action. But this cautious strategy also has its risks because poor communication and misperception of the incoming signal by the opponent can lead to an overreaction and an accompanying escalation.

MOTIVES DRIVING THE ADOPTION OF SPECIFIC NEGOTIATION CODES

In addition to understanding the tactical combinations that create the potential for escalation, it is useful to understand the motives that drive the pursuit of certain tactics. Such knowledge can provide project managers with a greater understanding of the rationale underlying a dispute and thereby enable them to identify workable solutions. To this end, it is worth summarising the results of research that has indicated the factors that determine a party's negotiation code.

Uncertainty of financial responsibility

When the location of responsibility for a disputed matter is blurred, uncertainty becomes a tool that can be exploited in attempts to avoid financial responsibility. Parties may attempt to take advantage of the inherent ambiguity of a dispute and the responsibility patterns that surround it by trying to redefine them in their favour. For example, people may attempt to distort the interpretation of contractual conditions that are in their favour in the hope that other parties may not notice it. The popularity of this tactic is related to its low risk of escalation. But such manipulative behaviour is not necessary, especially when there is a general willingness to accept contractual responsibilities.

Preconceived belief structures

People's initial behavioural response to a dispute is influenced by their preconceived beliefs about their opponent's interests and motives. If they are negative, then tactics tend to become coercive quickly, but when they are positive, people are far more willing to adopt accommodative tactics in their negotiations.

Personal interests

People's actions are often designed to protect their own interests and it is common for people to exacerbate the conflict by gathering into groups of similar interests to reinforce their power base. This is a serious problem since it results in pockets of information within the organisation with no communication between them. The inevitable result is that the dispute deepens rapidly.

Organisational policies

People's tactics are often guided by the policies of their employing organisations. For example, the policy of some contracting organisations may be to encourage their staff to pursue claims, one of which is likely to result in coercive tactics. Similarly, some clients may have strict policies on budgetary control, which might cause their consultants to adopt inflexible tactics in their negotiations. Such combinations of policies are highly dangerous since they may lead to the unintentional and uncontrollable escalation of a dispute.

An opponent's tactics

It would be reasonable to expect that a party's negotiation code would, in part, be a response to their opponent's code. Indeed, research indicates that the negotiation codes of opponents in a construction dispute typically follow parallel paths. That is, when one escalates, so does the other. This has important managerial implications since it indicates little imagination and courage in negotiation and suggests that a dispute reaches a point where its momentum towards escalation cannot be stopped by seeking compromise. This is the dangerous stage at which emotion and its accompanying irrationality guide behaviour. It appears that once the seeds of escalation are sown, it is too late for managerial intervention to have any significant effect on a dispute. At this point, the manager is largely powerless and is left with no option other than to firefight, preventing further accumulation of momentum and mitigating damage, letting the dispute take its natural course. A dispute will resolve itself only when all the pent-up frustrations of the disputing parties are dissipated, something that may take several escalations. This means that ad hoc solutions coercively imposed by managers or by third parties, which do not release the frustrations and tensions underlying a dispute, will merely create the superficial appearance of a solution. In the long term, underlying tensions will burst through in a re-emergence of the same dispute or in an illogical escalation of an unrelated and seemingly harmless dispute. It appears that disputes are only capable of being truly resolved by those whose interests are dependent on the outcome. The managerial implication is that managers should aim to facilitate this process rather than to forcibly impose quick solutions.

NEGOTIATION SKILLS

Effective negotiators achieve their results partly by understanding the factors that influence people's attitudes and behaviour and partly by exercising a range of skills that can be described in general terms as persuasion. All managers need to develop these skills because much of a manager's job is concerned with resolving issues on which people have different views but need to agree about solutions. The better a manager is at convincing other people of the need to accept or support a particular course of action, the less frequently serious disagreements will get in the way of progress, or decisions have to be imposed by senior management. The effective manager is a persuasive manager.

There is no single characteristic of persuasiveness: it is an amalgam of skills that are discussed below.

Your approach

The quality of any sort of negotiation is determined primarily by the tone set by whoever starts the meeting. There are two extremes: confrontational or collaborative. Inexperienced negotiators sometimes become defensive and adopt an abrupt, aggressive stance — though often this arises from fears about weaknesses in their own case or fears about having to compromise. Occasionally, too, an experienced but manipulative negotiator will adopt this confrontational approach when dealing with someone they recognise as inexperienced or nervous. A common ploy, designed to 'knock an opponent off balance', is to open the meeting with a wholly unexpected complaint or demand. For example: 'I know we are here to talk about unit costs but I'm not willing to get into this until you explain why you failed to supply me with accurate, quality data yesterday!' Note the personalised nature of this kind of attack ('I', 'you', 'you', 'me'). Whether conscious or not, this personalisation tends to turn the aim of the discussion into a win-or-lose battle between two people rather than an attempt to find a solution to the problem under discussion. The confrontational approach cannot be recommended, even if it appears to succeed in a few cases. The problem is that such cases generate resentment and undermine any later requirement for the two people concerned to work in collaboration.

The most productive style is the collaborative or problem-solving approach. This sets aside any personal feelings about winning or losing and concentrates on the issue to be resolved. In this approach the fact that the other person holds a different view or seeks a different outcome is not seen as a personal challenge. It simply creates a situation that both parties need to resolve. In essence, your message to the other person is: 'We have a problem that we have to discuss in order to find a solution that satisfies both of us', and not 'If it wasn't for you, there wouldn't be a problem, so I want you to back off and accept the solution!' This does not mean that there should be no targeted outcome, simply that the most effective way to work towards a goal is for the solution to emerge as a jointly agreed one. This is often described in the literature on negotiation as a 'win–win' approach.

Being considerate

One of the most important factors in developing persuasive skills is to continually look at your own negotiation objectives (and how you are trying to achieve them) from the other person's viewpoint. People cannot be ordered to agree with you, and it is unrealistic to expect people to adopt an altruistic attitude towards negotiations. In reality, most people are concerned with how your proposals might benefit them, not you. The fact that you have a problem is, by definition, your problem, not theirs. Making it seem like their problem as well as yours is the key to securing meaningful persuasion in negotiations, and the way to achieve this is to show them either that if they fail to address the issue constructively they may experience a problem themselves or that if they help you find a solution they may gain some benefit. The more you can stress a mutuality of interests, the more readily you will be able to persuade.

Listening

The two components of any discussion are talking and listening. Saying the right things in the right way at the right time, knowing when not to speak, and listening very carefully to what the other person says, are the essential characteristics of the skilled persuader. Inexperienced negotiators often feel under a compulsion not just to initiate the discussion but also to set out their whole case as quickly and comprehensively as possible. One symptom of a lack of confidence is compulsive talking, which includes the frequent interruption of the other person whenever they say something with which the negotiator disagrees. The wise approach is to develop a dialogue with the other person, giving them every opportunity to say what they wish to, and building up your case as the discussion proceeds. Attentive listening is essential if you are to discover the way the other person's mind is working. It is also important to listen not just to the words but to how they are said. Are statements being made confidently, or with some hesitation? Are there signs of irritation or impatience? Is there an immediate rejection of your proposals, or can you detect some willingness to consider them? These are the types of question you should be asking yourself while the discussion proceeds. Listening is as important as talking. So the general rules are:

- Do not state the whole of your case at the beginning: encourage a dialogue with the other person and develop your case as the discussion proceeds.
- Do not interrupt when the other person makes statements that you disagree with or that are incorrect: wait for them to finish and then use impersonal terminology to achieve corrections, rather than making personal criticisms.
- Be an attentive listener to the tone, as well as the substance, of what is said.

Investigating

Just as you are trying to persuade the other person in a negotiation to accept your views or proposals, so they will try to persuade you. If they can show that the ideas they put forward achieve better results than your solution, it is not a matter of weakness to be persuaded. What is important, however, is that their case should be very thoroughly examined. This is best done by probing questions and a thorough investigation of the problem before negotiation begins, so that you can be confident of making a sound assessment of its strengths and weaknesses. There are three common flaws in the way negotiators argue their cases:

- making factually incorrect statements
- omitting relevant factual information
- misusing statistics.

Each should be looked for and tested, if necessary, by your asking questions.

Questioning is generally much more effective than simply making statements or asserting that the other person is wrong. An even more common fault is to argue on factual grounds but omit facts that, although relevant, happen not to support one's case. The 10 per cent reduction in productivity may be an accurate figure, but it is misleading because no mention is made of a lengthy dispute during the period concerned. A calmly placed question such as 'How many of those problems were affected by the dispute?' can be much more effective than a belligerent 'You are conveniently forgetting the dispute!' It is equally important to be wary of accepting the validity of certain statistics, particularly averages and percentages when very small numbers of cases are involved. Thus, in a negotiation about office rentals, one person may state that the average new rental for the area is $200 per square metre. An appropriate probing question would be 'How many new rentals are included in that figure, and what were the lowest and highest rentals?' This may reveal that the average is almost meaningless, as only six cases were involved and the figure was distorted by one very high rental for a building of unusual prestige.

Faced with questions of this kind, the other person may realise for themselves that their argument is flawed, rather than having you tell them so. Exposing flaws tactfully is a powerfully persuasive tool.

Time out

Experienced negotiators frequently make effective use of breaks or adjournments in a discussion. Adjournments can be used for two main purposes:

- to give those concerned an opportunity to consider new points or proposals before making any commitments
- to bring a halt to a discussion that has become too emotionally charged.

Inexperienced negotiators can be over-influenced by pressure to agree quickly to proposals they have not thought about enough. It is important to resist this pressure, and one way of doing so is to be quite open about your need for time to think. Just say, 'that's an interesting idea but I need a little time to consider it. Let's have a break, and resume in half an hour'. Another reason for adjournments is that discussions do sometimes become heated, and good solutions are rarely possible when tempers are raised. Even a very short break at such a time can result in a greatly improved atmosphere when the discussion resumes. It just needs someone to take the initiative and say something like, 'We seem to be getting bogged down. I think it might be a good idea if we have a quick break for a cup of tea at this point and resume in 10 minutes'.

Compromising

The best outcome of a discussion may appear to be when you persuade the other person to accept your unaltered proposal or point of view, but this is not always desirable. For example, the other person's input to a discussion may reveal factors that you have failed to take into account. Attitudinally, you need to focus on the issue under discussion and remain determined to reach an agreed outcome, rather than be determined to hold to your original decision or to beat your opponent. An opponent's input may make them feel more involved and motivated in a change and may also improve the solution arrived at. It is generally a mistake to rush into making clear-cut proposals. People need time to think about possible changes to their position and often react badly to any apparent attempt to rush them into making a decision. A key point to remember when discussing concessions and compromise is the importance of encouraging the other person to feel positive about changing their position, that is, not making them feel they are losing. A change may be of great benefit to you, but that is not the point to emphasise. You need to help the other person recognise the personal benefits of any compromise they might agree to. So it is good to applaud their helpfulness and wisdom and not gloat about getting your own way.

Summarising

To prevent decisions being forgotten, it is often useful to pause in a discussion at appropriate moments, taking stock of where it has reached, confirming with a written note anything agreed so far. This can be initiated by a suggestion such as: 'I would now find it helpful if we could just summarise where we have reached. If we also make a note of what we have agreed so far it will help us avoid going over ground we have already covered. As I understand it, we have dealt with a, b and c, and agreed x, y and z. If you're happy with that I'll make a brief note and we can then move on to d, e and f'.

Reaching agreement

After a period in which various options have been considered, it can be quite difficult to bring a discussion to a firm and mutually satisfactory conclusion. But there comes a point when ideas and possibilities for a solution can no longer be discussed hypothetically or on a no-commitment basis: someone has to take the initiative and put forward a clearly defined proposal. Timing is again important. Most discussions have highs and lows, and the time to conclude an agreement is obviously when there is a collaborative, not confrontational, atmosphere. As with the floating of possible solutions, so the persuasive approach to making a final proposal is to emphasise its benefits to the other person. In addition, it is necessary to be very firm about the finality of a proposal that genuinely represents the limit to any compromise. Inexperienced negotiators often weaken their ability to take this stance because they have described earlier positions as final, only to agree to further compromise. So, if you make a final proposal, be sure it is final, and explain why. Also ensure that the final agreement includes all necessary points, is clearly understood, and is not expressed in ambiguous terms.

Body language

All the preceding headings in this section have dealt with verbal interaction. But the quality of communication between two people in a discussion is also influenced or made evident by body language. In verbal communication attentive and perceptive listening provides valuable clues about the other person's attitude. Observation of their posture, the way they sit and gesticulate, can provide further clues. For example, you make a statement and the other person sits back and folds his arms. This is probably a sign that he is unimpressed, or even offended. Leaning forward generally indicates interest. Face-touching is often connected with doubt; eyebrow-raising is a classic indication of surprise. Experienced negotiators also use body language to communicate the way they feel. For example, one way of indicating that the exploratory part of a discussion is at an end is to gather up one's papers or to close an open file. Alternatively, the finality of a decision can be reinforced by a firm hand gesture: placing it palm down on the table. Body language needs to be interpreted with caution — it is not an exact science. But to overlook the signals that can be read from posture, gesture and facial expression is to miss a potentially rich source of indications of attitudes and intentions.

CONCLUSION

This chapter has discussed the process of negotiation and the range of conditions that can lead a dispute into a self-perpetuating cycle of unwanted escalation, which becomes increasingly difficult to break. These discussions lead to a series of recommendations that may help construction project managers reduce the potential for inadvertent dispute escalation.

1   *Become skilled in negotiation.*
2   *Prepare yourself thoroughly for negotiations* and approach them in a planned, sensitive, responsive and considerate manner. Listen to others, emphasise mutuality of interests in resolution, do not rush, be prepared to compromise, understand how different negotiation tactics can lead to an unintentional escalation, and record any agreements made.
3   *Information management.* Few people deliberately go into negotiations wishing to precipitate a war and most disputes arise from tactical miscalculations by people in opposition. That is, people have problems in communicating their intentions to opponents and in predicting their opponent's response. It follows that the efficient management of communication between adversaries is critical to dispute management, so that the misunderstandings that lead to unintentional escalations are avoided.
4   *Avoid the coercive imposition of solutions.* The adoption of aggressive tactics by one party is likely eventually to produce a similar response from an opponent and lead to an escalation of the situation. The potential for this tandem escalation seems to be related to the equitable balance of power that exists in construction projects. While the legitimate power that derives from contractual conditions may enable one party to suppress another's efforts to protect their interests temporally, parties to a construction project are also able to wield a considerable amount of illegitimate power, which most seem prepared to use, to break a stalemate. It is evident that to prevent an escalating situation, one party needs to break the mould and show a willingness for conciliation and compromise. It is likely that this will induce a similar response from an opponent. However, such a move to a conciliatory negotiation code must be a genuine one and must be perceived as such by an opponent. If it is perceived as not genuine, due to a lack of trust between parties, then it will have no alleviating impact on an opponent's tactical stance. In light of the apparent inability of coercive power tactics to bring about the successful long-term resolution of a dispute, it seems that project managers should encourage innovative and

courageous negotiation strategies among stakeholders and attempt to move parties towards type C to D negotiation codes. However, complacency in having done so is dangerous, since it is particularly easy for a dispute to escalate out of a type C mode. The reason is that when parties are operating in such a mode they are inherently uncertain and therefore nervous about an opponent's motives. In this sense, any peace is a fragile one in which opponents need little temptation to move towards a more aggressive negotiation stance.

5   *Strive to achieve a common definition of the disputed matter.* Project managers should work at reaching a common definition of a dispute and minimising the uncertainty that is likely to arise around responsibility structures. Any uncertainty tends to get exploited by parties as a way of furthering their interests. This is particularly so when competitive tendering is abused or, in times of recession, when low margins magnify the impact of a problem and make parties more reticent to accept their contractual responsibilities. Much can be done to avoid this potential problem before the project starts, by insisting on all-party risk analyses and by clarifying perceptions about risk distribution through open discussion. In this respect, managers should beware of standard forms of contract since they may create the delusion among project participants that there is a common understanding of risk distribution patterns. Standard contracts must not become a substitute for effective communication in the early stages of a construction project.

6   *Beware of allocating blame.* Project managers should also be wary of disputes that imply that one or other party is to blame. In such situations, the tendency is for those implicated to adopt inflexible and aggressive negotiation codes in the defence of their interests.

7   *Beware of temporary coalitions.* Those with common interests tend to collect into temporary coalitions to increase their power-base in negotiations and thereby further their interests. This is dangerous and tends to intensify the ferocity of disputes by clarifying and strengthening differences. It also breaks down communications and makes people more fearless and reckless in their negotiation tactics.

8   *Consider the structure of project teams.* Project managers can contribute to the avoidance of conflict escalation by carefully considering the structure of their project teams at the start of a project. In particular, they should avoid combining companies or individuals who have preconceived negative perceptions of each other from recently completed projects. Furthermore, the compatibility of managerial styles and information systems should be considered. During a dispute, information systems, leadership qualities and personal relationships are tested to the extreme and any weaknesses exacerbated. Ironically, it is during such times that strong relationships, good leadership and efficient communications are most important.

9   *Beware of goal inflexibility.* Finally, project managers should be aware of the dangers of inflexibility in client goals and in those of the contracting organisations they employ. There is a difficult balance to maintain between controlling project goals and providing the flexibility required to avoid conflict. The danger

of inflexibility is that it engenders a sense of defensiveness and rigidity in the negotiation process, which in turn leads to an atmosphere of escalation. In contrast, what is needed during a dispute is flexibility and the freedom to compromise on all sides. It is a scenario that will only be achieved when participating organisations are permitted to make the necessary contingency allowances for their risks.

EXERCISES

1   Why is it important to develop negotiation skills in the construction industry?
2   Through what mechanisms can disputes accidentally escalate?
3   What are the nine key elements of successful negotiation?

PART 4

RISK AND SAFETY
IN PROJECT
MANAGEMENT

CHAPTER 15

INTRODUCTION TO RISK MANAGEMENT

INTRODUCTION

Most decisions and events in life have a wide range of possible outcomes, some expected, some hoped for, and a few unforeseen. The possibility of unsatisfactory or undesirable outcomes occurring can never be ruled out. However, if the presence of uncertainty is predicted and its likely impact estimated with a reasonable degree of confidence, appropriate management action may be taken to mitigate the likely impact.

Risk and uncertainty are inherent in day-to-day activities, irrespective of the type of industry, the size of a project or the nature of the environment. The presence of risk is desirable because opportunity can arise from risk, so there is a natural balance between them. Trading in shares is the obvious example: with more risk, an investor expects to derive greater profits, while low returns are usually associated with low risk.

Construction projects, whether large or small, are surrounded with risk and uncertainty throughout their lifecycle. The project manager is responsible for identifying risky events, assessing their magnitude and developing appropriate responses.

Uncertainty exists where there is an absence of information about future events, conditions or values. The probability of uncertainty occurring cannot be quantified in order to express its economic loss. Uncertainty commonly gives rise to risk, which is defined as the potential for realisation of unwanted negative consequences of an event. It may also be defined as exposure to economic loss or gain that has a known probability of occurrence.

The precise distinction between risk and uncertainty has been the subject of many inconclusive arguments. In this text the terms are regarded as synonymous and no attempt is made to distinguish between them.

The purpose of this chapter is to review the concept of risk management and to demonstrate its benefits in managing projects. The text will explain how risk management assists decision-making by carefully analysing the likelihood and the consequences of identified risk events. Information will be presented from the perspective of a project manager who, in liaison with an expert risk manager and other team members, assumes the overall responsibility for risk management.

RISK MANAGEMENT

Risk management is commonly defined as 'the identification, measurement and economic control of risks that threaten the assets and earnings of a business or other enterprise' (Spence 1980: 22).

Risk management is a systematic approach to identifying sources of risk and uncertainty, assessing their impact, and developing appropriate management responses. These are the main three steps of risk management. A more comprehensive risk management process was defined by the Australian/New Zealand Risk Management Standard (AS/NZS 1999: 7–8) in the following points:

- establishing the context
- identifying risk
- analysing risk
- treating risk
- monitoring and reviewing decisions.

This chapter expands the process defined by AS/NZS by including important elements of risk classification and data elicitation. The expanded model of risk management is illustrated in Figure 15.1.

Risk management was originally formulated as a useful addition to a range of established techniques used in financial analysis, taking account of risk in financial forecasts. It has now become a necessary and useful adjunct to strategic planning by encouraging constructive dialogue across a wide range of tasks in project management. Accepting that an initial assessment of risk is no more than an attempt at understanding problems, the main benefit of risk management is in developing a team-based decision-making process

Figure 15.1 The risk management process

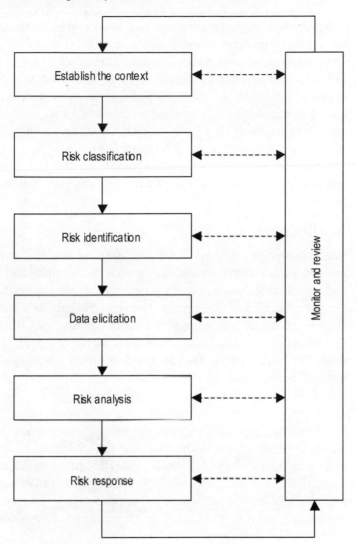

SOURCE Adapted from AS/NZS 1999.

based on an open and frank examination and re-examination of issues and events that show uncertainty.

Risk management is being widely applied in most industries, including construction. The construction industry benefits from its applications in hazard assessment, financial analysis, feasibility, strategic planning, lifecycle costing, scheduling, cost estimating, design appraisal, contingency planning and the like. It has also

Introduction to Risk Management

benefited from predicting the likely impact of a range of external political, economic, social and environmental factors on projects.

Since 1993, risk management has been compulsory on all NSW Government projects in excess of A$5 million in value. The NSW Government prepared comprehensive guidelines (NSW Government 1993) that defined risk management policy and process. These guidelines are similar to the AS/NZS with the notable addition of a risk management plan which organisations bidding for government work must prepare for PFI projects, projects classified as 'significant', and all other projects in excess of A$20 million.

Let's now examine individual steps in the risk management process when assessing the likelihood and the consequence of project risk.

Establishing the context

Establishing the context is the first step in the risk management process. Its aim is to examine the project, its strengths and weakness, and the environment in which it operates. It also identifies the project participants and attempts to interpret the project's business plan and objectives. It sets a broad scope and boundaries for the implementation of risk management. It also establishes operational, technical, financial, legal, and social and any other criteria against which risk will be assessed.

Risk classification

A process of risk identification can greatly benefit from establishing an appropriate risk classification model. Numerous such models exist (see Erikson 1979; Porter 1981; Cooper & Chapman 1987; Kangari & Farid 1987; Uher 1993). These models map categories of risk relevant to the activities of organisations. The classification model of project risk adopted in this chapter comprises 'External risks' and 'Internal risks'.

External risk refers to events that are outside the control of the project, for example regulations, market competitiveness, weather, interest rates, national strikes, availability of resources, and project location.

Internal risk refers to events that are generated by the project itself, for example changes in scope and design, errors in tender documentation, the use of onerous contract conditions and bid shopping, and the lack co-ordination of subcontractors.

Depending on the type and size of project, external and internal risk groups could further be aggregated into subcategories to make the risk identification process more effective. The external risk group could, for example, comprise economic, social, political, industrial, legal and environmental subcategories, while the internal risk group could be aggregated according to the project lifecycle stages or individual project participants.

Risk identification

The principal benefits of risk management are usually derived from the process of identifying risk. This is because identifying risk involves detailed examination of the project, its components and its strategy, which helps the project manager and the project team to understand better the complexity of the project, its design, the site on which it will be erected, and the likely influence of a range of external environmental factors. The project manager thus becomes aware of potential weaknesses for which treatment responses must be developed. The manager may also become aware of opportunities that may present themselves. But because the process of risk identification is usually a subjective one relying on the manager's ability to recognise potential risks, it may over-emphasise or under-emphasise, or even overlook, the importance of some risks.

In identifying risk, the project manager is often guided by past history and personal experience. But reliance on the past has inherent problems. Retrospectively, it is fairly easy to work out what decision the project manager should have made in consideration of a set of events that had actually taken place. Just because something happened on one or more past projects as a result of a particular event doesn't mean that it will happen on the current project. Another potential problem is related to the project manager's reliance on a small set of personal experiences gathered in the past, on which his opinion has been formed. This opinion may guide him towards making a particular decision that may be totally inappropriate in light of risk events to which the current project is actually exposed. One other danger in the use of history when identifying risk is the inability of the past to prepare adequately for new events not encountered before. Ashley (1989) warns against 'hindsight biases' or journeys back into the past because they may lead the project manager to identify the wrong risks.

Several methods are available for risk identification. In spite of the dangers pointed out just above, to a greater or lesser extent all the methods rely on history. They generally fall into two distinct approaches: bottom-up and top-down.

BOTTOM-UP RISK IDENTIFICATION

Bottom-up risk identification works with the pieces and tries to link them in a meaningful, logical manner. Examples of this approach are numerous:

A checklist approach A checklist is a database of risks to which projects have been exposed in the past. Depending on the volume of past data, it may be aggregated according to the type or size of project, the industry sector, the procurement type and the like. According to Mason (1973), this method offers the most useable risk identification method for construction contracting by allowing the project manager to identify, in a rational manner, risks to which a project is exposed. Reliance on historical data in a checklist may be both the strength and the weakness of this approach. For as long as it is used only as a guide that helps the project manager to identify risks thoroughly and systematically, a checklist approach may be highly effective. But when too much emphasis is placed on risks that are either irrelevant as far as a current project is concerned, or their real impact is likely to be too small to worry about, incorrect decisions could be made.

A financial statement method This method of risk identification is based on the premise that financial statements serve as reminders of exposure to economic loss. Since they show both budgeted and actual costs, the analysis of negative variances may reveal specific areas of projects where the exposure to risk was greatest. This information assists the project manager in searching for specific 'risky events'. The weakness of the method is that it provides little help in identifying construction-related risks.

A flow chart approach This approach attempts to construct a visual chart of the actual production process showing important components of the process and the flow of work. The project manager has an opportunity to focus on each element of the chart at a time and simultaneously consider the possibility of something going wrong with such an element. This approach may lead to identifying a series of risk events that could have significant impact on the project.

A brainstorming approach Brainstorming is probably the most popular approach to risk identification. It involves project participants taking part in a structured workshop where they systematically examine every part of the project under the guidance of the workshop facilitator with a view to identifying likely risk events.

Scenario-building for risk identification This approach relies on the development of two scenarios: the most optimistic scenario where everything occurs as expected; and the most pessimistic scenario where everything goes wrong. The aim of this approach is to assess the two opposed scenarios in order to identify the factors or risks that might influence project performance. Scenario-builders are often seriously constrained by their own experience. It has long been recognised that they assign a higher probability of occurrence to scenarios that they build from their own personal histories than to those they have not actually experienced.

An influence diagram approach An influence diagram presents a more comprehensive view of the likely project risks. It helps to identify risk by a detailed assessment of cause–effect relationships among project variables. The project manager would first examine a particular variable outcome (effect) and by working backwards attempt to define the causes of variation. Once identified, these causes then become effects for which causes are sought in turn, and so on. For example, the cost escalation arising from variation orders may be caused by errors in design documentation, the client's changes to the scope, or new regulations imposed by the local authority. The errors in design documentation may be caused by an incomplete brief, a lack of co-ordination, or insufficient time set aside for design and documentation. The causes of the incomplete brief will further be examined, and so on. This simple example is illustrated in Figure 15.2, where circles or nodes represent high-risk events in the production process and arcs or lines illustrate cause–effect links between these events. When all possible cause–effect relationships have been identified, the project manager is presented with a highly detailed graphic map of possible risk events.

Figure 15.2 Example of influence diagram

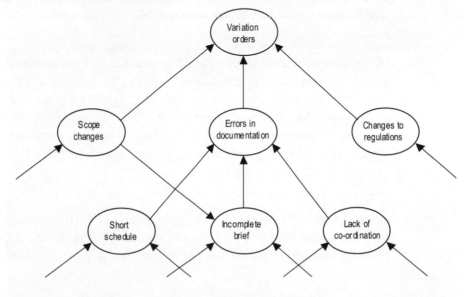

Introduction to Risk Management

TOP-DOWN RISK IDENTIFICATION

A top-down approach generates a holistic view of a project from which risk variables that are likely to affect the project are deduced. The two most commonly used approaches are:

A case-based approach A case-base approach provides an opportunity to examine a specific case in its entirety. Most commonly, a past project for which a wealth of information is available would be selected as a case study. This examination of a case serves as an excellent training ground for new managers who gain an overall perspective of a typical project situation and its associated risks.

An aggregate or bottom-line approach A better-known term for an aggregate or bottom-line approach is a contingency allowance. It is a global perception of the volume of risk and its impact on the project. It is expressed as a percentage of a certain performance measure, such as contract period or project cost. Project contingency is commonly formulated by the project manager in liaison with the client and other team members. It reflects subjective perception of the likely impact of risk on the project. While attractive for its simplicity, it lacks the ability to explain in any meaningful way the basis for developing a risk management response to anticipated problems.

Data elicitation

Before identified risks can be analysed, their likely magnitudes need first to be determined. This process is referred to as 'data elicitation'. Eliciting data for the identified risks is perhaps the most difficult task in risk management. Data is usually derived from databases, random experiments or the knowledge of experts.

ELICITING OBJECTIVE DATA

Information extracted from a database or a random experiment provides objective data that can be expressed in the form of a probability distribution. In this form, data is highly suitable for the quantitative risk assessment. Betts (1991) believes that objective data is preferred because of its consistency and perceived accuracy. But the use of databases in the construction industry is rare and the cost of random experiment high.

ELICITING SUBJECTIVE DATA

Subjective data is elicited by brainstorming, which accesses the knowledge and experience of the project participants. It may appear to be an arbitrary procedure or guesswork, but it is not so. If carried out by a properly structured brainstorming process, managed by an experienced facilitator, it provides collective expert knowledge of high quality. The best-known brainstorming technique is the Delphi method, which seeks information from a group of experts by means of an iterative questionnaire technique (though while highly effective, the Delphi method can be rather time-consuming). Other techniques suitable for data elicitation are the fractile and the histogram methods.

A fractile method breaks subjective probability into a series of discrete assessments from answers to a structured questionnaire. This approach to data elicitation, also referred to as the cumulative distribution fractile, attempts to relate possible values X of the uncertain variable to the probability that the true value of the uncertain variable is X or less. From this a certain number of discrete points or fractiles can be drawn. The application of the fractile method can be found in Hertz and Thomas (1983), Franke (1987) and Phillis (1987).

An alternative to the fractile method is to fix the values rather than probabilities corresponding to the different fractiles. The manager is asked to assess the probabilities of X being between each respective fractile. This is known as the histogram method or the fixed value method in recognition of the way that discrete probability distributions are usually displayed (Bunn 1984). Brotherton (1993) considered the fractile method to be more appropriate for eliciting subjective data because of its simplicity.

The most common source of subjective risk data is the knowledge and experience of the project manager and other team members. The problem is that no two decision-makers are likely to have the same amount of knowledge and experience, and consequently are unlikely to perceive risk the same way. To prevent inconsistencies, elicitation of subjective risk data should be approached as a group task.

Encoding risk data

Objective risk data derived from databases or through a random experiment can easily be fitted into one of many standard probability distributions. In this format, risk data can be assessed quantitatively.

Subjective data may be expressed as likelihood and consequence of risks in 'high' or 'low' terms, or as estimates of specific values of risk variables that fit simple probability distributions, such as uniform, triangular, trapezoidal and discrete. The simplest way of transforming subjective data into probability distributions is to arbitrarily determine a two-point estimate of the highest and lowest value of the random variable. These two points describe the uniform probability distribution, which is the simplest of the four. A three-point estimate of the highest, most likely and lowest values describes the triangular probability distribution. An extension of the triangular distribution is the trapezoidal distribution, which is characterised by two estimates of the most likely values together with the highest and the lowest estimates. Shapes of simple probability distributions are given in Figure 15.3.

Figure 15.3 Examples of simple probability distributions

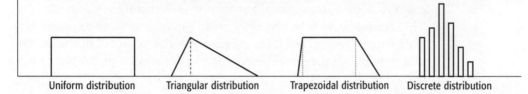

| Uniform distribution | Triangular distribution | Trapezoidal distribution | Discrete distribution |

Uniform distribution assumes that a range of possible values for a given risk variable can only be expressed between its minimum and maximum limits. While the use of this distribution may be acceptable in some applications, for example in predicting the exchange rate of the currency, it is doubtful that it would accurately model the cost and time of construction.

Triangular distribution is viewed as being adequate for most applications, particularly in estimating cost and time. It is characterised, in addition to its minimum and maximum limits of values, by the most likely value. Low and high probabilities will be assigned to the values close to the minimum and maximum limits of the range respectively, while values in between the minimum, maximum and most likely limits are distributed in a linear fashion.

Trapezoidal distribution is described by the minimum and maximum limits, and by two estimates of the most likely values.

Discrete distribution shows frequency of occurrence of various outcomes of a given risk variable. Such frequencies are in fact probabilities of occurrences that describe the range of possible outcomes.

When using simple probability distributions, care is required in determining their lower and upper limits. There are no precise rules governing the determination of these limits, but they should be set on the assumption that there is only a small chance, for example 1–2 per cent, that their values would be exceeded.

Chau (1993, 1995) showed that triangular distribution cannot accurately represent the distribution of cost and time values because such distribution is usually skewed with a long right-pointing tail.

This results in an overestimation of the risk exposure of the most likely estimate. To overcome this problem, the triangular distribution should be transferred to the log-triangular distribution.

A concept of fuzzy sets may also be adopted for encoding subjective data. Fuzzy set theory was originated in 1965 as a mathematical theory of vagueness. The theory helps to transfer a linguistic model of the human thinking process to a fuzzy algorithm which emphasises the human ability to extract information from masses of inexact data. The implementation of fuzzy strategies in encoding subjective data for applications such as forecasting and cost estimating would seem to have considerable merit, but much remains to be done to convert this idea into a working system. More information can be found in Kaufmann (1975), Zeleny (1982) and Schmucker (1984).

Risk analysis

The purpose of risk analysis is to measure the impact of the identified risks on a project. Depending on the available data and the manager's experience and expertise in risk management, analysis could be performed qualitatively or quantitatively.

QUALITATIVE RISK ANALYSIS

Qualitative assessment of risk is popular for its simpler and more participative approach. It involves subjective assessment of the derived data in terms of the likelihood and consequence of risk. When appropriately structured and systematically applied, qualitative risk analysis serves as a powerful decision-making tool. Let's briefly review examples of qualitative risk analysis techniques.

The Risk Management Guidelines of the NSW Government (1993) and the Australian and New Zealand Risk Management Standard (AS/NZS 1999) These describe a qualitative risk assessment process in which risk events are expressed in terms of likelihood and consequence. The aim of this form of risk assessment is to isolate the major risks and exclude those that are regarded as minor or have low impact.

An estimate of the likelihood of each risk may be expressed on a simple scale from low to high, or on a more detailed scale from rare, unlikely, moderate, likely, to almost certain.

An estimate of the consequences of each risk may also be expressed as either low or high, or as suggested in AS/NZS (1999) in terms of being insignificant, minor, moderate, major or catastrophic.

Introduction to Risk Management

A risk management matrix can then be formed using estimates of likelihood and consequence. An example of two such matrices is given in Figure 15.4.

In the simple matrix (a) in Figure 15.4, minor risks will commonly be ignored. Moderate risk will be carefully analysed and appropriate management responses formulated, while major risk will be given the utmost attention.

In the complex matrix (b) in Figure 15.4, low risks (L) will most likely be ignored while high risks (H) will be given top priority. Significant risks (S) and moderate risks (M) will be prioritised and treated accordingly.

Figure 15.4 Examples of risk management matrices

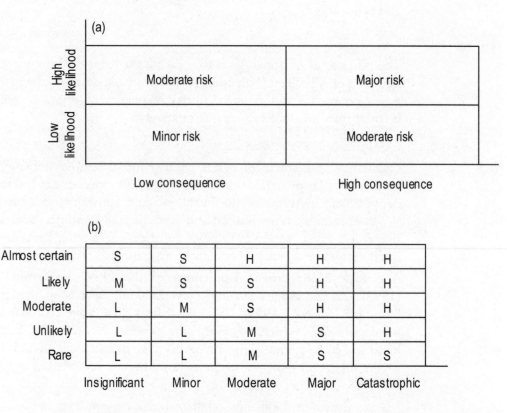

NOTE H, high risk; S, significant risk; M, moderate risk; L, low risk.
SOURCE Adapted from NSW Government 1993 and AS/NZS 1999.

A control hazard study method Control hazard study is a project-based or plant-based hazard assessment technique for identifying, analysing and responding to those hazards (risks) that affect production, safety, health and environmental requirements. It examines various hazards and operability of processes at each major project stage. This form of risk analysis is suitable for the following applications (Walters 1995):

- reviewing safety, health and environmental issues
- identifying and documenting potential hazardous events, their causes and consequences
- checking that plant buildings comply with the design
- assessing compliance of employee health, site safety and environmental protection with the project's policy, and laws and regulations.

The hazard and operability study (HAZOP) The main benefit of the HAZOP method is in a systematic review of the design and the construction strategy which identifies potential hazards to people, property and the environment (Neowhouse 1993). It can also be used to identify operability problems that affect efficiency of production. HAZOP is commonly performed as a group brainstorming exercise involving a multidisciplinary team of relevant experts.

The management confidence technique (MCT) This was developed by Jaafari (1990). The MCT enables the evaluation of the probability at which a given project will fail to meet its original objectives because of the combined effects of all the likely constraints acting on the project. An additional objective is to allow the project manager to test, on an interactive basis, whether or not a project is likely to succeed or fail on the supposition that appropriately formulated strategies will minimise the effect of perceived constraints and thus reduce the probability of failure. It is assumed that the combined effect of the constraints can be subjectively evaluated. However, the process of subjective evaluation must take into account the weight that a given constraint carries plus a score that will denote its intensity as it specifically affects the project under consideration. Such weight and score values may be set by the project manager for each constraint, taking into account the factual information available and the project manager's own perception of the presence of constraints and weights.

QUANTITATIVE RISK ANALYSIS

Quantitative risk analysis techniques are stochastic or random in nature. The most important feature of a stochastic process is that the outcome cannot be predicted with certainty. The application of any such technique requires that risk data be expressed as a probability distribution.

The choice of technique usually depends on the type of problem, the available experience and expertise, and the capability of the computer software and hardware. A wide range of techniques is available including sensitivity analysis, probability analysis using Monte Carlo simulation, controlled interval and memory (CIM) analysis, decision trees and utility theory. A brief review of quantitative risk analysis methods will now be presented.

Sensitivity analysis Sensitivity analysis seeks to place a value on the effect of change of a single risk variable within a particular risk assessment model by analysing that effect on the model output. A likely range of variations is defined for selected components of the risk assessment model and the effect of change of each of these risk variables on the model outcome is then assessed in turn across the assumed ranges of values. Each risk is considered individually and independently, with no attempt to quantify probability of occurrence.

The importance of sensitivity analysis is that often the effect of a single change in one variable can produce a marked difference in the model outcome. Sometimes the size of this effect may be very significant indeed.

In practice, a sensitivity analysis will be performed for a large number of risks and uncertainties in order to identify those that have a high impact on the project outcome and to which the project will be most sensitive. If the project manager is interested in reducing uncertainty or risk exposure, then sensitivity analysis will identify those areas on which effort should be concentrated.

The outcome of sensitivity analysis is a list of selected variables ranked in terms of their impact. This is best presented graphically, with a spider diagram being the most preferred form of representation (Figure 15.5).

Sensitivity analysis may also be plotted in the form of a series of cumulative density functions representing the cumulative project cost. The extent of separation of individual functions is caused by the presence of risk. This is illustrated in Figure 15.6.

A spider diagram is an X–Y graph with the horizontal and vertical axes showing a percentage change in the outcome variable and the percentage of change in the level of risk respectively. A spider diagram in Figure 15.5 illustrates assessment of the impact of four risk variables:

Figure 15.5 A spider diagram of the four risk variables

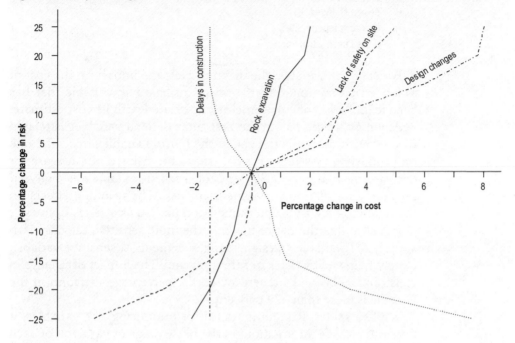

Figure 15.6 Sensitivity analysis expressed as cumulative density functions

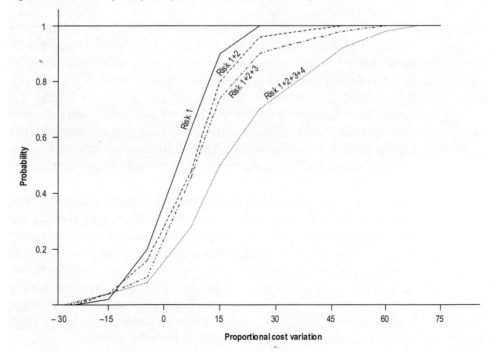

Introduction to Risk Management

- problems associated with rock excavation
- design changes
- lack of safety on site
- delays in construction.

For each risk variable, the diagram plots the impact on the cost of a defined proportionate variation in a single risk variable that has been identified as having some risk associated with its cost estimate.

Assume that the risk in the first three defined variables is related to cost overruns, while the risk in the fourth variable affects the rate of production. Assume further that the risk could increase or decrease by 5, 10, 15 or 20 per cent. Each risk variable is analysed separately for different increments of risk. The resulting effects on the total project cost is determined and plotted as a series of lines.

It is clear that the flatter the line, the more sensitive the cost variation in that variable. 'Design changes' is the most sensitive variable, closely followed by 'Lack of safety on site'. The impact of 'Delays in construction' becomes severe when the percentage variation in the variable is more than −20 per cent.

While a spider diagram is useful for prioritising risk variables, it does not provide information on the likely range of variation of each risk variable. This may be overcome by introducing probability contours, which subjectively define ranges within which a particular variable is expected to lie.

Probability contours are constructed by subjectively identifying the range of values such as time and cost within which a particular variable is expected to lie at each level of probability. For example, two probability levels considered are 70 per cent and 90 per cent (Figure 15.7). The manager then attempts to determine in turn the impact that 70 per cent and 90 per cent probability of variation will have on each risk variable. The impact of each of those probabilities on the risk variables are estimated and marked on the spider diagram. When such marks are connected, they form 70 per cent and 90 per cent probability contours.

It is then possible to deduce that there is 70 per cent probability that the range of risk variation will be between +18 per cent and −22 per cent, and that there is 90 per cent probability that it will be between +21 and −25 per cent. Similarly, the variation in cost for the 70 per cent probability contour is between −4.5 and +5.9 per cent and between −2.1 and +7.5 per cent for the 90 per cent probability contour. It may further be deduced that within the 70 per cent probability contour, 'Delays in construction' is the most sensitive parameter, while 'Design changes' is more sensitive within the 90 per cent contour.

Figure 15.7 A spider diagram with the probability contours

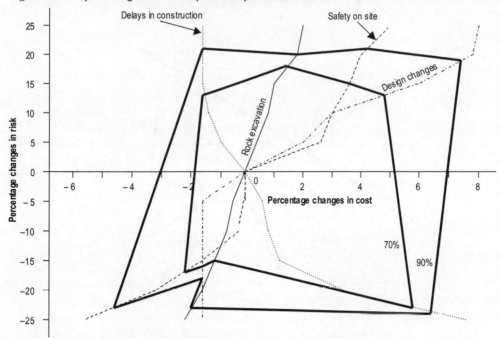

SOURCE Adapted from Flanagan & Norman 1993: 100.

Flanagan and Norman (1993) explained the use of spider diagrams in comparing and ranking outcomes of decision alternatives. This is illustrated in Figure 15.8, which shows two development options, X and Y. Each option is characterised by two critical risk variables.

From the cost point of view, option X is preferred to option Y. However, option X is more sensitive to the variation in the risk variables A and B than option Y at the 80 per cent probability contour. Option Y seems to offer a greater degree of cost certainty.

The main weakness of sensitivity analysis is that selected risk variables are treated individually as independent variables, and interdependence is not considered. The outcomes of sensitivity analysis thus need to be treated with caution where the effects of combinations of variables are being assessed.

Probability analysis The weakness of sensitivity analysis is that it looks at risks in isolation. Probability analysis overcomes this. It is a statistical method that assesses a multitude of risks that may vary simultaneously. In combination with Monte Carlo simulation, it provides a powerful means of assessing project uncertainty.

Introduction to Risk Management

Figure 15.8 Sensitivity analysis of two development options

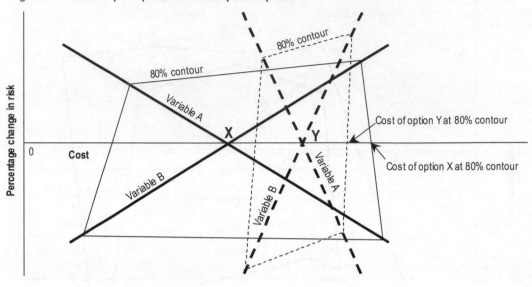

SOURCE Adapted from Flanagan & Norman 1993: 100.

The key to probability analysis is the development of a risk analysis model. The model should include variables affecting the outcome, taking into account the interrelationships and interdependence between the variables. Each risk variable in the model is expressed as a probability distribution. The model must permit assessment of risks at the desired level of detail and accuracy. This may require breaking down risk variables into their subcomponents in order to accurately assess the impact of risk on such subcomponents.

Probability analysis is commonly performed using Monte Carlo simulation. The Monte Carlo technique generates random numbers that are related to probability distributions of individual risk variables in the model. After each iteration, the model calculates one specific outcome from the generated risk values from individual probability distributions. After a large number (at least 100) of iterations have been performed, the Central Limit Theorem, an important concept in statistics, ensures that the outcomes fall on a normal curve, which is fully described by its mean and standard deviation (SD). A typical probability analysis process using Monte Carlo simulation involves the following steps:

- developing a risk analysis model (Figure 15.9)
- eliciting data and assigning appropriate probability distributions to the identified risk variables

Figure 15.9 Example of a probability analysis model

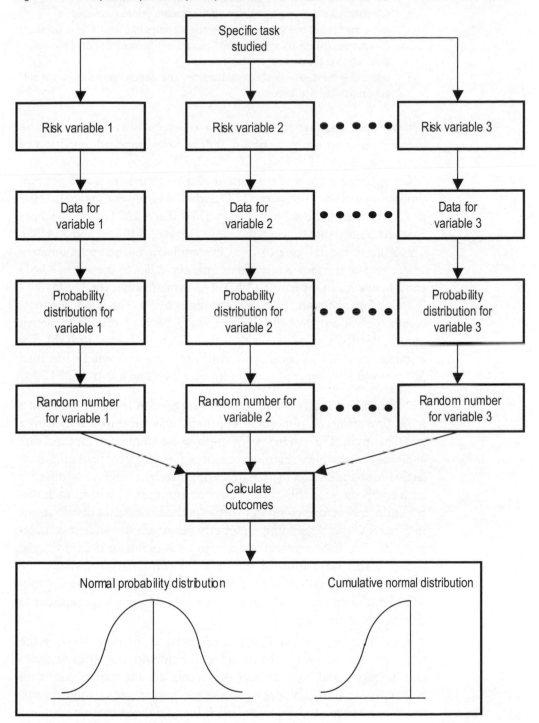

Introduction to Risk Management

- selecting a random value from each of the probability distributions
- calculating the outcome by combining the random values generated
- repeating the last two steps a large number of times to obtain a large number of outcomes, which according to the Central Limit Theorem fall on a normal distribution curve
- calculating the measures of central tendency and dispersion of the normal output probability distribution.

For example, assume that the result of probability analysis is the mean of the cost estimate of \$22 000 and the standard deviation of the cost estimate of \$1120.

Given the properties of the normal distribution, there is a 68 per cent probability (the area between +1 SD and −1 SD from the mean) that the project cost lies in the range between \$20 880 and \$23 120, and a 95 per cent probability that it lies in the range between \$19 760 and \$24 240. This is illustrated in Figure 15.10(a). In cumulative terms, approximately 16 per cent of the area under the normal curve lies to the left of −1 SD and 84 per cent to the left of +1 SD. The similar values for −2 SD to +2 SD are 2.5 per cent and 97.5 per cent respectively. The likely cost outcomes for various probabilities can easily be read off a cumulative normal distribution curve such as that given in Figure 15.10(b). For example, there is 84 per cent probability that the cost will be less than \$23 120 and 97.5 per cent probability that it will be less than \$24 240.

The controlled interval and memory method A less complex and a more powerful computational approach was proposed by Cooper and Chapman (1987). Their CIM method is a numerical method that combines probability distributions. It is based on representing subjective probabilities as histograms with different interval widths for each component distribution but a constant interval within each distribution. Interactive computer software then combines distributions in various ways, depending on whether they are dependent or independent. The CIM method appears to be superior to the traditional Monte Carlo simulation in the areas of computational simplicity, greater accuracy, and flexible treatment of dependence among probability distributions. This is true particularly where a large number of risk variables are involved.

A decision tree method This is a geometrical method showing the structure of a decision problem. A tree is made up of a series of nodes and branches and each branch represents an alternative course of action or decision. The decision tree method is not covered in this book. See Moore and Thomas (1984) for more information.

Introduction to Risk Management

Figure 15.10 The normal distribution and the cumulative distribution functions

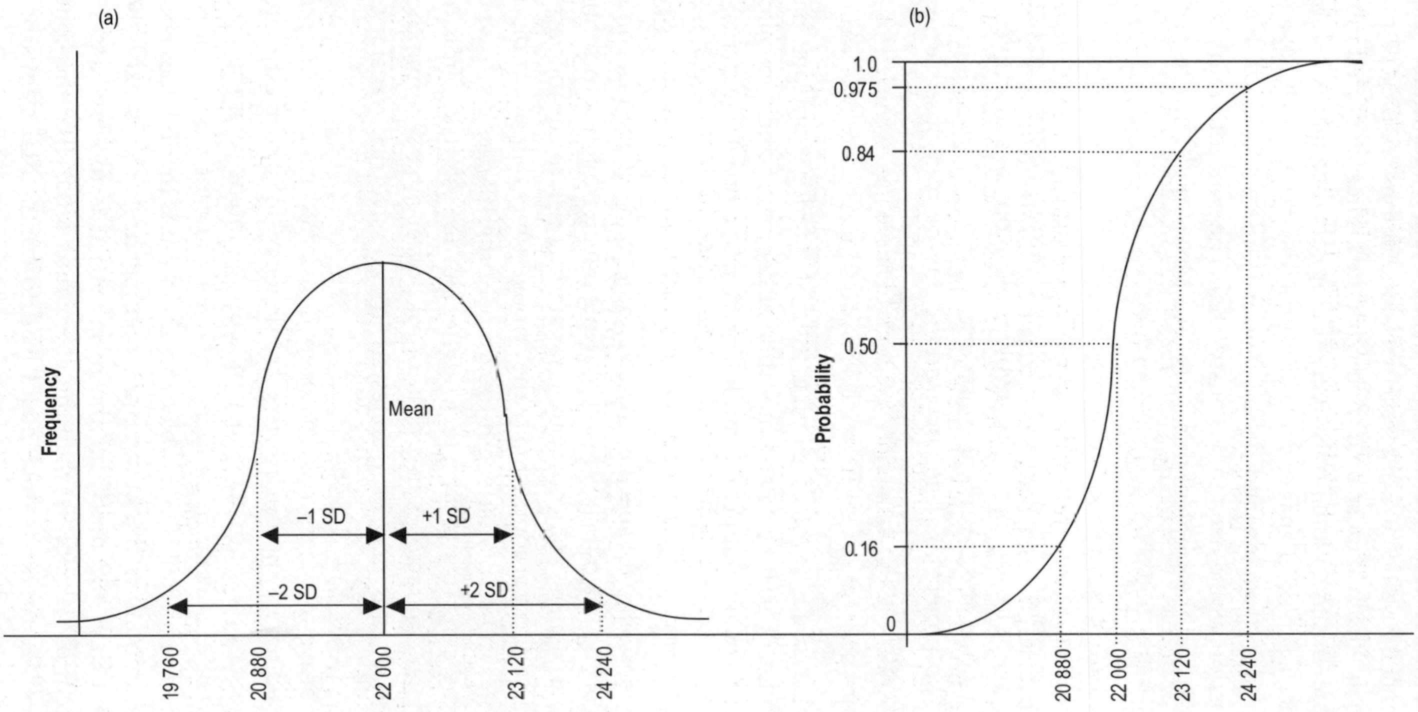

The utility theory method Decision theory determines 'expected monetary values' of different decision options. A project manager would ideally choose an option with the highest expected monetary value. But this would only apply if a project manager is risk-neutral. Since a project manager could also be a 'risk-taker' or a 'risk-avoider', in order to understand his behaviour in risky situations, his utility function would need to be constructed. The concept of utility theory is beyond the scope of this book. See Moore and Thomas (1984) for more information.

Risk response

Risk response is an action or a series of actions designed to deal with the presence of risk. This involves developing and implementing mitigation and treatment strategies.

The project manager may adopt one of two possible response strategies: transfer the risk and/or control it.

RISK TRANSFER

Risk transfer involves shifting the risk burden from one party to another. This may be accomplished either contractually by allocating risk through contract conditions or by insurance. Contractual transfer is a popular form of risk transfer. It is used extensively in construction contracts, where one party with power transfers the responsibility for specific risks to another party. The most commonly occurring risk transfers involve clients transferring risk to contractors and designers, and contractors transferring risk to subcontractors.

The client is contractually the strongest member of a project team and may use that strength to compel other parties with whom he forms contracts to absorb a higher portion of risk. Similarly, the general contractor is in a position of power after securing the main contract; this may be manifested in excessive risk transfer to subcontractors and suppliers.

Parties to whom risk has been transferred generally respond by including an appropriate risk contingency in a cost estimate. The amount of a contingency depends on the intensity of the risk transferred. It is likely to be high when a significant amount of risk has been transferred. The problem for the client is that he does not know the amount of the contractor's contingency. If it is excessive, the client pays too much for the project. If it is too low and the full extent of risk occurs, then the client runs a risk that the contractor may lose money, which could lead to contractual claims or lower quality standards or could even bankrupt the contractor.

The essential principle of risk transfer is that risks should be equitably shared among the parties to a contract on the basis of their ability to control and their capacity to sustain such risks. This is the philosophy of risk allocation promoted by Abrahamson (1973).

Risk transfer by insurance is highly desirable in those situations where insurance policies exist. The purpose of insurance is to convert the risk into a fixed cost. This approach assigns a dollar value to the risk.

RISK CONTROL

When risk can neither be transferred nor insured, management action is required to manage it. This is commonly achieved through the processes of risk avoidance and risk retention.

Risk avoidance or reduction A simple approach to managing risk is to avoid it in the first place. For example, the risk of an unreliable concrete supply from a single supplier can be avoided by engaging two suppliers. Similarly, if the contractor believes that the client imposes an excessive burden of risk through contract conditions, the contractors may avoid the risk altogether by not bidding for the job. Not all risks can be avoided but their impact can be reduced, for example by developing alternative solutions or even redesigning parts of the project. In collaboration with the project participants, the project manager needs to develop appropriate treatment strategies and to assign responsibility for their implementation.

Risk retention While risk avoidance and risk minimisation can help reduce the overall level of risk, some residual risk will always remain. It is this risk that requires close attention. The most common approach to controlling residual risk is to convert it into a contingency allowance. Contingency may be expressed as a single-value estimate of a particular measure of performance such as time, cost, hours of work, or as a probability distribution of a particular measure of performance.

Perhaps the most serious weakness of a single-value contingency approach is its inflationary impact on the base estimate. When a percentage contingency is added to the base estimate, every item in that estimate will be inflated by the percentage figure of the contingency, irrespective of whether such items represent 'risk' or not. For example, a 10 per cent contingency added to the cost estimate inflates the cost of each item in the estimate by 10 per cent. If the Pareto Principle holds, then only about 20 per cent of such cost items in the cost estimate represent the real risk. They should then account for approximately 80 per cent of the uncertainty.

When a risk contingency is expressed in the form of a probability distribution, it is possible to determine its value at a specific level of probability. For example, a prudent manager may prefer to accept a certain level of risk but at the probability level of 85 per cent or higher. It is then a simple task to determine the actual value of the risk at that level of probability.

Monitoring and review

Although listed last, monitoring and review do not constitute the final step of the risk management process. In fact monitoring and review start almost from the beginning and are maintained throughout the entire process. Since no risk is static, monitoring and review ensure that changing circumstances are identified and reassessed. This is necessary to ensure the relevance of the risk management process.

Risk management plan

A risk management plan is a written statement describing the risk management process. It specifies the processes employed and summarises the results obtained in each risk assessment step. It aggregates identified risks into low–high impact and likelihood types. It lists elicited values of risk variables and assigned probability distributions, where appropriate. It defines treatment strategies, particularly for high impact or likelihood risks and assigns responsibilities. It lists discarded risks. It also reports on resource requirements, timing of actions, and monitoring and reporting mechanisms.

A typical risk management plan might adopt the following structure (NSW Government 1993):

1   proposal familiarisation
2   risk analysis
3   management of risks
4   implementation monitoring.

Proposal familiarisation This section of a risk management plan describes the purpose of the plan, issues and objectives, criteria for assessment and the key elements to be analysed. It also provides details of the risk management policy and defines the project organisation structure.

Risk analysis In this section identified risks are listed, together with information about their likelihood and consequences and assessment of their impact, and priority listing of major, moderate and minor risks.

Management of risks Under this heading the project manager will describe in detail management's response to risk. It will include a summary of a management action in response to major and moderate risks, and a schedule of discarded minor risks.

Implementation monitoring This section describes who will have the authority and responsibility for implementation and reporting an action plan.

CORRELATION OF RISK VARIABLES

Risk assessment using probability analysis by way of simulation is based on a model that facilitates computation of the output probability distribution by combining values of each risk variable in a particular manner. Risk variables could be related to each other in one of three possible ways:

- they may be totally independent of each other
- they may be fully dependent on each other
- they may be partially dependent on each other.

In other words, risk variables are correlated and the extent of correlation varies from −1 to +1. The correlation value of −1 indicates 100 per cent indirect dependence (for example, if the level of interest rate increases by 30 per cent, the demand for property would decline by 30 per cent), whereas +1 correlation shows 100 per cent direct dependence. The correlation value of 0 represents independence between the risk variables. The values between −1 and +1 indicate partial correlation.

In constructing a probability analysis model, it is essential to replicate as accurately as possible correlations among risk variables if the outcome of simulation is to be meaningful. Chau (1995) argued that the correct assessment of correlation among risk variables is far more important in terms of accuracy of the result than the choice of input probability distributions of risk variables.

Defining correlations among risk variables is an arduous task because of the lack of objective data. The difficulty in assigning correct values of correlation often leads project managers to adopt an assumption of independence among risk variables, the side effect of which is an underestimation of the real value of the risk.

COMPUTER MODELLING

The advent of powerful PC computers and the availability of appropriate simulation software have accelerated application of a computer-based risk assessment at a very low cost. Computer software such as Cyclone, Disco, I Think and spreadsheets enhanced with risk analysis add-in packages, such as @Risk and Crystal Ball, are excellent low cost/high performance risk analysis tools.

Spreadsheets in particular are a highly effective yet inexpensive means of performing simulation. A typical spreadsheet offers fast computation of data across columns and rows, easy access to numerous mathematical and statistical functions, and excellent graphics. These capabilities can be further enhanced when linked with compatible risk analysis add-in software. In this configuration a spreadsheet becomes a powerful risk analysis tool. It provides access to the whole range of functions that control the simulation process. The manager is able to select an appropriate probability distribution for a particular data set from a wide range of available standard probability distributions and control the extent of correlation among risk variables. The length of a simulation run and the choice of a risk analysis method can also be varied to suit different applications. The output is both statistical and graphical.

RISK MANAGEMENT APPLICATIONS

The construction industry is diverse and complex. It requires expertise form a wide range of experts who come together to conceive, design and produce a one-off project. In view of the size and nature of construction projects, poor decision-making can have disastrous financial consequences for the client and other participants in the project.

Much has already been said and written about risk and uncertainty. The vast volume of published research on this subject suggests that a theoretical framework of risk management has been developed sufficiently to facilitate its widespread use. While high-risk industries such as the petrochemical industry, oil exploration, defence and airspace have for long benefited from the concept of risk management, the construction industry has been slow to adopt it. Examples of applications are few and are largely related to the feasibility studies of large developments. This is despite the vast knowledge of risk management that has been accumulated since the 1960s, the emergence of excellent models of risk assessment including those defined by the NSW Government Risk Management Guidelines (NSW Government 1993) and the AS/NZS (1999), and the availability of powerful computers and sophisticated software for a range of risk management applications.

Uher and Toakley (1997: 80) identified a number of obstacles preventing the application of risk management in the construction industry. These were:

- inadequate knowledge
- inadequate skill
- lack of understanding of potential benefits
- ignorance
- fear of working with probability and statistics
- negative attitudes.

The purpose of the next section is to examine a range of possible applications of risk management in the construction industry. Let's now explore some of them.

Qualitative risk assessment

Qualitative risk assessment is generally preferred to quantitative assessment. In recent years, the frequency of qualitative assessment has increased as a result of the availability of structured and systematic qualitative methods such as those defined by the NSW Government (1993) and AS/NZS (1999).

These methods are based on a simple yet powerful process of determining the likelihood and the consequences of identified risk. By tabulating this information in a simple matrix, the impact of risk can be assessed on a scale from high to low. It is then possible to categorise risk events as being either major, moderate or minor. For each category of risk, the project manager develops appropriate mitigation and treatment strategies.

Qualitative risk assessment is suitable for most applications, but it should not replace quantitative assessment where objective data is readily available. The NSW Government Risk Management Guidelines give worked examples of qualitative risk assessment of applications such as the feasibility of a new road construction, the extension of an existing road, the development of a commercial budget and a business plan, the development of a new health facility and tender risk assessment.

Quantitative risk assessment

A number of specific quantitative risk assessment applications have been developed in the past for use in high-risk environments. They were primarily targeted at large engineering, petrochemical and aeronautical projects. While effective, their use has been sporadic beyond those industries. The construction industry in particular would benefit from their implementation. Let's examine some of these applications.

PROBABILITY ANALYSIS OF FEASIBILITY

The Synergistic Contingency Evaluation and Review Technique (SCERT) was one of the first methods that applied probability analysis to feasibility. The SCERT was developed by Chapman (1979) in the late 1970s for BP International to assess risks associated with North Sea petroleum projects. The emphasis of this risk analysis technique was to provide a systematic approach to the planning and financial evaluation of large engineering projects that involved significant risks. The SCERT models projects on the basis of a decision tree/semi-Markov process coupled with qualitative risk assessment procedures.

Only BP International has used the SCERT in its most comprehensive form. However, a simplified version of the SCERT was successfully adopted by Gulf Canada Resources Incorporated in the early 1980s (see Ward & Chapman 1991).

In recent years, probability analysis of feasibility is being performed by Monte Carlo simulation. An example of a feasibility study based on Monte Carlo simulation is presented later in this section.

PROBABILITY COST ESTIMATING

A cost estimate of a project is an instrument intended to predict the total expected expenditure. The starting point is the measurement of quantities of work followed up by their pricing. Measuring is a mechanical process that should produce few errors as long as the correct information is available, whereas pricing is more intuitive and discriminating, particularly in determining duration of activities, resource demand levels and productivity rates. It introduces the element of risk into estimating.

Traditionally, project cost plans have been prepared in the form of a single-value estimate. For example, the contractor's tender was calculated from estimates of materials, labour, plant, equipment and preliminary items, which were summed to form a deterministic base-cost estimate. Thereafter, a risk contingency, an estimate of the contractor's overhead cost and a profit margin were added, each expressed separately as a percentage of the base estimate. The final project cost estimate was thus formulated as a single number. But how accurate is this estimate? If contractors are able to assess accurately the value of project risks, the estimate is likely to be reliable. But there is little evidence of contractors undertaking a systematic and detailed assessment of project risks (Perry & Hayes 1985; Uher 1996). In most cases, their expression of a risk contingency is at best an educated guess. The traditional single-value estimating method fails to deal adequately with risk and uncertainty. Since it does not accurately predict the final project cost, a more reliable cost-estimating approach is needed. Probability estimating based on Monte Carlo simulation provides a suitable framework for a systematic and disciplined assessment of project risks that creates a more robust and reliable cost estimate.

Spooner (1974) described the first probability cost-estimating model based on the Central Limit Theorem. Diekmann (1983) examined a number of probability-estimating approaches and found Monte Carlo simulation to be the most promising method. Jaafari (1990) and Uher (1996) further demonstrated the suitability of the Monte Carlo method in probability estimating in practice.

However, the implementation of probability estimating using Monte Carlo simulation is hindered by the reliance of contractors on bills of quantities. Bills of quantities are used widely in Australia as tender documents and contractors use them extensively in preparing cost estimates. They are popular because their structure is aggregated for trades, which facilitates a more effective elicitation and formulation of subcontract bid prices. The problem is that when compiled in this manner, a bill contains thousands of cost items, which makes it unsuitable for probability estimating. If its structure could be simplified or converted into a more suitable format it would be more suitable for probability estimating.

One possible solution is to convert a bill of quantities into an elemental cost plan. However, the groupings of materials, resources and plant in a bill rarely follow the structure of a standard elemental cost plan and the effort and time in transforming its structure into a cost plan by encoding every item in the bill could not be justified.

The most promising alternative is to use a trade summary in a bill of quantities as a model for probability cost estimating. This concept was pioneered by Brotherton (1993) and Uher (1996). Electronically compiled bills of quantities can automatically generate a trade summary. For example, the trade 'Excavation' would normally contain dozens of individual items of quantities and costs associated with excavation and filling activities. In a trade summary format, those individual cost items will be summarised into one 'Excavation' cost.

Although each trade item is expressed in money terms and does not give a break-up of quantities and unit rates, it can nevertheless be 'expanded' to include risk components within each trade together with their accompanying quantities and unit rates. This approach to probability estimating will be demonstrated in the second specific example of risk management applications presented later in this chapter.

Introduction to Risk Management

PROBABILITY COST PLANNING

Jaafari (1990) proposed an elemental cost planning structure as a base for probability estimating using the Monte Carlo method. Because such a structure has a small number of elements, a model of the probability cost estimate would be fairly small. This approach is highly suitable for estimating a project cost at the conceptual stage when only sketchy information is available. However, since most contractors prefer to use a bill of quantities as a framework for cost estimating, the Jaafari method has been less popular in cost estimating in pre-construction and construction stages.

PROBABILITY LIFECYCLE COSTING

Hall (1986) and Flanagan and colleagues (1987) expanded the concept of probability cost estimating to lifecycle costing using Monte Carlo simulation.

PROBABILITY SCHEDULING

PERT was the first example of probability scheduling. It was developed in 1958, in parallel to the critical path method, by Malcomb and colleagues (1959) to assist the US Navy in the development of the Polaris submarine/ballistic missile system. It was primarily conceived as a project control technique to ensure that the highly complex and strategically important Polaris project would be delivered on time. The fact that the project was completed and commissioned well ahead of its schedule is being largely attributed to PERT.

The fundamental aim of PERT is to track the progress of a project and show, at different intervals, the probability of completing that project on time. PERT relies on statistics, particularly the Central Limit Theorem, in calculating the probability of project outcomes.

Throughout the 1960s PERT remained the principal planning and control technique in the United States, particularly on large military and aerospace projects such as Atlas, Titan I, Titan II, the Minuteman rocket systems and the Apollo space program. But by the late 1960s the critical path method gained in popularity and has since become the most widely used planning and control technique. More information on PERT can be found in Uher (2003).

In recent years, the technique of Monte Carlo simulation has become popular in probability scheduling. The heart of probability scheduling is a computer-generated critical path schedule, which either has its own simulation capabilities or could be enhanced with such capabilities through add-in simulation software. The I Think software is an example of the former case and the Monte Carlo for Primavera of the latter.

The development of a probability critical path schedule requires the activity's period to be expressed as a distribution of possible durations. Triangular function is the one most commonly adopted. A risk analysis is then performed on a schedule. The output distribution of a simulated schedule is, according to the Central Limit Theorem, a normal distribution. When expressed as a cumulative density function, the planner is able to determine the probability of completing the project by the contract date.

In the 1980s BP International developed the Probabilistic Schedule Appraisal Program (PAN) and the Cost and Time Risk Analysis Program (CATRAP) (Baker 1986). The PAN method was used for the appraisal of schedules and the CATRAP method for both the time and cost risk analyses. CATRAP was particularly useful for generating forecasts of the rate of project expenditure in the form of S-curves. It automatically generates probability distributions of project expenditures from historical data and projected forecasts.

Specific examples of risk management application

Two fully worked out cases illustrating the application of risk management will now be examined. The first case is a feasibility study of a proposed construction development and the second case is a probability estimate of a medium-sized commercial building.

FEASIBILITY STUDY USING MONTE CARLO SIMULATION

Problem: The property developer considers building a block of sixteen villa houses in two possible locations: Location A is in Sydney (south) and Location B is in Gosford (about 100 km north of Sydney). The developer wants to know which of the two possible locations offers better return on investment. The information known to the developer is summarised in Table 15.1.

TABLE 15.1 DATA FOR THE DEVELOPMENT PROJECT

	Location A ($)	Location B ($)
The cost of land	1 200 000	800 000
Real estate agent's commission fee	60 000	50 000
Holding charges 12 months after the commencement of the development (will cease when all the villas have been sold)	300 000	144 000
The discount rate	10 per cent	10 per cent
The maximum period of sale of the villas	6 months	6 months

Solution: The developer's project manager has identified a number of risk variables that were likely to have a significant impact on the project:

- the development approval cost
- the design and documentation cost
- the construction cost
- the number of villa houses sold within the first three months
- the level of interest rate
- the sale price
- the length of the conceptual and design stages
- the period of construction.

The project manager then elicited data for the identified risk variables and assigned probability distributions. These are given in Tables 15.2 to 15.9.

TABLE 15.2 THE DEVELOPMENT APPROVAL COST EXPRESSED AS DISCRETE DISTRIBUTION

	Location A ($)	Probability A (%)	Location B ($)	Probability B (%)
Development approved	5 000	20	2 000	30
Conditional approval	35 000	60	20 000	60
Development rejected	40 000	20	30 000	10

TABLE 15.3 THE DESIGN AND DOCUMENTATION COST EXPRESSED AS TRIANGULAR DISTRIBUTION

	Location A ($)	Location B ($)
Minimum cost	40 000	35 000
Most likely cost	45 000	40 000
Maximum cost	50 000	45 000

The cost of design and documentation will be affected by the cost of construction. The project manager expects 90 per cent positive correlation between these two variables.

TABLE 15.4 THE CONSTRUCTION COST EXPRESSED AS TRIANGULAR DISTRIBUTION

	Location A ($)	Location B ($)
Minimum cost	1 500 000	1 360 000
Most likely cost	1 600 000	1 400 000
Maximum cost	1 780 000	1 600 000

The construction cost will be influenced by the period of construction. The project manager expects that these two variables will have about 90 per cent positive correlation.

TABLE 15.5 THE NUMBER OF HOUSES SOLD WITHIN THE FIRST THREE MONTHS EXPRESSED AS TRIANGULAR DISTRIBUTION

	Location A.	Location B
Minimum number	8	6
Most likely number	13	11
Maximum number	16	16

The speed of sales of the houses is likely to be influenced by the level of interest rate. The project manager expects 75 per cent negative correlation between these two variables.

TABLE 15.6 THE LEVEL OF INTEREST RATE EXPRESSED AS UNIFORM DISTRIBUTION

	Location A (%)	Location B (%)
Minimum level	8	8
Maximum level	13	13

TABLE 15.7 THE SALE PRICE PER HOUSE EXPRESSED AS TRIANGULAR DISTRIBUTION

	Location A ($)	Location B ($)
Minimum cost	240 000	180 000
Most likely cost	275 000	210 000
Maximum cost	310 000	225 000

TABLE 15.8 THE PERIOD OF THE CONCEPTUAL AND DESIGN STAGES EXPRESSED AS TRIANGULAR DISTRIBUTION

	Location A (months)	Location B (months)
Minimum period	5	3
Most likely period	6	4
Maximum period	8	6

Introduction to Risk Management

The period of the conceptual and design stages is likely to be influenced by the outcome of the development application. The project manager expects about 90 per cent positive correlation between these two variables.

TABLE 15.9 THE PERIOD OF CONSTRUCTION EXPRESSED AS TRIANGULAR DISTRIBUTION

	Location A (months)	Location B (months)
Minimum period	8	8
Most likely period	10	10
Maximum period	11	11

In the next phase of risk assessment, the project manager constructed a simulation model of a feasibility study for the two development options in a spreadsheet (Table 15.10). The model lists income-generating activities first, followed by activities generating costs. The model was then assessed in terms of a net present value of the two development options.

The model was built in the Excel spreadsheet and was analysed by Monte Carlo simulation using the add-in software @Risk. Probability distributions of the identified risk variables were entered into the model using the standard @Risk functions. Correlations between the risk variables were assigned using the @Risk correlation matrix.

For this exercise, the number of iterations was set to 1000. The result of simulation in the form of statistics is given in Table 15.11.

Clearly, both development options are feasible since they offer positive NPV. Of the two options considered, location A provides better financial return than location B. For example, there is a 90 per cent probability of NPV to be at least $125 647 if the project is built in location A, whereas in location B NPV would only be $28 934 at the same level of probability. There is a 98 per cent probability of a profitable investment in location A, while in location B the probability is slightly lower at 92 per cent. But it is worth noting that the standard deviation of location B is substantially smaller than that of location A, indicating that the output normal distribution of location B is narrower than that of location A. It means that location B will be less sensitive to variation in the level of risk.

The graphical output in the form of an output normal distribution is also useful in decision-making (Figure 15.11). The output distributions are presented as cumulative density functions. In this format, it is possible to read from the graph the expected value of NPV for a specific level of probability.

TABLE 15.10 THE SIMULATION MODEL OF A FEASIBILITY STUDY

	Location A	Location B
Income from sales		
Sale price per house	$275 000	$205 000
No. of sales in 3 months	12	11
No. of sales in 6 months	16	16
Interest rate	10.5%	10.5%
Income in 6 months	$4 400 000	$3 280 000
Development cost		
Land cost	$1 200 000	$800 000
Development approval cost	$35 000	$20 000
Design and documentation cost	$45 000	$40 000
Construction cost	$1 626 667	$1 453 333
Real estate agent's commission fee	$60 000	$50 000
Subtotal cost	$2 966 667	$2 363 333
Period of design and approvals	6	4
Period of construction	10	10
Period of sales	6	6
Holding charges	$300 000	$144 000
Total development cost	$3 266 667	$2 507 333
NPV at 10 per cent over 2 years	$371 361	$209 991

NOTE The table shows the expected or mean values of the risk variables.

TABLE 15.11 THE SIMULATION STATISTICS OF THE FEASIBILITY EXAMPLE

	Location A	Location B
Minimum value	-$138 729	-$182 695
Maximum value	$913 231	$501 485
Mean value	$386 610	$207 269
Standard deviation	$197 505	$132 842
90% probability	$125 647	$28 934
Probability of positive NPV	98%	92%

In conclusion, the developer has at least 92 per cent probability of a profitable investment in either location. Location A is likely to realise a higher level of return than location B, but location A will be more sensitive to variation in the level of risk exposure.

PROBABILITY ESTIMATING USING MONTE CARLO SIMULATION

Problem The project in question is a small commercial office building built in the early 1990s in Sydney. The project was fully designed and documented before tendering and was procured using the traditional method. The electronically compiled bill of quantities was available for tendering. There were eight contractors bidding for the project. Contractor X, who provided information for this case study, was one of the bidders. Using the traditional, single-value estimating method, contractor X calculated the cost of construction as $9 018 119. This cost is inclusive of a risk contingency of 4.8 per cent. The contractor's tender was rejected as being too high. A contract was eventually awarded to another contractor for around $8 800 000.

The purpose of this case study was to investigate whether or not a probability cost-estimating method would have improved contractor's X chance of winning the project.

Solution The first task in developing a probability cost-estimating model was to convert the bill of quantities created in the Buildsoft Estimating System into a suitable trade summary format (Table 15.12).

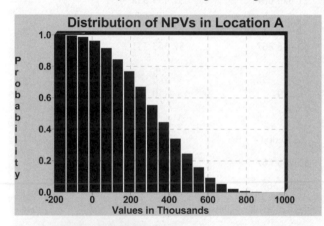

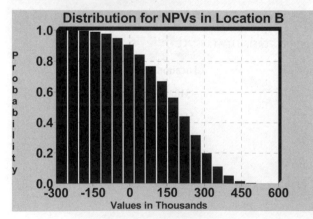

Figure 15.11 Cumulative density functions of the output normal distributions. These figures were generated by @Risk software.

TABLE 15.12 A TRADE SUMMARY OF THE COST ESTIMATE

No.	Trade summary items	(%)	Cost ($)
1	Demolition	0.6	53 650
2	Groundworks	3.6	324 470
3	Piling	0.6	57 230
4	Concrete supply	3.9	355 200
5	Concrete place	1.4	121 880
6	Reinforcement supply	2.0	181 390
7	Reinforcement place	0.8	71 060
8	Formwork	5.8	525 620
9	Structural steel	0.9	85 090
10	Metalwork	6.8	610 350
11	Prestressing	1.9	169 710
12	Masonry	4.0	365 190
13	Stonework	0.4	33 470
14	Carpentry	0.5	47 440
15	Roofing and roof plumbing	1.7	155 530
16	Suspended ceiling	5.7	512 020
17	Plastering	2.7	247 730
18	Wall and floor tiling	1.3	112 770
19	Doors and frames	2.1	189 610
20	Windows	6.1	550 630
21	Glazing	0.2	21 390
22	Wall and floor finishes	3.0	268 160
23	Painting	1.8	160 380
24	Hardware	1.0	93 240
25	Hydraulics	4.6	418 780
26	Mechanical services	9.1	823 690
27	Electrical services	10.8	970 430
28	Lift services	5.7	510 230
29	Miscellaneous services	2.7	246 970
30	Fixtures and fittings	1.1	103 360
31	Provisional sum	2.2	202 015
32	Contingency	4.8	429 434
	Total	100.0	9 018 119

Introduction to Risk Management

The next task was to identify specific risks related to each trade summary item. The chief estimator of contractor X identified a number of risks and expressed them in terms of quantity, unit cost rate and cost. In this detailed format, risk could be related to quantities, unit cost rates and cost items. The trade summary was then expanded by the inclusion of specific risk variables into a format illustrated in Table 15.13.

TABLE 15.13 THE EXPANDED TRADE SUMMARY OF THE COST ESTIMATE

NO	ITEM DESCRIPTION	UNIT	QUANTITY	RATE ($)	ITEM (TOTAL $)	TRADE (TOTAL $)
1	DEMOLITION	item				53 650
2	GROUNDWORKS					324 470
	Site clearing	m²	3 960	6.50	25 740	
	Excavation	m³	2 310	33.00	76 230	
	Fill	m³	1 870	33.00	61 710	
	Surface preparation	item			22 320	
	Pathways and pavements	item			57 790	
	Landscaping	item			63 270	
	Miscellaneous	item			17 410	
3	PILING					57 230
	Site establishment	item			3 780	
	Excavation	m³	1 170	19.00	22 230	
	Dewatering	item			5 340	
	Reinforcement	t	17	650.00	11 050	
	Concrete	m³	146	18.00	2 628	
	Miscellaneous	item			12 202	
4	CONCRETE SUPPLY					355 200
	25 Mpa concrete	m³	2 220	160.00	355 200	
5	CONCRETE PLACE					121 880
	Pumping	m³	2 160	8.00	17 280	
	Placement	m³	1 990	26.00	51 740	
	Curing	m²	3 010	1.00	3 010	
	Finishes	item			26 870	
	Miscellaneous	item			22 980	

TABLE 15.13 THE EXPANDED TRADE SUMMARY OF THE COST ESTIMATE *(CONTINUED)*

NO	ITEM DESCRIPTION	UNIT	QUANTITY	RATE ($)	ITEM (TOTAL $)	TRADE (TOTAL $)
6	REINFORCEMENT SUPPLY					181 390
	Bar	t	170	885.00	150 450	
	Mesh	m²	2 240	5.40	12 096	
	Scheduling and transport	item			550	
	Miscellaneous	item			18 294	
7	REINFORCEMENT PLACE					71 060
	Reinforcement	t	187	380.00	71 060	
8	FORMWORK					525 620
	Formwork	m²	12 670	37.00	468 790	
	Narrow widths	m	2 740	17.00	46 580	
	Miscellaneous	item			10 250	
9	STRUCTURAL STEEL					85 090
	Supply and fix	item			85 090	
10	METALWORK					610 350
11	PRESTRESSING					169 710
12	MASONRY					365 190
	Common bricks	m²	2 010	50.00	100 500	
	Face bricks	m²	2 670	72.00	192 240	
	Labour and sundries	item			72 450	
13	STONEWORK					33 470
14	CARPENTRY					47 440
15	ROOFING AND ROOF PLUMBING					155 530
	Membrane	m²	120	75.00	9 000	
	Metal roofing	item			88 700	
	Insulation	m²	1 510	17.00	25 670	
	Roof plumbing	item			32 160	
16	SUSPENDED CEILING					512 020
	Plasterboard	m²	4 480	42.00	188 160	
	Proprietary systems	m²	5 950	50.00	297 500	
	Miscellaneous	item			26 360	
17	Plastering					247 730
	Operable wall	item			72 450	

TABLE 15.13 THE EXPANDED TRADE SUMMARY OF THE COST ESTIMATE (*CONTINUED*)

NO	ITEM DESCRIPTION	UNIT	QUANTITY	RATE ($)	ITEM (TOTAL $)	TRADE (TOTAL $)
	Stud wall frames	m²	3 240	13.00	42 120	
	Insulation	m²	3 140	8.00	25 120	
	Plasterboard lining	item			86 630	
	Miscellaneous	item			21 410	
18	WALL AND FLOOR TILING					112 770
	Wall tiles	m²	260	75.00	19 500	
	Floor tiles	m²	1 130	78.00	88 140	
	Miscellaneous	item			5 130	
19	DOORS AND FRAMES					189 610
20	WINDOWS					550 630
21	GLAZING					21 390
22	WALL AND FLOOR FINISHES					268 160
	Resilient finishes	item			68 900	
	Carpet	m²	5 130	32	164 160	
	Underlay	m²	5 130	6	30 780	
	Edge strips	m	160	27	4 320	
23	PAINTING					160 380
24	HARDWARE					93 240
25	HYDRAULICS					418 780
26	MECHANICAL SERVICES					823 690
27	ELECTRICAL SERVICES					970 430
28	LIFT SERVICES					510 230
29	MISCELLANEOUS SERVICES					246 970
30	FIXTURES AND FITTINGS					103 360
31	PROVISIONAL SUM					202 015
32	CONTINGENCY					429 434
	TOTAL ESTIMATE					9 018 119

The expanded trade summary became a preliminary model of a probability cost estimate, which was created by the Excel spreadsheet. Risk data was then elicited for the identified risk variable from the chief estimator to fit triangular distributions. From this information the final cost estimating model was prepared (Table 15.14).

TABLE 15.14 THE FINAL MODEL OF THE COST ESTIMATE

NO.	DESCRIPTION	QUANTITY			UNIT RATE			COST		
		MINIMUM	MOST LIKELY	MAXIMUM	MINIMUM	MOST LIKELY	MAXIMUM	MINIMUM	MOST LIKELY	MAXIMUM
1	DEMOLITION							42 918	53 650	64 377
2	GROUNDWORKS									
	Site clearing	3 760	3 960	4 355						
	Excavation	2 189	2 310	2 65?	27	33	50			
	Fill	1 776	1 870	2 150	26	33	40			
	Surface preparation							17 858	22 320	26 781
	Pathways and pavements							46 232	57 790	69 348
	Landscaping									
	Miscellaneous									
3	PILING									
	Site establishment									
	Excavation	1 111	1 170	1 40?	16	19	25			
	Dewatering							5 073	5 340	7 476
	Reinforcement									
	Concrete									
	Miscellaneous									
4	CONCRETE SUPPLY									
	25 Mpa concrete	2 164	2 220	2 553	152	160	168			
5	CONCRETE PLACE									
	Pumping	2 091	2 200	2 340						
	Placement	1 892	1 990	2 188						

TABLE 15.14 THE FINAL MODEL OF THE COST ESTIMATE (*CONTINUED*)

NO.	DESCRIPTION	QUANTITY			UNIT RATE			COST		
		MINIMUM	MOST LIKELY	MAXIMUM	MINIMUM	MOST LIKELY	MAXIMUM	MINIMUM	MOST LIKELY	MAXIMUM
	Curing									
	Finishes									
	Miscellaneous									
6	REINFORCEMENT SUPPLY									
	Bar	164	170	187	840	885	930			
	Mesh									
	Scheduling and transport									
	Miscellaneous							14 635	18 294	21 952
7	REINFORCEMENT PLACE									
	Reinforcement				310	370	428.9			
8	FORMWORK									
	Formwork	12 037	12 670	13 939	31	37	44			
	Narrow widths									
	Miscellaneous									
9	STRUCTURAL STEEL									
	Supply and fix									
10	METALWORK							488 275	610 350	732 417
11	PRESTRESSING							135 768	169 710	203 652
12	MASONRY									
	Common bricks	1 909	2 010	2 309	41	50	60			

TABLE 15.14 THE FINAL MODEL OF THE COST ESTIMATE (*CONTINUED*)

NO.	DESCRIPTION	QUANTITY			UNIT RATE			COST		
		MINIMUM	MOST LIKELY	MAXIMUM	MINIMUM	MOST LIKELY	MAXIMUM	MINIMUM	MOST LIKELY	MAXIMUM
	Face bricks	2 535	2 670	2 936	60	72	83			
	Labour and sundries									
13	STONEWORK									
14	CARPENTRY									
15	ROOF, ROOF PLUMBING									
	Membrane				61	72	83			
	Metal roofing									
	Insulation				14	16	18			
	Roof plumbing									
16	SUSPENDED CEILING									
	Plasterboard				35	42	49			
	Proprietary systems				41	50	60			
	Miscellaneous							21 678	27 100	32 517
17	PLASTERING									
	Operable wall									
	Stud wall frames	3 174	3 300	3 426						
	Insulation									
	Plasterboard lining									
	Miscellaneous			18 199	21 413	24 623				
18	WALL AND FLOOR TILING									
	Wall tiles									

TABLE 15.14 THE FINAL MODEL OF THE COST ESTIMATE (CONTINUED)

NO.	DESCRIPTION	QUANTITY			UNIT RATE			COST		
		MINIMUM	MOST LIKELY	MAXIMUM	MINIMUM	MOST LIKELY	MAXIMUM	MINIMUM	MOST LIKELY	MAXIMUM
	Floor tiles									
	Miscellaneous									
19	DOORS AND FRAMES		151 689	189 610	227 531					
20	WINDOWS		440 504	550 630	660 760					
21	GLAZING									
22	WALL AND FLOOR FINISH									
	Resilient finishes									
	Carpet									
	Underlay									
	Edge strips									
23	PAINTING									
24	HARDWARE									
25	Hydraulics			397 841	418 780	439 719				
26	MECHANICAL SERVICES									
27	ELECTRICAL SERVICES		921 912	970 430	1 018 948					
28	LIFT SERVICES									
29	MISCELLANEOUS SERVICES		228 963	255 900	282 837					
30	FIXTURES AND FITTINGS			82 688	103 360	124 031				
31	PROVISIONAL SUM									
32	Contingency									

@Risk software was used for simulation. The chief estimator identified numerous dependencies among the risk variables and expressed their values based on his personal experience. The simulation was then performed using 5000 iterations. The stability of the output distribution was monitored during simulation using convergence statistics and acceptable stability was attained after around 1000 iterations. The output was generated both statistically (Table 15.15) and graphically (Figure 15.12).

TABLE 15.15 THE SIMULATION STATISTICS OF THE COST ESTIMATING EXAMPLE

STATISTICAL REPORT
No. of Iterations: 5000

Minimum	$8 389 495
Maximum	$8 947 206
Mean	$8 654 646
Standard deviation	$94 16
Variance	8.87E+09
Skewness	−2.24E-02
Kurtosis	3.0835
Errors calculated	0
70%	$8 699 372
80%	$8 735 809
90%	$8 775 69
100%	$8 947 206

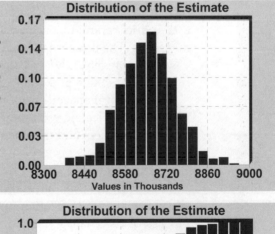

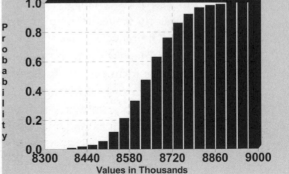

Figure 15.12 The output normal distribution in the form of a frequency and a cumulative density function. These figures were generated by @Risk software.

The expected (mean) value of the cost estimate is given by the statistics as $8 654 646, with minimum value of $8 389 495 and maximum value of $8 947 206. The distribution has a relatively narrow range of 6.2 per cent due to its small value of standard deviation. What the statistics clearly show is that the original single-value cost estimate of $9 018 119 was too high. Had it been reduced to just under $8 800 000, which would have made it the cheapest bid, contractor X would still have had over 90 per cent probability of building the project within that price.

The contractor's contingency was clearly set too high. The magnitude of the identified risks was well below the level of this contingency. Had the contractor developed a probability cost estimate, he would have gained a detailed understanding of the problem which would have helped him to make more informative judgments about the presence of risk and the value of the contingency.

CONCLUSION

Risk management is an important component of the project management body of knowledge. Risk is always present in construction projects and its careful management is essential for achieving successful outcomes. The presence of risk in construction projects should not be viewed negatively, however, since a high level of risk can be successfully converted to a potentially high level of financial reward.

The project manager's responsibility is to develop a risk management plan for dealing with project risk. Such a plan details outcomes of risk assessment and treatment strategies, and assigns responsibilities for action.

Risk management should be viewed as an integral part of good management practice. It generates a realistic perspective of business situations for which it attempts to develop innovative solutions. A process of risk management should be applied to a project from its very early stages to help in ongoing decision-making throughout a project's lifecycle.

Risk management offers many benefits. Perhaps the most important is its focus on details. By identifying and assessing potential project risks, a project manager is able to explore potential problems in depth. Simply knowing more about a potential problem removes some uncertainty and makes the project manager more confident about outcomes. Moreover, with more detailed knowledge the quality of decision-making is enhanced. This also contributes to better scheduling and cost budgeting, and reduced dependency on contingencies.

EXERCISES

1   Do you seek risk in decision-making or do you avoid it? Do you treat risk consistently in decision-making? Do you respond to risk in your personal life differently from that in your business life? If you do, why? What is the reason for your attitude towards risk?

2   What is an appropriate structure of a risk management plan? What specific issues should be covered?

3   Explain in detail the operation of at least four different risk identification methods.

4   What is the Delphi technique used for? How does it work?

5   Describe a qualitative approach of risk assessment.

6   Explain in detail the use of sensitivity analysis.

7   Describe probability risk analysis.

8   What is Monte Carlo simulation?

9   Describe management response to risk.

10   What options are available to a project manager for controlling residual risk?

CHAPTER 16

MANAGING HEALTH AND SAFETY IN CONSTRUCTION PROJECTS

INTRODUCTION

The construction industry is characterised by a hostile and uncontrollable production environment, a complex product, a fragmented structure, and increasing economic constraints. It is not surprising, therefore, that the industry has more accidents of greater severity than any other industrial sector (Gyi et al. 1996). For example, a recent survey (April 2000) of the impact of the Construction (Design and Management) (CDM) regulations in the United Kingdom concluded that while fatalities have dropped from 88 to 70 per year over the four years since the introduction of these regulations, major accidents have risen from 2627 to 4619. European statistics are similarly depressing, with the construction sector employing less than 10 per cent of the working population but accounting for 15 per cent of all accidents at work and 30 per cent of all fatal accidents (Bishop 1993). In Australia the situation is no different, with the rate of compensated injuries and disease in the construction industry being 37.4 per 1000 workers, 63 per cent higher than the all-industry average of 22.9 (RCBCI 2002). This record and a perceived inability to improve performance through self-regulation has led to the introduction of new health and safety regulations (OHS 2001) which impose far more stringent requirements on project participants, particularly contractors, to manage the risks associated with occupational health and safety (OHS) in the construction workplace. Although international comparisons are difficult because of differing legislation, definitions and reporting systems, the construction industry's safety record is also depressing in many other areas such as Hong Kong, the United States and Singapore (Kisner & Fosbroke 1994; Lingard & Rowlinson 1994; Mattila et al. 1994; Martin et al. 1996). It appears that the problem of OHS is a global one. There is therefore no justification for complacency and the aim of this chapter is to investigate the attitudinal, behavioural, economic and ethical reasons underlying the construction industry's fairly poor safety record.

SAFETY ATTITUDES AND THE CONCEPT OF VALUE

Despite its poor record, the construction industry's attitude towards OHS appears to be apathetic. Managers display complacency and a lack of concern for the safety of their workforce, which is perpetuated by the industry's masculine values and macho image (Eves 1989; Bergman 1990; Everly 1997; Lingard & Rowlinson 1997). This attitude extends beyond an individual's personal sense of safety to their treatment of others. For example, in a survey of over a hundred companies, Nabarrow & Nathanson (1996) found that 60 per cent of the respondents did not consider the risks to the general public arising from their activities. It was also found that 40 per cent of firms failed to carry out risk assessments on the activities of independent subcontractors. Similarly, while 75 per cent produced written emergency procedures, very few communicated them to subcontractors, visitors or the public. This recklessness is likely to be, in part, a product of client failures to identify OHS issues as a priority. For example, in the lists of client requirements produced by Bennett and Flanagan (1983), Hewitt (1985) and Masterman (1994), OHS does not feature once, the focus being exclusively on cost, time, quality, functional and aesthetic criteria.

Surprisingly, despite a growing legal awareness of OHS responsibilities, Grey's (1996) report, which focused wholly on value for money, only mentioned the issue of OHS once. This is particularly surprising in light of the Health and Safety Executive (HSE) report of 1993, which estimated that accidents on construction sites might account for up to 8.5 per cent of tender prices. It appears that greater attention to OHS issues could contribute significantly to securing greater value for money from the construction industry.

But it seems that the dominant attitude within the industry towards OHS continues to be a cavalier one. Decisions on OHS provisions are not based on ethical considerations about people's basic rights to health and safety at work, but on economics. This attitude towards decision-making, which is centred around the concept of value, was noted by Moore (1991), who argued that social requirements such as health and safety, maximum working hours, pensions and working conditions are being dictated by the marketplace. In terms of management attitudes, organisational decisions are too often based on the view that expenditure on safety provisions reduces profits and undermines a firm's efficiency. As Hunter (1997: 6) recalls, 'we are constantly faced with the economic argument, in other words, safety provisions are said to cost money without producing tangible results'. Whittington and colleagues (1992: v) also recognised the economics driving OHS decision-making and cited other problematic constraints such as time pressures and the short-term nature of contracting relationships. They argued that these restrictions adversely affected safety performance by:

- diffusing responsibility for the co-ordination and control of critical safety issues
- making it more difficult to establish a system in which contractors are held accountable for poor safety performance
- reducing the quality and quantity of onsite supervision
- increasing the difficulty of maintaining safety standards within one site and across different projects
- reducing the industry's infrastructure, in particular the provision of training
- reducing the likelihood that safety and risk management will be dealt with systematically and at an early stage in the project lifecycle.

Whittington and colleagues (1992: 6) summarised it this way:

a number of fundamental flaws were identified in the way that both the industry and potential clients are currently responding to these demands. Unless project management is driven by a client with a high safety profile, decisions on organisation, planning and subcontracting of work were found to be primarily based on meeting commercial goals. This is not to say that safety issues are ignored but they tend to be dealt with late in the project life cycle and have little impact on early and often critical decision making. There was also undue emphasis on the failure of individual workers. This results in a reliance on short term solutions rather on any attempt to uncover more fundamental management and organisational problems.

INVESTING IN SAFETY

Few would debate that in a construction environment, absolute safety is a utopian ideal. Irrespective of the precautions taken or finance expended, no project can be totally free of risk. Expenditure on safety in all its forms (systems, management, training and protective equipment) does not guarantee safety, although it is accepted that it can reduce the likelihood and severity of accidents. This persistent belief has meant that there has never been any commercial inducement to invest significant resources into safety management. Another disincentive to investment is that there is no obvious positive commercial advantage to giving safety a high priority in business planning. Firms that have a high safety profile are not necessarily placed at any obvious commercial advantage (Preece & Male 1997). Furthermore, while ill health costs the UK industry an estimated £12 billion per year (CBI 1997), many of these costs are indirect and therefore not obviously accountable. This makes cost–benefit analysis difficult (Moore 1991). Finally, although statutory instruments require risks to be as low as reasonably practicable, fines and punishments remain relatively small-scale in the area of safety (Jarvis 1997). Although these penalties are increasing in many countries to include corporate manslaughter and corporate killing, prosecution is often difficult and it seems that many managers are still prepared to risk the penalties in order to economise on their budgets.

In summary, ethical principles in relation to safety are being dislodged in favour of increased efficiency and minimising costs, which are seen as justifiable in light of economic pressures on industry. Similar views are reflected in a discussion paper from the Institute of Directors (IOD 1996), which has described OHS as a 'prosaic rather boring subject', a burden to business that has been implemented by an 'over zealous inspectorate'. These attitudes are of grave concern as the Institute represents the senior management within the British construction industry. This prejudice can percolate through organisations and help create an erroneous culture that is unsympathetic to the needs and protection of the worker.

ERRONEOUS ASSUMPTIONS THAT UNDERPIN OHS ATTITUDES

Many of the erroneous assumptions that are used as an excuse to avoid the issue of quality (Crosby 1979) are reflected in the OHS literature.

Managing Health and Safety in Construction Projects

The first is that safety standards are subjective, that they mean different things to different people, and that their intangible nature makes them difficult to measure (Hogh 1987). Perhaps it is a consequence of this assumption that a reliable and widely acceptable measure of safety performance has never been developed. Current measures are uncoordinated and rely on a range of different proactive and reactive data including safety auditing, attitudinal measures, accident statistics, near misses reporting and the financial cost of accidents (Budworth 1996).

A second assumption is that all accidents take place on site and are therefore the result of workforce error (Shimmin et al. 1981; Salminen 1995; Falconer & Hoel 1996). This perception has been compounded by the rapid growth in subcontracting and self-employment, which has in turn led to a situation where responsibility for OHS is difficult to identify (HSE 1994). However, an analysis of fatal accidents by Bishop (1993) has shown that 35 per cent of all accidents have their root cause in design, with a further 28 per cent being attributable to managerial decisions. This type of research led directly to the introduction of the CDM 1994 Regulations in the UK and the new *Occupational Health and Safety Regulation 2001* in Australia, both of which place collective responsibility on all members of the construction team, including clients, for OHS matters. Bennett and colleagues (1994) have pointed to a lack of understanding of how decisions made in design influence other aspects of the construction process. Accepting that approximately 80 per cent of the costs of a building are committed by sketch design stage, it may be that a similarly large proportion of 'designed in' accidents are in place by this stage (Brandon 1978; Ferry & Brandon 1991).

The third assumption is that OHS legislation is overly burdensome for employers (IOD 1996). While Workcover in Australia, the Health and Safety Executive (1994) and the Confederation of British Industry (1990) in the United Kingdom, accept that the laws relating to OHS are seen as voluminous, complicated and fragmented, they do not support the overall assertion that they are unduly burdensome. This assertion has been qualified by Jarvis (1997: 5), who found differences in opinion on how it should be enforced: 'Small firms encourage a more prescriptive approach; they want to be told what to do. While larger firms prefer deregulation which they perceive increases flexibility and efficiency'.

The fourth assumption is that safety is the sole responsibility of the safety officer or department (Waring 1992; Whittington et al. 1992). It is an assumption that is reinforced by OHS officers, who have a tendency to over-centralise their responsibilities (Deacon 1994). In short, there is a general lack of collective responsibility for health and safety within the construction industry. If Heinrich and colleagues' (1980) assertion that most accidents are caused by a chain of events across a spectrum of organisational levels is true, then this attitude is a dangerous one.

HEALTH AND SAFETY CULTURE

Thus it seems that the economic, macho, adversarial and selfish values that guide managerial decision-making in the construction industry lie at the heart of its poor safety record. It follows that a change in OHS performance would require a change in managerial values and culture within the construction industry.

Seymour and Rooke (1995) point out that the research community's values reflect those of the construction industry, and in this sense it is not surprising that OHS research has been neglected in favour of more economically oriented research. With some notable exceptions (Whittington et al. 1992; Duff et al. 1993; Sawacha 1993), the attention that has been given to OHS has been generally restricted to the exploration of causal relationships between organisational characteristics and OHS performance (Hinz & Roboud 1988; Niskanen & Lauttalammi 1989; Jaselskis et al. 1996). Notwithstanding the difficulties of developing reliable performance measures, such an emphasis may have been justifiable in the early stages of OHS research. However, having built a basic understanding of influences on OHS performance, it is important to understand the underlying attitudinal and behavioural issues affecting OHS in the early stages of construction.

LEGISLATION AND SAFETY

In recent years there has been greater emphasis given to the role of legislation in managing OHS in the construction industry, largely brought about by the industry's seeming inability to control its accident rates voluntarily. In Europe, Australia, America and much of Asia, the demands of risk-related OHS legislation are increasing. Furthermore, construction clients are becoming more risk-averse, which is reflected in their preference for design and construct and BOOT procurement systems and for the private financing of projects. Contractors in particular are facing greater levels of risk for longer periods and this has focused their attention on risk management techniques.

Risk management has been used extensively in industries such as mining, petrochemicals and pharmaceuticals, and there appears to be an increasing political agenda within the Australian construction industry to follow a similar path, particularly in the area of safety. For example, the new *Occupational Health and Safety Regulation 2001* has a particularly strong emphasis on risk management. This is not surprising because without exception, the considerable amount of literature on risk management has been uncritical. The problems of risk management are seldom discussed and it is invariably presented as a panacea for improving the reliability and profitability of construction. Indeed, it is difficult to question the basic premise of risk management, which is to predict and plan for the future, and there is considerable evidence to indicate that organisations that manage their risks effectively hold a significant commercial advantage over those that do not. However, there is less evidence that risk management has had benefits in more qualitative areas such as safety. While it is easy to develop a risk management system, many companies have found that it is far more difficult to implement it effectively so that it becomes a key aspect of organisational culture.

In Australia the New South Wales construction industry pays one of the highest worker's compensation premium rates due to its poor OHS performance. These high rates of insurance result from the fact that the building construction industry has far more injuries (57 per 1000) than the New South Wales State average (20 per 1000). In fact the building construction industry has consistently performed worse than other New South Wales industries in terms of OHS, other than the coalmining industry. As part of its overall strategy to substantially lift the construction industry's productivity, the NSW Government is insisting on improved management of occupational health, safety and rehabilitation (OHS&R) with the aim of:

- reducing the level of accidents
- curtailing lost time due to industrial disputation
- greatly improving productivity for the industry.

The thrust of the initiative is directed at making OHS&R management an integral part of the organisational culture of a company or enterprise. In this way, sound OHS&R practices will become an automatic component of everyday work practice. As a part of this process, the WorkCover Authority of NSW (WorkCover) has implemented a new regulation, the *Occupational Health and Safety Regulation 2001*. The new Regulation is made under the *Occupational Health and Safety Act 2000* and replaces all 36 existing regulations made under the *Occupational Health and Safety Act 1983*, the *Construction Safety Act 1971*, and Part 3 of the *Factories, Shops and Industries Act 1962*. The *Occupational Health and Safety Regulation 2001* places obligations on employers to adopt a systematic risk management, performance-based approach to managing workplace health and safety. It also establishes an important legislative framework that places a duty on employers to consult employees in every part of the process. The purpose of consultation (which is a major part of the regulation) is to ensure that employees can contribute to all decisions affecting their health and safety. This is done through OHS committees or OHS representatives or simply by sharing information. An employer must also provide the necessary amenities, accommodation, first aid facilities, equipment, information, instruction, training and supervision of those working on or in any premises that are under its control.

COMMUNICATION AND BEHAVIOUR

In addition to the attitudinal and value problems of the construction industry being a contributor to poor OHS performance, it is likely that communication problems are a contributing factor. These have been well documented and appear to be an inherent characteristic of the construction industry (Crichton 1966; Crawshaw 1976; NEDO 1983). More recently, Loosemore (1996) identified communication as essential to the detection and communication of potential problems, the process of defining potential problems, the provision of information to arrive at solutions, the implementation of solutions, and the provision of feedback. He found that the structure of communication between project participants influences the extent to which a simple problem escalates into a crisis and an organisation's ability to mitigate its impact on project goals. Unfortunately, he also found the paradox that during problematical times, when good communication is particularly important, it is less likely to happen due to people's tendency to protect their own interests. It is reasonable to assume that the efficiency and structure of communications between those responsible for OHS decisions would be a factor influencing performance.

DESIGN IN HEALTH AND SAFETY

The historical emphasis on the construction stage and the important influence of design decisions on OHS performance justify a focus on the design process. Design entails the combination and balance of ideas generated by the design team, in a way that should reflect the needs of the client's brief. Much design involves the use of basic components and materials in new and different ways. There is no single method or system underlying the creation of a design. Most design strategies are reiterative and consist of the generation of several potential solutions or hypotheses, which are evaluated, refined and combined until an acceptable solution is created (Gray et al. 1993). It is a complex process that involves a large number of culturally, technically and geographically diverse but interdependent specialists making their contributions at different times. As a result of the various parties involved in the design process, a design is generated and evolves in reaction to change and differing problems that may be encountered because of lack of clarity of the client's brief, financial constraints, resource availability, multifunctional end users, necessary changes to the design, and the involvement of specialist contractors. Within this environment of evolution and change, health and safety can often become of secondary importance to more immediately demanding requirements.

THE PSYCHOLOGY OF ACCIDENT PREVENTION

The Challenger space disaster and the Zeebruge ferry disaster are among the most potent examples of a disturbing behavioural phenomenon associated with OHS issues which characterises all organisations: the reluctance of people to prevent accidents, despite being aware of their advance signs and being in possession of the necessary knowledge to prevent them (Jarman & Kouzmin 1990; Pijnenburg & Van Duin 1990). This tendency has been noted in an OHS context by HSE (1996), who found that on many construction projects, accidents are often under-reported by over 60 per cent. Organisations appear to create conditions that reduce diligence in the detection of potential accidents by encouraging people to subconsciously ignore their advance signs rather than communicating them to the wider organisation. Over time, such behaviour increases the cost of correcting potential problems and reduces further their likelihood of being reported, a cycle that continues until a crisis develops to force an organisation to respond. This self-perpetuating cycle of decline occurs because unreported potential problems become obscured and magnified by subsequent dependent activities. It is with this dangerous phenomenon that this chapter is concerned. More precisely, its aim is to investigate the conditions and psychological mechanisms that fuel it.

PREVENTIVE MANAGEMENT OF ACCIDENTS

It is possible to classify accidents along a continuum from 'sudden' to 'creeping', depending on the length of the chain of events that cause them. From this point onwards, this chain of events will be referred to as an 'accident chain'. Sudden accidents have relatively short accident chains and normally demand an element of reactive management. In contrast, creeping accidents present numerous opportunities for prevention along their accident chain. For a manager, it is more desirable to prevent accidents by tackling the events in an accident chain rather than the resultant accident, and the first step in being able to do this is to develop an understanding of accident chains.

Accident chains can vary in two ways: by their length and conspicuousness. A good example of a site accident with a relatively long and inconspicuous accident chain is one that originates at the design phase by the incorporation of a dangerously inappropriate material or component into the plans and specifications. The causal link between such an event and an accident is so indirect that such a danger is likely to remain undetected until it manifests itself on site. In contrast, a decision to use dangerous machinery within a congested working environment is more conspicuous and more likely to be detected as a potential problem. Thus while accidents are infinite in their variety, they can be classified along two continuums, which are associated with the character of their accident chains. For managers, this classification is useful because the greatest challenges are posed by short, rapidly evolving and inconspicuous accident chains. This relationship is illustrated in Figure 16.1.

Unfortunately the poor OHS record of the construction industry (Bishop 1993; Gyi et al. 1996) suggests that preventive 'accident chain management' is conducted inadequately. While it is tempting to conclude that the construction industry is incompetent in comparison with other industries or attracts more accident-prone personnel, it may be that the nature of construction activity produces more accident chains with the problematical characteristics of inconspicuousness and shortness. This hypothesis has never been tested, but whatever the reasons for its poor record, it is reasonable to assume that the industry would benefit from a greater understanding of the accident prevention process and the psychological mechanisms of its personnel which prevent its effective application.

Figure 16.1 The relationship between the length and consciousness of accident chains and coping difficulty for organisations

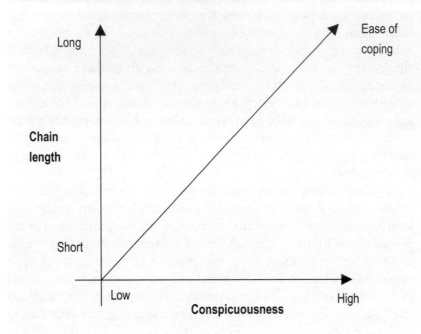

A MODEL OF ACCIDENT PREVENTION

The prevention of accidents requires a process of control whereby an organisation monitors and corrects events that have the potential to cause accidents. Effective control systems are therefore essential to this process and their importance is directly related to the level of risk an organisation faces. In construction projects, effective control systems take on particular importance because risk levels are high, there are few implicit controls, and the production process is small-batch and unmechanised in nature (Miller & Rice 1967; Hofstede 1981; Morris & Hough 1987; Skitmore et al. 1989; Winch 1989). Where there are low levels of risk and implicit controls built into the tasks of a continuous, repetitive and highly mechanised production process, less control is needed. A model of accident prevention based on the principles of control is depicted in Figure 16.2 and it is around this model that the remainder of this chapter is based.

Figure 16.2 A model of accident prevention

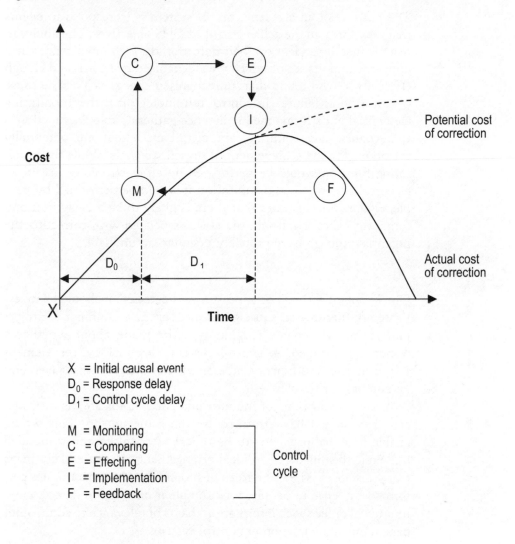

X = Initial causal event
D_0 = Response delay
D_1 = Control cycle delay

M = Monitoring
C = Comparing
E = Effecting
I = Implementation
F = Feedback

Control
cycle

EXPLAINING THE MODEL OF ACCIDENT PREVENTION

The value of the model presented in Figure 16.2 is that it identifies five distinct but interrelated phases of activity that have to be conducted effectively if accidents are to be prevented. These are monitoring, comparing, effecting, implementation and feedback.

Monitoring

The purpose of an accident control system is to respond to events that may threaten the achievement of OHS objectives. The ability to react is first dependent on their detection through intelligence gathering, the importance of which was recognised by Morris and Hough (1987: 212), who found that 'unrecognised change is a classic cause of catastrophic failure'. In control terminology it is the 'monitoring element' that performs the intelligence-gathering role. Its function is to recognise and communicate disturbances that are potentially threatening to the achievement of prescribed goals. While the monitoring function would be performed by an electronic device in an electrical system, in an organisation it would be performed by people. Figure 16.1 illustrates that there is likely to be a delay in monitoring (D_o) and, as a result, the costs associated with correcting the potential problem have probably begun to accumulate.

Comparing

Once a potential threat or opportunity is detected, it should be investigated further to assess its significance for the attainment of project goals. This will enable judgments to be made about whether a response is justified. In control systems terminology, the element responsible for this function is referred to as the 'comparing element' or 'comparator'.

While the functions of monitor and comparator are distinct, ideally they should be performed by the same person. This would enable potential problems to be quickly recognised and responded to. While this might be an ideal arrangement, there are likely to be many instances where the roles of monitor and comparator are performed by separate people, for example if one person detects environmental or feedback information that is of relevance to some other person operating in another control system.

A comparator operates on the basis of environmental and feedback information supplied to it by monitors and is usually a specialist with formal responsibility for the protection of one specific goal. There will probably be a number of specialist comparators in an organisation because one person is unlikely to have broad enough expertise to make an assessment of one event's impact on a project's full range of goals. Typically, the feedback and environmental data that converge on a comparator are in a raw state and of little immediate relevance to project goals. The skill of comparators is in recognising the impact of this data on the specific goal that they protect, and it is for this reason that they are likely to be well educated, professionally qualified specialists.

Specialist knowledge and feedback information alone are not enough to enable a comparator to make judgments about whether a particular event is a threat or opportunity that demands a response. Rather, each specialist comparator also needs to have a clear idea of the standard of performance that is required in his particular area of expertise. That is, the comparators need a clear grasp of project goals.

Effecting

If a comparator considers a detected event not to be a threat or an opportunity, then the control process terminates at that point. The emphasis of the control system then returns to one of monitoring, which has in any event been continuing in parallel. However, if a comparator does perceive a threat or opportunity that justifies a response, then the control process progresses to the next stage, which involves formulating an appropriate response. In control terminology this is the function of the 'final controlling element' or 'effector', which is responsible for acting directly on the controlled body. In organisational terms it would be a person with the necessary decision making authority who will choose from a range of alternative courses of action (Simon 1960; Albers 1969; Cleland & King 1983). Although these courses of action may be independently generated by an effector, they are more likely to be generated by collaboration with comparators, who are more qualified to make recommendations. The skill of an effector is to synthesise the varied and possibly conflicting advice provided by the range of comparators, who have an interest in the nature of the final decision.

Effectors exist at all organisational levels and it is the relative significance of a threat or opportunity that determines the level of decision-making invoked. Normally, the more significant the threat or opportunity, the higher the level of decision-making. As well as being determined by the seriousness of a threat, the level invoked is also likely to depend on the degree to which decision-making authority is decentralised — a decentralised structure facilitates a lower level of decision-making. This is an important point because greater proximity of decision-makers to threats or opportunities permits a faster response by reducing the need for communication between hierarchical levels. Highly centralised structures are notoriously slow at adapting to change (Wright et al. 1992). The fastest response to a threat would arise from a situation where the monitor, comparator and effector were the same person. In such a situation

Managing Health and Safety in Construction Projects

there would be no communication problems and no delays in responding to a threat. However, in the absence of this ideal responsibility structure, effective teamwork is essential to minimise potential communication problems and damaging delays.

Implementation

Figure 16.2 shows that implementation occurs some time (D_1) after the initial detection of a potential problem by a monitor, and it is the speed of the control cycle that determines the extent of this delay. The figure also shows that during this delay, the costs of correcting a problem have accumulated further and that they would continue to accumulate along the dotted line if the control response was slower. However, once implementation has commenced they should accumulate at an ever-decreasing rate until the problem is completely resolved. The value of Figure 16.2 is that it illustrates the importance of an effective accident prevention system to the mitigation of the costs associated with unexpected problems.

By definition, the implementation of an effector's decision is likely to involve widespread social, technical and monetary change. Technical change would relate to modifications in the physical routines of the job and social change to the way in which project members have to alter their behaviour and established patterns of relationships (Lawrence & Seiler 1965). Monetary change relates to the extra resources that are required to implement those changes and to the redistribution of those resources between project participants. In a highly interdependent organisation such as a construction project, such change is likely to lead to the emergence of a network of communications. The structure of this network will be determined by the nature of the effector's decision in terms of the change it requires and by the interdependencies between those affected. The recognition of interdependencies by organisational members and the smooth flow of information between them will determine how efficiently any change will be brought about.

Feedback

Throughout the implementation phase, knowledge of results is crucial to ensure the efficiency of the control process in realigning planned and actual performance. This information is known as feedback and is acquired by continued monitoring of the organisation to collect performance data. This feedback should be continuously assessed in relation to project goals and, if necessary, further adjustments made until planned performance is in line with actual performance. In this way the control process should be a continuous, cyclical and converging one.

BARRIERS TO EFFECTIVE SAFETY CONTROL SYSTEMS

Monitoring

There are numerous factors which can adversely influence the efficiency of monitoring.

NATURAL RESISTANCE TO CHANGE

The process of responding to a potential problem involves change to the social and reward structure of the organisation and often to the physical nature of its product. Those who are potential losers in the new order generally exhibit the lowest sensitivity to the advanced warning signs of that problem. Therefore, to minimise resistance to change, managers must seek to identify and protect potential losers.

AMBIGUOUS GOALS

Monitoring problems also relate to the clarity of performance standards, to goal priorities, and to a poor understanding of the factors that affect their attainment. This would seem to be particularly relevant to the detection of accidents since safety performance standards are rarely made a priority, are difficult to quantify, and are rarely elucidated on construction projects (Bennett & Flanagan 1983; Hewitt 1985; Masterman 1994; Grey 1996). Furthermore, research has a long way to go in developing a full understanding of the factors that cause accidents.

CONFLICTS OF INTEREST AND CLOUDED RESPONSIBILITY STRUCTURES

Another problem relates to the conflicts of interest within construction projects, which means that diligence in the detection of problems will not always be in a person's own interests. This is partly a product of the matrix structure of construction project organisations (Bresnen 1990; Walker 1996) but also the result of complex and ambiguous construction contracts which confuse responsibilities for accident reporting. It is evident that much could be done in the construction industry's contractual and organisational structure to enhance the reporting of potential accidents. Confusion over responsibilities for hazard reporting is also a function of the complexity of a problem. Complex multidisciplinary problems appear to have a paradoxical impact in that confusion over responsibilities can result in the problem being over-monitored or under-monitored in a collective assumption that someone else is taking the responsibility.

INCONSPICUOUS ACCIDENT DATA

Information about potential problems is fed through dedicated management information systems (MISs), which run vertically through an organisation. These systems provide routes through which specialist information flows relating to specific goals. Not only do problems exist in construction projects in the collection, storage and retrieval of information relating to potential safety problems, but there is also a lack of lateral communication between different systems. This is problematic because of the interdependency of different goals. For example, a requirement to increase the speed of construction to meet a reduced program deadline may have safety implications.

INFORMAL REPORTING SYSTEMS

The main problem in relation to health and safety is likely to be the absence of a formal MIS for accident data existing on many projects. This would mean that managers have to draw their information from the range of well-established MISs relating to cost, time and quality goals. Since the purpose of such information is not to report accidents, the manager is faced with having to deduce the possibility of potential accidents from a synthesis of non-specific information. It is evident that attention should be given to the creation of specific MISs for safety data and that information relating to potential accidents should be presented to managers in a more conspicuous, easily digestible and prominent way for quicker and easier assimilation. The process of deduction that currently faces construction project managers is an arduous one which is unlikely to be undertaken within the time pressures of a construction project. Competitive bidding, which results in overly low margins, is also likely to discourage such efforts. Indeed, within the current economic climate and structure of construction projects, attention to safety issues demands a certain degree of self-sacrifice, since a manager's performance is likely to be measured by other criteria.

INCONSISTENT TEAM MEMBERSHIP

The consistency of membership within a project team has an impact on receptivity to warning. For example, where team membership is not consistent throughout a project, people tended to suppress problems in the hope that they would be dealt with by other people, later in the project. The result of this 'buck-passing' behaviour can be disproportionately damaging, particularly when the problem has a high

interdependence with other activities. The widespread consequences of such behaviour ensure that a simple problem will more
readily escalate into a serious accident.

UNSTABLE SENSITIVITY

Receptivity to warning may vary throughout the life of a construction project. For example, consider the dormant period between the
phases of design and construction. During this period of relative
inactivity, people's attention is temporarily focused elsewhere and
the team structure is in a state of transition. It is evident that project
managers must seek to identify these dangerous transition periods
and maintain the attention of the project team during them.

INEPTITUDE

In addition to the influences of organisational structure on accident
prevention, it is important to mention ineptitude as a relatively simple yet, due to its delicacy, easily forgotten factor in accident prevention. While reasonable professional standards are notoriously
difficult to measure, a number of instances have occurred where the
advance signs of a potential problem were plainly evident yet overlooked. The problem of ineptitude is worthy of mention because it
is likely to be a particular problem in an OHS context due to a historical lack of concern with OHS issues, a widespread acceptance of
low safety standards, and a lack of adequate training in this area.

Comparing

It is at the comparing stage that potential hazards are assimilated in
terms of their ability to cause an accident and that responsibilities for
mitigation are established. There are numerous instances of problems detected by monitors but covered up at the comparing stage
where they should have been recognised and communicated to decision-makers for action. The following influences have been found to
be instrumental in inducing this problem.

A LACK OF COLLECTIVE RESPONSIBILITY

One reason for communication breakdown is the disturbance to the
status quo that reporting would bring, a disturbance which, for some,
could involve personal liability. Every problem that arises on a construction project brings about a redistribution of resources which creates both winners and losers, winners encouraging the change and
losers resisting it. While there exist opposing forces to reveal and
suppress problems, potential losers are found to be more forceful
than potential winners. This contributes to a tendency to ignore

communicated problems. This inconsistent behaviour, which results in disjointed reporting chains and a lack of collective responsibility, is partly a function of unequal risk distribution patterns. To avoid it, managers must seek to develop a culture of collective responsibility by negotiating contracts that share, rather than separate, project risks.

OPPOSING PROJECT GOALS

A further factor that encourages the suppression of problems is the diametrical nature of some project goals. For example, to avoid delays it is often necessary to commit more resources, which means that cost targets are threatened. Or one can ask people to work faster, which has safety implications. The resolution of one problem may therefore create further problems of control elsewhere in an organisation. It has been found that if the person who is informed of a problem is responsible for an opposing goal then he will be less likely to deal with it. This can be a particular problem in OHS issues because dealing with a safety problem will probably cause delays and involve the commitment of resources. It is a conflict of interests that is exacerbated because time and cost goals are likely to be of higher priority than safety goals. This is a factor that will encourage people to subject safety warnings to stricter tests than information that supports policy aspirations. The managerial implication of these findings is that managers must seek to reduce inappropriate and unnecessary pressures on people and be more tolerant of justifiable changes that are inconsistent with their own expectations. A project manager must also be aware of the impact on other aspects of organisational performance when considering any one goal as an absolute priority.

DIVERSE INFORMATION

There are problems related to people's ability to assimilate the information for identifying potential problems. This is partly related to the need to assimilate information from diverse sources and different MISs but also to a lack of diligence on the part of comparators. Both problems are likely to be exacerbated in OHS issues because there are few people who understand well enough the impact of events on OHS performance in the construction industry. A lack of clear and measurable safety goals compounds this problem.

DISPUTES

During the comparing process, responsibility for problem resolution is decided. Disputes are therefore common. When handled poorly, disputes escalate into conflict, magnifying the consequences of minor problems by creating a stalemate that damages interpersonal relations. Because disputes become increasingly difficult to resolve as they escalate, it is crucial that project managers are sensitive to their advance signs and act on them early. It is evident that conflict management is an important aspect of safety management.

Effecting

All of the aforesaid problems of self-interest and ambiguity, identified in relation to monitoring and comparing, also adversely influence the effecting process. But there are also unique problems and these are discussed below.

CENTRALISATION OF DECISION-MAKING RESPONSIBILITY

One problem relates to the hierarchical distance of the decision-making authority from the point of problem detection. During a crisis, when reference to high authority is justified or in highly centralised organisations where the necessary decision-making authority exists at higher organisational levels, this is a particular problem. This is because it slows down the speed of response. Rapid response can be vital in situations where an accident chain is short. Therefore managers should aim to structure their organisation so that the roles of monitor, comparator and effector are as autonomous as possible. Autonomy minimises communication interfaces and is one way of achieving responsiveness.

LACK OF RESOLVE

A lack of resolve is sometimes reflected in people's tendency to lower performance standards by accepting damage to goals rather than attempting to mitigate it. For example, when there is a cost problem on site, managers might accept a reduction in safety standards to reduce costs rather than thinking about ways of achieving both. The matrix structure of construction project organisations is a likely contributory factor in producing this behaviour because it separates staff personal interests from those of the client. This means that they are often defending goals that are not necessarily in their interests, even though they may have had formal responsibility for them. This behaviour is more likely to occur in relation to OHS standards since they are less overt than those relating to costs, time and quality. This means that they can be lowered less conspicuously. To resolve this problem, project managers must make safety a high-profile priority and avoid matrix structures by pursuing procurement arrangements that minimise conflicting objectives. Partnering arrangements are a possible solution to this problem, as is a shortening of the supply chain by reducing the number of firms involved in the procurement process. While partnering seeks to build trust and converge interests, shortening the supply chain is a structural measure that reduces fragmentation and thereby the number of problematical interfaces within a construction project organisation.

Implementation

Poorly considered decisions inevitably lead to problems in implementation. But there are also instances of implementation problems that are independent of decision quality. These problems are largely related to the social, technical and monetary change that decision implementation involves. The high interdependence of construction project organisations ensures that even the most minor changes have widespread implications throughout the organisational structure. The efficiency with which the necessary change is brought about is compromised by resistance to it, by people failing to recognise their interdependencies and by poor communication. Project managers should therefore give attention to ensuring effective communication and to the recognition of interdependencies between project members if they are to ensure high levels of safety performance in the construction industry.

Feedback

Having implemented a decision, feedback is essential to ensure that a decision has had its intended effect. The problems encountered in this phase of the process relate to conflicting interests, poor communication, and difficulties in measuring project goals. The last is likely to be a particular problem in relation to OHS issues because of the immense problems in measuring OHS performance when compared to cost, time and quality performance.

CONCLUSION

This chapter has used a model of accident prevention as a vehicle to investigate the psychological mechanisms that underpin the poor OHS performance of the construction industry. This model has been useful in identifying a range of organisational conditions that appear to cause behaviour and that reduce the efficiency of accident prevention systems within construction projects. By identifying the causes of these conditions it has been possible to suggest managerial strategies to mitigate them. These are summarised in Table 16.1 and will allow construction project managers to achieve higher levels of safety performance.

TABLE 16.1 PROBLEMS AND SOLUTIONS IN THE ACCIDENT PREVENTION CYCLE

PROBLEM	M	C	E	I	F	SOLUTION
Resistance to change	*	*	*	*	*	Identify and protect potential losers.
Lack of concern for safety	*	*	*	*	*	Make safety a priority; clarify safety performance standards and make them realistic; appraise managers of safety record; train to enhance understanding of influences on performance; reduce time pressures.
Conflicts of interest; clouded responsibility structures	*	*	*	*	*	Avoid matrix structures; keep contractual arrangements simple; avoid abuses of competitive bidding.
Diverse and inconspicuous accident data	*	*	*			Create a specific MIS relating to health and safety to ensure that safety information is presented in a more conspicuous and easily digestible form.
Unstable team structure	*	*		*	*	Avoid unnecessary changes in team membership.
Unstable sensitivity	*	*				Be aware of slack periods and continually re-emphasise the project's importance.
Ineptitude	*	*	*	*	*	Raise awareness by providing safety training; employ those with a concern for safety; make safety a priority; reduce time and financial pressures.
Fear of blame	*	*	*	*		Create a supportive rather than a penal environment.
Opposing goals		*		*		Tolerate justifiable changes in goal priorities; do not make one goal an absolute priority.
Disputes		*	*			Be sensitive to tensions that pre-empt conflict; manage conflict positively.
Difficulty in measuring safety performance		*	*		*	Set clear and measurable safety performance standards.

PROBLEM	M	C	E	I	F	SOLUTION
Lack of responsiveness			*		*	Make monitors, comparators and effectors as autonomous as possible; decentralise decision-making authority.
Lack of resolve			*			Resist lowering performance standards; avoid matrix structures; make safety a priority; encourage partnering; shorten supply chains.
Poorly made decisions				*		Pay attention to all of those factors influencing decision-making in this table.
Poor information flow	*	*	*	*	*	Identify those with conflicting interests and focus on the information flow between them; emphasise the pattern of interdependencies between people.

NOTE M = Monitoring; C = Comparing; E = Effecting; I = Implementation; F = Feedback.

EXERCISES

1   What are the main causes of the construction industry's poor attitude towards safety?

2   How can health and safety performance be improved in the construction industry?

3   What are the main impediments to improving the construction industry's poor attitude towards safety?

APPENDIX

OCCUPATIONAL HEALTH AND SAFETY

Occupational Health and Safety Regulation 2001

Summary of consultative risk-management provisions

Hazard identification

Controlling risk starts with identifying the hazards in your workplace.

The OHS Regulation requires that the employer must identify workplace hazards (Reg: 200(1)).

Employees are likely to be aware of, or readily able to help identify, hazards in their place of work.

The employer should encourage employees to notify their supervisor of hazards in the workplace whenever they become aware of them and consult employees when seeking to identify hazards.

The OHS Regulation requires that employees must take reasonable steps to prevent risks to health and safety at work by notifying the employer or supervisor of any matter that may affect OHS (Reg: 307(1))

Risk assessment

After identifying workplace hazards the next step is to assess the risks to health and safety that they pose.

The employer must assess the risks such hazards pose to the health and safety of their employees and any other person potentially at risk (Reg: 201(1)).

A risk assessment is the process of determining the level of risk involved and the likelihood of injury, illness or death occurring. This means evaluating the probability and consequences of injury or illness from exposure to an identified hazard or hazards.

The OHS Act requires that the employer must consult their employees when risks to their health and safety arising from work are assessed or when the assessment of those risks is reviewed (Act: 15(a)).

Risk control

The next step is to implement appropriate measures to control risk.

The OHS Regulation requires that an employer must eliminate risks, or if this is not reasonably practical, must minimise the risk to the fullest extent possible (Reg: 105; 203(1) & (2)).

To ensure the health, safety and welfare of employees it is preferable to eliminate the risk wherever possible. Sometimes elimination of the risk may not be possible or reasonably practical. But remember it is the employer's duty of care to ensure the health, safety and welfare of their employees. The following list will help in determining measures for controlling risk.

Level 1 *Eliminate the risk (e.g. discontinue the activity or not use the plan).*

Level 2 *Minimise the risk by:*
- Substituting the system of work or plant (with something safer).
- Modifying the system of work or plant (to make it safer).
- Isolating the hazard (e.g. introduce a restricted work area).
- Introducing engineering controls (e.g. guardrails or scaffolding).

Level 3 *Other controls:*
- Using Personal Protective Equipment (e.g. eye protection).
- Adopting administrative controls such as hazard warning signs (e.g. 'persons working above' and specific training and work instructions (e.g. brittle roof).

Note that the control measures at Level 1 give the best results and should be adopted wherever practicable. The measures at the lower levels are less effective and require more frequent reviews of the hazards and systems of work. In some situations a combination of control measures may need to be used.

Because of their specific knowledge about the work processes, plant or substances with which they work, employees may often be able to identify very practical and effective risk control measures.

The OHS Act requires that the employer must consult employees when decisions are made about the measures to be taken to eliminate or control those risks (Act: 15(b)).

The employer should establish mechanisms and/or procedures to ensure all employees exposed to a risk are consulted about how the risk should be controlled.

Various WorkCover Codes of Practice provide practical guidance about how specific hazards may be controlled. Where applicable, such Codes of Practice may form a useful basis of consultation in determining appropriate control measures.

The employer should consider the views of employees for controlling risk as part of their risk assessment. Following consultation, the employer should implement appropriate control measures based on their risk assessment.

When controlling risks for certain hazards, an employer must comply with any specific risk controls set out in the OHS Regulation.

Review of risk assessment and risk control measures

The OHS Act requires that the employer consult their employees when introducing or altering the procedures for monitoring risks (including health surveillance procedures) (Act: 15c).

A review of risk control measures will be particularly relevant if the employer is considering changes to the workplace or to the way in which work is done.

The OHS Act requires that the employer must consult their employees when changes that may affect health and safety are proposed to the premises where persons work, to the systems or methods of work or to the plant or substances used for work (Act: 15(e)).

The employer should consider the OHS implications of all changes he may wish to make to the workplace or the way in which work is done. Where there may be OHS implications the employer must consult with his employees.

This includes consultation with employees before the purchase of new plant and substances to be used for work. It makes sense, for example, to consult employees in assessing the risks posed by any new piece of plant that the employer may wish to purchase. In this way, the employer might identify a piece of plant for purchase with the appropriate guarding for doing the work safely already affixed, rather than having to modify the plant after its installation in the workplace.

The employer should also consult his employees after the occurrence of an injury, illness or incident to review why the control measures may have failed and to identify any necessary changes to the measures in place for controlling risk. Possible questions might include:

- What factor(s) contributed to the injury, illness or incident?
- Could the injury, illness or incident have been prevented?
- Should the employer's safety procedures and systems have prevented the injury, illness or incident?
- If no, what needs to be changed?
- If yes, why didn't the OHS system prevent the injury, illness or incident?
- What can be done or has been done to prevent the injury, illness or incident from recurring, and when?

Measuring workplace safety performance

Ongoing monitoring of risk control measures and measuring the effectiveness of OHS programs contributes to safer workplaces and facilitates continuous improvement in OHS outcomes.

The employer should consider ways of involving employees in the development and application of OHS performance indicators as a means of raising OHS awareness, giving employer and employees alike a clear sense of common purpose and a focus on continuous improvement of OHS outcomes.

Training and information

The OHS Act provides that the employer must provide appropriate supervision, instruction and training to ensure the health, safety and welfare of employees (Act. 8(1)(d)).

The employer should consult his employees about the information and training necessary to undertake their work safely.

The employer should consult with his employees to ensure that such information and training is in a form that is accessible and easily understood by such employees.

In identifying training needs, it is important to consider a range of factors including the nature of the task, the age and experience of the worker, and the plant and substances used.

Training procedures should be documented, produced and well displayed. Employers and employees can work together in the development of such procedures and these should be jointly reviewed and monitored.

The employer should review training needs and consult employees about training arising from changes proposed to the systems or methods of work, or to the plant or substances used for work.

REFERENCES

Abrahamson, MW (1973) Contractual risks in tunnelling: how they should be shared, *Tunnels and Tunnelling*, November: 587–98.

ABS (1999) *1999 Year Book Australia*, Australian Bureau of Statistics, Canberra.

Ackoff, RL (1969) *Systems, Organisation and Interdisciplinary Research in Systems Thinking* (ed. FE Emery), Penguin Books, London.

Adams, JS (1963) Towards an understanding of inequity, *Journal of Abnormal and Social Psychology*, 67: 422–36.

AFR (2000) Lifting the lid on immigration, *Australian Financial Review*, March: 20–1.

AGC (1995) *Partnering: Changing Attitudes in Construction*, Associated General Contractors of America, Washington DC.

Ajzen, I and Fishbein, M (1980) *Understanding Attitudes and Predicting Behaviour*, Prentice Hall, Upper Saddle River NJ.

Albers, HH (1969) *Principles of Management*, 3rd edn, John Wiley & Sons, New York.

Allen, JG (1976) *Migrants in Industry*, Health Commission of NSW, Division of Occupational Health and Safety, Sydney.

Altmeyer, R (1988) *The Enemies of Freedom: Understanding Right-wing Authoritarianism*, Jossey-Bass, San-Francisco CA.

Anon (1992) Margerison and McCann discuss the Team Management Wheel, *Industrial and Commercial Training*, 24(1): 289–32.

Ansoff, I (1984) *Implanting Strategic Management*, Prentice Hall International, Englewood Cliffs NJ.

Antony, RN (1988) *The Management Control Function*, Harvard Business School, Cambridge MA.

Archer, R (1996) *On Teams*, Irwin, London.

Argyris, C (1962) *Inter-personal Competence and Organisational Effectiveness*, Homewood Publishing, IL.

Argyris, C (1970) *Intervention Theory and Method*, Addison-Wesley, Reading MA.

Argyris, C (1990) *Overcoming Organisational Defences*, Allwyn & Bacon, London.

AS (1986) *Australian Standard 2124-1986: General Conditions of Contract*, Standards Australia, Sydney.

AS/NZS (1999) *Australian/New Zealand Standard on Risk Management AS/NZS 4360:1995*, Standards Australia and Standards New Zealand.

Ashley, DB (1989) Project risk identification using inference subjective expert assessment and historical data, *Proceedings of the Internet International Expert Seminar in Connection with the PMI/Internet Joint Symposium*, Atlanta, 12–13 October: 9–28.

Atkinson, J (1984) *Manpower Strategies for Flexible Organisations*, Institute of Personnel Management, London.

Augoustinos, M and Sale, L (1997) Constructions of racism in student talk, unpublished manuscript, University of Adelaide.

Azar, B (1997) Prejudice is a habit that can be broken, In MH Davis (ed.) *Social Psychology Annual Editions, 97/98*: 148–9.

Bagilhole, BM, Dainty, ARJ and Neale, RH (1995) Innovative personnel practices for improving women's careers in construction companies: methodology and discussion of preliminary findings, In *Proceedings of 11th Annual ARCOM Conference*, University of York: 686–95.

Baker, RW (1986) Handling uncertainty, *International Journal of Project Management*, 4(4): 205–10.

Barbara, S (1997) Team roles and team performance: is there really a link? *Journal of Occupational and Organisational Psychology*, 70(3): 241–58.

Barnes, M (1989) The role of contracts in management, In *Construction Contract Policy: Improved Procedures and Practice*, Centre for Construction Law and management, Kings College, London: 119–38.

Barnes, M (1991) Risk sharing in contracts, In *Civil Engineering project procedure in the EC*, Proceedings of the Conference Organised by the Institution of Civil Engineers, Heathrow, London, January.

Barrett, P and Stanley, C (1999) *Better Construction Briefing*, Blackwell Science, Oxford.

Barrick, MR and Mount, MK (1993) Autonomy as a moderator of the relationship between the big five personality dimensions and job performance, *Journal of Applied Psychology*, February: 111–18.

Beardwell, I and Holden, L (eds) (1994) *Human Resource Management: A Contemporary Perspective*, Pitman Publishing, London.

Becker, F (1990) *The Total Workplace*, Van Nostrand Reinhold, New York.

Belbin, RM (1984) *Management Teams: Why They Succeed or Fail*, Heinemann, London.

Belbin, RM (1993) *Team Roles at Work*, Butterworth-Heinemann, London.

Benne, KD and Sheats, P (1948) Functional roles of group members, *Journal of Social Issues*, 4(1): 41–9.

Bennett, J and Flanagan, R (1983) For the good of the client, *Building*, 1, April: 26–7.

Bennett, J (1991) *International Construction Project Management: General Theory and Practice*, Butterworth-Heinemann, Oxford.

Bennett, L, Grey, C and Hughes, W (1994) *The Successful Management of Design*, Centre for Strategic Studies in Construction, University of Reading.

Bergman, D (1990) A killing in the boardroom, *New Statesman and Society*, 1, June: 15–16.

Bertalanffy, L (1972) *General Systems Theory: Foundations, Development, Applications*, G. Braziller, New York.

Betts, M (1991) Achieving and measuring flexibility in project information retrieval, *Construction Management and Economics*, 9(3): 231–45.

Betts, M and Lansley, P (1993) Construction Management and Economics: A review of the first ten years, *Construction Management and Economics*, 11(2): 221–45.

Bishop, D (1993) The Professionals' view, paper given to the joint Health and Safety Construction Industry Council Conference, February, HSE/CIC.

Blake, R and Mouton, JF (1964) *The Management Grid*, Gulf Publishing Co, Houston TX.

Brandon, P (1978) A framework for cost exploration and strategic cost planning in design, *Chartered Surveyor, Building and QS Quarterly*, 5(4): 60–5.

Braveman, H (1974) *Labour and Monopoly Capital. The Degradation of Work in the Twentieth Century*, Monthly Review Press, New York.

Bresnen, M (1990) *Organising Construction: Project Organisation and Matrix Management*, Routledge, London.

Bresnen, MJ, Bryman, AE, Ford, JR, Beardsworth, AD and Keil, ET (1986) Leadership orientation of construction site managers, *Journal of Construction Engineering and Management*, ASCE, 112(3): 370–86.

Brislin, R (1993) *Understanding Culture's Influence on Behaviour*, Harcourt Brace, Jovanovich College Publishers, Fort Worth TX.

Bromilow, FJ (1969) Contract time performance: expectations and the reality, *Building Forum*, 1(3): 70–80.

Bromilow, FJ (1970) The nature and extent of variations to building contracts, *Building Economics*, 9(3): 93–104.

Bromilow, FJ (1971) Building contract cost performance, *Building Economics*, 9(4): 126–38.

Brooke, PP, Russell, DW and Price, JL (1988) Discriminant validation of measures of job satisfaction, job involvement and organisational commitment, *Journal of Applied Psychology*, May: 139–45.

Brooker, P and Lavers, A (1995) Perceptions of the role of alternative dispute resolution in the settlement of construction disputes: lessons for the UK from the US experience, In *Proceedings of TG 15 Conference on Construction Conflict: Management and Resolution*, CIB Publication 171, October, Lexington KY, 49–69.

Brotherton, R (1993) Risk Analysis of Construction Cost Estimates, B.Build. thesis, School of Building, UNSW, Sydney.

Bryson, JM and Alston, FK (1996) *Creating and Implementing your Strategic Plan*, Jossey-Bass, San Francisco CA.

BS 1523 (1967) *Glossary of Terms used in Automatic Controlling and Regulating Systems, Part 1: Process and Kinetic Control*, British Standards Institution, London.

Budworth, N (1996) Indicators of performance in safety management, *Health and Safety Practitioner*, 15(11): 23–29.

393

Bunn, DW (1984) *Applied Decision Analysis*, McGraw-Hill, New York.

Burke, R (1999) *Project Management: Planning and Control Techniques*, 3rd edn, John Wiley & Sons, Chichester.

Burns, T and Stalker, G (1961) *The Management of Innovation*, Tavistock, London.

Carnall, CA (1982) *The Evaluation of Organisational Change*, Gower, Aldershot.

Cavill, N (1999) Purging the industry of racism, *Building Magazine*, May: 20–2.

CBI (1990) *Developing a Safety Culture: Business for Safety*, Confederation of British Industry, London.

CBI (1997) *Managing Absence — In Sickness and in Health*, Confederation of British Industry, London.

CDM (1994) Construction design and management regulations, HMSO, London.

Chapman, CB (1979) Large engineering project risk analysis, *IEEE Transactions of Engineering Management*, EM-26(3): 78–85.

Chau, KW (1993) On the assumption of triangular distribution in risk analysis of construction cost, *Proceedings of the 1993 W55 CIB International Symposium on Building Economics*, Lisbon, 7: 33–71.

Chau, KW (1995) Monte Carlo simulation of construction costs using subjective data, *Construction Management and Economics*, 13(5): 369–83.

Cherns, AB and Bryant, DT (1984) Studying the client's role in construction management, *Construction Management and Economics*, 2: 177–84.

Chitkara, KK (1998) *Construction Project Management: Planning, Scheduling and Controlling*, Tata McGraw-Hill Publishing, New Delhi.

CIDA (1993) *Best Practice Project Management Guide*, Construction Industry Development Agency, Sydney.

CIDB (1998) *Construction Economics Report: Third Quarter 1998*, Construction Industry Development Board, Singapore.

CII (1989) *Partnering: Meeting the Challenges of the Future*, Construction Industry Institute, Austin TX.

CII (1996) *Partnering: Models for Success*, Research Report No 8, Construction Industry Institute, Australia.

CII Australia (1997) *Improving Project Performance through People*, Construction Industry Institute, Australia.

CIOB (1988) *Project Management in Building*, Chartered Institute of Building, Berkshire, UK.

CIOB (2002) *Code of Practice for Project Management for Construction and Development*, 3rd edn, Chartered Institute of Building, Blackwell Publishing, Oxford.

Cleland, DI (1995) Leadership and the project management body of knowledge, *International Journal of Project Management*, 13(2): 83–8.

Cleland, DI and King, WR (1983) *Systems Analysis and Project Management*, 3rd edn, McGraw-Hill, New York.

Collier, K (1969) *Construction Contracts*, Reston, London.

Connaughton, JN and Green, SD (1996) *Value Management in Construction: A Client's Guide*, Construction Industry Research and Information Association, London.

Cooper, D and Chapman, C (1987) *Risk Analysis for Large Projects*, John Wiley & Sons, London.

Covey, SR (1989) *The Seven Habits of Highly Effective People*, The Business Library, Melbourne.

Covey, SR (1994) *Daily Reflections for Highly Effective People*, Fireside, Simon & Schuster, New York.

Crawshaw, DT (1976) *Coordinating Working Drawings*, BRE Current Paper, 60/76, Department of the Environment, London.

Crichton, C (1996) *Interdependence and Uncertainty: A Study of the Building Industry*, Tavistock, London.

Crosby, P (1979) *Quality is Free: The Art of Making Quality Certain*, McGraw-Hill, New York.

Crozier, M (1964) *The Bureaucratic Phenomenon*, University of Chicago Press.

CSSC (1988) *Building Britain 2001*, University of Reading.

Cullen, A (1997) The conditions of tender: a separate contract, *ACLN*, 56: 54–7.

Daft, RL (1983) *Organisational Theory and Design*, West Publishing Co., St Paul MN.

Davenport, P (2000) *Adjudication in the NSW Construction Industry*, Federation Press, Sydney.

Essentials of Construction Project Management

Dawson, P and Palmer, G (1995) *Quality Management: The Theory and Practice of Implementing Change*, Longman, Melbourne.

Deacon, A (1994) The role of safety in total quality management, *Safety and Health Practitioner*, 12(1): 18–21.

De Bono, E (1991) *Conflicts: A Better Way to Resolve Them*, Penguin Books, London.

Deresky, H (1997) *International Management: Managing across Borders and Cultures*, Addison-Wesley Educational Publishers Inc, USA.

Devine, PG (1989) Stereotypes and prejudice: their automatic and controlled components, *Journal of Personality and Social Psychology*, 56(1): 5–18.

Diekmann, M (1983) Probabilistic estimating: mathematics and applications, *Journal of Construction Engineering and Management*, ASCE, 109(3): 297–308.

DIMA (1998) *Major Source Countries (July 1997 to June 1998 settler arrivals, by country of birth)*, Department of Immigration and Multicultural Affairs, Canberra.

Donnelly, R and Kezsbom, DS (1994) Overcoming the responsibility–authority gap: an investigation of effective project team leadership for a new decade, *Cost Engineering*, 36(5): 33–41.

Duff, AR, Robertson, IT, Cooper, MD, Phillips, RA (1993) *Improving Safety on Construction Sites by Changing Personnel Behaviour*, HSE Contract Research Report No 51/1993, HMSO, London.

Dunphy, DC (1981) *Organisational Change by Choice*, McGraw-Hill, Sydney.

Eagly, AH and Chaiken, S (1993) *The Psychology of Attitudes*, Harcourt Brace, Sydney.

Erikson, CA (1979) Risk Sharing in Construction Contracts, PhD thesis, University of Illinois.

Everly, M (1997) Demolishing a paper tiger, *Health and Safety at Work*, 19(2): 20–2.

Eves, D (1989) Builders condemned for indifference to safety, *Guardian*, 4 October, London.

Falconer, L and Heol, H (1996) The perceptions of line and senior managers in relation to occupational health issues, *Occupational Medicine*, 46: 151–6.

Fehlig, C (1995) Project partnering/team building in the public sector, *Proceedings of the Team Building/Partnering: Public Sector Conference*, CII, Texas A&M University, 10–15.

Fenn, P and Speck, C (1995) The occurrence of disputes on United Kingdom construction projects, In *Proceedings of ARCOM 11th Annual Conference*, University of York, September, 581–91.

Ferry, DJ and Brandon, PS (1991) *Cost Planning of Buildings*, 6th edn, BSP Publications, Oxford.

Fiedler, F.E. (1967) *A Theory of Leadership Effectiveness*, McGraw-Hill, New York.

Fisher, G (1988) *Mind-sets: The Role of Perception and Culture in International Relations*, Intercultural Press, Yarmouth ME.

Flanagan, R (1980) Tender Price and Time Prediction for Construction Work, PhD thesis, Aston University, UK.

Flanagan, R (1986) *Patterns of Competitive Tendering*, Section 7, Department of Construction Management, University of Reading.

Flanagan, R and Norman, G (1993) *Risk Management and Construction*, Blackwell Scientific Publications, Oxford.

Flanagan, R, Kendell, A, Norman, G and Robinson, GD (1987) Life cycle costing and risk management, *Construction Management and Economics*, 5: S53–S71.

Flanagan, R, Norman, G, Meadows, J and Robinson, G (1989) *Life Cycle Costing: Theory and Practice*, BSP Professional Books, Oxford.

Franke, A (1987) Risk analysis in project management, *International Journal of Project Management*, 5(1): 29–34.

Fry, LW and Smith, DA (1987) Congruence, contingency and theory building, *Academy of Management Review*, January: 117–32.

Fryer, B (1990) *The Practice of Construction Management*, 2nd edn, BSP Professional Books, Oxford.

Fulmer, RM (1983) *The New Management*, 3rd edn, Macmillan, New York.

Galbraith, JR (1973) *Designing Complex Organisations*, Addison-Wesley, New York.

Galbraith, JR (1977) *Organizational Design*, Addison-Wesley, Reading MA.

Gale, AW (1992) The construction industry's male culture must feminise if conflict is to be reduced: the role of education as a gate-keeper to a male construction industry, In P Fenn and R Gameson (eds) *Construction Conflict Management and Resolution*, E & FN Spon, London, 416–27.

Gardiner PD and Simmons JEL (1992) Analysis of conflict and change in construction projects, *Construction Management and Economics*, 10: 459–78.

Gardiner, PD and Simmons, JEL (1995) Case explorations in construction conflict management, *Construction Management and Economics*, 13: 219–34.

George, AL (1991) Strategies for crisis management, In George (ed.) *Avoiding War*, 216–29.

George AL (ed.) (1991) *Avoiding War: Problems of Crisis Management*, Westview, San Francisco CA.

Gibb, JR (1984) Defensive communication, In Kolb, DA, Rubin, IM and McIntyre, M (eds) *Organisational Psychology: Readings on Human Behaviour in Organisations*, Prentice Hall, Englewood Cliffs NJ, 279–84.

Gilbert, GP (1983) Styles of project management, *International Journal of Project Management*, 1(4): 189–93.

Gilmour, P and Hunt RA (1995) *Total Quality Management: Integrating Quality into Design, Operations and Strategy*, Longman, Melbourne.

Glackin, M (1999) Some kind of refuge, *Building Magazine*, April: 23–7.

Gray, C (1996) *Value for Money: Helping the UK Afford the Buildings it Likes*, Reading Construction Forum, University of Reading.

Gray, C and Flanagan, R (1989) *The Changing Role of Specialist and Trade Contractors*, Chartered Institute of Building, Ascot, UK.

Gray, C, Hughes, WP and Bennett, J (1994) *The Successful Management of Design*, University of Reading.

Gyi, D, Gibb, A and Haslam, R (1996) A study to investigate the cause of accidents in the construction industry, *Proceedings of the 11th ARCOM Annual Conference*, 1: 11–16.

Gyles, R (1992) *Final Report of the Royal Commission into Productivity in the Building Industry in New South Wales*, Commonwealth Government, Canberra.

Hackman, JR and Morris, CG (1975) Group tasks, group interaction process and group performance effectiveness: A review and proposed integration, In L Berkowitz (ed.) *Advances in Experimental Social Psychology*, Academic Press, New York, 45–99.

Hackman, JR and Oldham, GR (1976) Motivation through the design of work: test of a theory, *Organisational Behaviour and Human Performance*, August: 250–79.

Hall, J (1971) Toward group effectiveness, *Teleometrics International*, Conroe TX.

Hall, JN (1986) Use of risk analysis in North Sea projects, *International Journal of Project Management*, 4(4): 217–22.

Hamilton, A (1997) *Management by Projects: Achieving Success in a Changing World*, Thomas Telford, London.

Handler, AB (1970) *Systems Approach to Architecture*, Elsevier, Amsterdam.

Handy, C (1994) *The Future of Work: A Guide to a Changing Society*, Blackwell, London.

Hannaford, I (1997) *Race: The History of an Idea in the West*, Johns Hopkins University Press, Baltimore MD.

Hatush, Z and Skitmore, M (1997a) Criteria for contractor selection, *Construction Management and Economics*, 15(1): 19–38.

Hatush, Z and Skitmore, M (1997b) Evaluating contractor prequalification data: selection criteria and project success factors, *Construction Management and Economics*, 15(3): 129–47.

Heinrich, HW, Petersen PE and Roos, N (1980) *Industrial Accident Prevention: A Safety Management Approach*, 5th edn, McGraw-Hill, New York.

Hersey, P and Blanchard, KH (2001) *Management of Organisational Behaviour: Leading Human Resources*, 8th edn, Prentice Hall, Englewood Cliffs NJ

Hertz, DB and Thomas, H (1983). Decision and risk analysis in a new product and facilities planning problem, *Sloan Management Review*, Winter: 17–31.

Hertzberg, F (1974) The wise old turk, *Harvard Business Review*, September–October: 70–80.

Hewitt, RA (1985) The Procurement of Buildings: Proposals to Improve the Performance of the Industry, unpublished report to the College of Estate Management for the RICS Diploma in Project Management, College of Estate Management, Reading, UK.

Hill, RC and Bowen, PA (1997) Sustainable construction: principles and a framework, *Construction Management and Economics*, 15(3): 223–39.

Hinze, J and Raboud, P (1988) Safety on large buildings, *Journal of Construction Engineering and Management*, 11(4): 286–93.

Hirsh, SK and Kummerow, JM (1990) *Introduction to Type in Organizations*, 2nd edn, Australian Psychologists Press, Melbourne.

Hoecklin, L (1994) *Managing Cultural Differences: Strategies for Competitive Advantage*, Addison-Wesley, London.

Essentials of Construction Project Management

Hofstede, G (1980) *Cultural Consequences: International Differences in Work-related Values*, Sage, Beverly Hills CA.

Hofstede, G (1981) Management control of public and not for profit organisations, *Accounting, Organisations and Society*, 6(3): 193–211.

Hofstede, G (1983) The cultural relativity of organisational practices and theories, *Journal of International Business Studies*, Fall: 82–9.

Hofstede, G (1992) *Cultures and Organisations: Software of the Mind*, McGraw-Hill, London.

Hogh, MS (1987) Cost effective safety, *Quarterly Bulletin on Health and Safety and the Environment*, 13(3): 1–3.

Hollingsworth, D, McConnochie, K and Pettman, J (1988) *Race and Racism in Australia*, Social Sciences Press, Wentworth Falls, NSW.

Homans, G (1950) *The Human Group*, Harcourt Brace Jovanovich, New York.

Horgan, MOC (1984) *Competitive Tendering for Engineering Contracts*, E & FN Spon, London.

Hornstein, HA (1986) *Managerial Courage*, John Wiley & Sons, New York.

Hoxie, R (1915) *Scientific Management and Labour*, Appleton, New York.

HSE (1993) *The Costs of Accidents at Work*, Health and Safety Executive, London.

HSE (1994) *Review of Health and Safety Regulations, Main Report,* Health and Safety Executive, London.

HSE (1996) *A guide to the reporting of injuries, diseases and dangerous occurrences regulations 1995*, Health and Safety Executive Books, Suffolk.

Hughes, WP (1994) Construction Management and Economics: abstracts and indexes 1983–94, *Construction Management and Economics*, University of Reading.

Hunter, JA, Stringer, M and Watson, RP (1991) Intergroup violence and intergroup attribution, *British Journal of Social Psychology*, 30(1): 261–6.

Hunter, W (1997) Safety needs to contribute to productivity, *Safety Management*, 13(4): 6.

IOD (1996) *Health and Safety, A Discussion Paper*, Institute of Directors, London.

Ireland, V (1985) The role of managerial actions in the cost, time and quality performance of high-rise commercial building projects, *Construction Management and Economics*, 3(3): 59–87.

Jaafari, A (1990) Probabilistic unit cost estimation for project configuration optimisation, *Transactions of the 1990 National Engineering Project Management Conference and Forum*, Sydney: 221–46.

Janis, IL (1971) Group think, *Psychology Today*, November: 43–6.

Jarman A and Kouzmin A (1990) Decision pathways from crisis: a contingency theory simulation heuristic for the Challenger space disaster (1983–1988), In A Block (ed.) *Contemporary Crisis: Law, Crime and Social Policy*, Kluwer Academic Press, Netherlands, 399–433.

Jarvis, R (1997) HSE in the spotlight, *Safety Management*, 13(3): 4–7.

Jaselskis, EJ, Anderson, SD and Russell, JS (1996) Strategies for achieving excellence in construction safety performance, *Journal of Construction Engineering and Management*, 122(1): 61–70.

John, OP (1990) The big five factor taxonomy: dimensions of personality in the natural language and in questionnaires, In LA Pervin (ed.) *Handbook of Personality Theory and Research*, Guildford, Press, New York, 66–100.

Johnson, DW and Johnson, FP (1994) *Joining Together: Group Theory and Group Skills*, 5th edn, Allyn & Bacon, Boston MA.

Jones, JC (1970) *Design Methods: Seeds of Human Futures*, John Wiley & Sons, New York.

Kahn, R and Katz, D (1960) Leadership practices in relation to productivity and moral, In Cartwright, D and Zander, A (eds) *Group Dynamics: Research and Theory*, 2nd edn, Row Paterson, Elmsford NY, 75–6.

Kallman, EA and Grillo, IP (1996) *Ethical Decision Making and Information Technology*, 2nd edn, McGraw-Hill, Singapore.

Kangari, R and Farid, F (1987) Construction risk management, *Proceedings CIB Symposium, Managing Construction Worldwide*, London, 138–47.

Katz, D and Kahn, RL (1978) *The Social Psychology of Organizations*, 2nd edn, John Wiley & Sons, New York.

Katzenbach, JR and Smith, DK (1993) The discipline of teams, *Harvard Business Review*, March–April: 111–20.

Kaufmann, A (1975) *Introduction to the Theory of Fuzzy Subsets*, Academic Press, New York.

Kavanagh, TC, Muller, F and O'Brien, JJ (1978) *Construction Management: A Professional Approach*, McGraw-Hill, New York.

Kelly, J, MacPherson, S, Male, S (1992) *The Briefing Process: A Review and Critique*, Paper No 12, Royal Institution of Chartered Surveyors, London.

Kerzner, H (1989) *Project Management: A Systems Approach to Planning, Scheduling and Controlling*, 3rd edn, Van Nostrand Reinhold, New York.

Kezsbom, DS, Schilling DL and Edward, KA (1989) *Dynamic Project Management: A Practical Guide for Managers and Engineers*, John Wiley & Sons, New York.

Kirkpatrick, SA and Locke, EA (1991) Leadership: do traits really matter? *Academy of Management Executive*, May: 48–60.

Kisner, S and Fosbroke, D (1994) Injury hazards in the construction industry, *Journal of Occupational Medicine*, 36(2): 137–43.

Kometa, ST, Olomolaiye, PO and Harris, FC (1994) Attributes of UK clients influencing project consultant's performance, *Construction Management and Economics*, 12(5): 433–43.

Kroeber, AL and Kluckholm, C (1954) *Culture: A Critical Review of Concepts and Definitions*, Random House, New York.

Latham, M (1994) *Constructing the Team*, Final Report of the Government/Industry Review of Procurement and Contractual Arrangements in the UK Construction Industry, HMSO, London.

Lawler, EE III (1973) *Motivation in Work Organizations*, Brooks/Cole, Monterey CA, 30–6.

Lawrence, PR and Lorsch, JW (1967) *Organisation and Environment: Managing Differentiation and Integration*, Harvard University Press, Cambridge MA.

Lawrence, PR and Seiler, JA (1965) *Organisational Behaviour and Administration*, Richard D Irwin Inc. and Dorsey Press, New York.

Lawson, B (1990) *How Designers Think*, 2nd edn, Butterworth Architecture, Oxford.

Leavitt, HJ (1951) Some effects of certain communication patterns on group performance, *Journal of Abnormal Social Psychology*, 46: 38–50.

Lendrum, T (1995) *The Strategic Partnering Handbook: A Practical Guide for Managers*, McGraw-Hill, Sydney.

Levido, GE (1990) Project management, unpublished lecture notes, School of Building, UNSW, Sydney.

Levido, GE, Green, JD, Bromilow, FJ and Toakley, AR (1981) *Contracting Performance with Various Non-Traditional Forms of Contract*, Research Report for the Project Managers Forum, School of Building, UNSW and the Division of Building Research, CSIRO, Sydney.

Lewin, K and Lippitt, R (1939) An experimental approach to the study of autocracy and democracy: a preliminary note, *Sociometry*, 1: 292–300.

Likert, R (1984) The nature of highly effective groups, In Kolb, DA, Rubin, IM and McIntyre, JM (eds) *Readings in Organisational Psychology: Readings on Human Behaviour in Organisations*, 4th edn, Prentice Hall, Englewood Cliffs NJ, 160–82.

Likert, R and Likert JG (1976) *New Ways of Managing Conflicts*, McGraw-Hill, New York.

Lingard, H and Rowlinson, S (1997) Construction site safety in Hong Kong, *Construction Management and Economics*, 12(6): 501–10.

Loosemore, M (1996) Crisis Management in Building Projects: A Longitudinal Investigation of Communication and Behaviour Patterns within a Grounded Theory Framework, PhD thesis, University of Reading.

Loosemore, M (1999a) A grounded theory of construction crisis management, *Construction Management and Economics*, 17(1): 9–19.

Loosemore, M (1999b) Responsibility, power and construction conflict, *Construction Management and Economics*, 17(3): 699–709.

Loosemore, M (2000) *Crisis Management in Construction Projects*, American Society of Civil Engineers Press, New York.

Loosemore, M and Chau, DW (2001) Racial discrimination towards Asian operatives in the Australian construction industry, *Construction Management and Economics*, in press.

Loosemore, M and Djebarni, R (1994) Tension, problems and conflict behaviour, In

S Rowlinson (ed.) *East Meets West, Proceedings of CIB W92 Procurement Systems Symposium*, University of Hong Kong, 187–95.

Loosemore, M and Lee, P (2001) An investigation of communication problems with ethnic minorities in the construction industry, *International Journal of Project Management*, in press.

Loosemore, M and Tan, CC (1999) Occupational stereotypes in the construction industry, *Construction Management and Economics*, 18(5): 51–6.

Loosemore, M, Nguyen, BT and Dennis, N (2000) An investigation into the merits of encouraging conflict in the construction industry, *Construction Management and Economics*, 18(4): 447–57.

Luck, RAC and Newcombe, R (1996) Integration of the project participants' activities within a construction project environment, In DA Langford and A Retik (eds) *The Organisation and Management of Construction: Shaping Theory and Practice*, vol. 2, E & FN Spon, London, 458–71.

Mahoney, TA, Jerdee, TH and Nash, AN (1960) Predicting managerial effectiveness, *Personnel Psychology*, Summer: 147–63.

Malcomb, DG, Roseboom, JH, Clark, CE and Fazar, W (1959) Applications of a technique for research and development program evaluation (PERT), *Operations Research*, 7: 646–9.

Margerison, CJ, and McCann, DJ (1990) *Team Management Index*, Team Management Resources, Brisbane.

Margerison, CJ, McCann, DJ and Davies, R (1986) Human resource management for TAA pilots and flight engineers, *Human Resource Management Australia*, May: 32–7.

Marsh, P, Roser, E and Harre, R (1978) *The Rules of Disorder*, Routledge, London.

Martin, AB, Linehan, A and Whitehouse, I (1996) *The Regulation of Health and Safety in Five European Countries*, HSE Contract Research Report No 84, Health and Safety Executive, London.

Maslow, A (1954) *Motivation and Personality*, McGraw-Hill, New York.

Mason, GE (1973) *A Quantitative Risk Management Approach to the Selection of Construction Contract Provisions*, Technical Report No 173, Construction Institute, Department of Civil Engineering, Stanford University.

Masterman, JC (1994) *A Study of the Basis upon which Clients of the Construction Industry Choose their Procurement Systems*, PhD thesis, UMIST, UK.

Mattila, M, Rantanen, E and Hitting, M (1994) The quality of work environment, supervision and safety in building construction, *Safety Science*, 17: 257–68.

Mayo, E (1933) *The Human Problems of an Industrial Civilization*, Macmillan, New York.

McAndrew, B (1993) *Changing Organisations*, Longman, London.

McClelland, DC (1961) *The Achieving Society*, Van Nostrand Reinhold, New York.

McGeorge, D and Palmer, A (1997) *Construction Management: New Directions*, Blackwell Science, Oxford.

McGregor, D (1960) *The Human Side of Enterprise*, McGraw-Hill, New York.

Meredith, JR and Mantel, SJ Jr (1989) *Project Management: A Managerial Approach*, 2nd edn, John Wiley & Sons, New York.

Migliorino, P, Miltenyi, G and Robertson, H (1994) *Best Practice in Managing a Culturally Diverse Workplace: A Manager's Manual*, Office of Multicultural Affairs, Canberra.

Miller, EJ and Rice, AK (1967) *Systems of Organisation*, Tavistock, London.

Miller, J (1971) Living systems: the group, *Behavioural Science*, 16: 302–98.

Mintzberg, H (1973) *The Nature of Managerial Work*, Harper & Row, New York.

Mintzberg, H (1976) Planning on the left side and managing on the right side, *Harvard Business Review*, 54(2): 49–58.

Mintzberg, H (1979) *The Structuring of Organisations*, Prentice-Hall, Englewood Cliffs NJ.

MLSA (1998) *The Annual Statistics Report of Labourers*, Ministry of Labour and Social Affairs, Riyadh.

Moore, PG and Thomas H (1984) *The Anatomy of Decisions*, Penguin Books, London.

Moore, R (1991) *The Price of Safety: The Market, Workers' Rights and the Law*, Institute of Employment Rights, London.

Morris, P and Hough, GH (1987) *The Anatomy of Major Projects: A Study of the Reality of Project Management*, Major Projects Association, John Wiley & Sons, Chichester.

Morris, PWG (1972) *A Study of Selected Building Project in the Context of Theories of Organization*, PhD thesis, Institute of Science and Technology, University of Manchester, UK.

Morris, PWG (1994) *The Management of Projects*, Thomas Telford, London.

Mullen, B and Copper, C (1994) The relation between group cohesiveness and performance, an integration, *Psychological Bulletin*, March: 210–27.

Murdoch, J and Hughes, W (1992) *Construction Contracts: Law and Management*, E & FN Spon, London.

Murray, HA (1938) *Explorations in Personality*, Oxford University Press, New York.

Nabarrow and Nathanson (1996) *The Prime Duty? Health and Safety in the UK*, Nabarrow & Nathanson, London.

Naoum, SG (1991) *Procurement and Project Performance*, Occasional Paper No 45, CIOB publications, Ascot, UK.

Napier, JA (1970) *A Systems Approach to the Swedish Building Industry*, National Swedish Institute for Building, Stockholm.

Napoli, J (1998) *Understanding Equal Employment Opportunity: A Guide for the Workplace*, Prentice Hall, Sydney.

NEDO (1983) *Faster Building for Industry*, Network Economic Development, HMSO, London.

NEDO (1988) *Faster Building for Commerce*, Network Economic Development, HMSO, London.

Neowhouse, MM (1993) The use of software based qualitative risk assessment methodologies in industry, In RE Melchers and MG Stewart (eds) *Proceedings of the Integrated Risk Assessment Conference*, Balkema, Rotterdam: 147–55.

Nesdale, AR (1997) *Ethnicity and Prejudice: Is There a Way out of the Labyrinth?* School of Applied Psychology, Gold Coast University, Qld.

Newcombe, B (1994) Procurement paths: a power paradigm, In S Rowlinson (ed.) *East Meets West: Proceedings of CIB W92 Procurement Systems Symposium*, University of Hong Kong, 243–51.

Niskanen, T and Lauttalammi, J (1989) Accident prevention in materials handling at building construction sites, *Construction Management and Economics*, 7: 263–79.

NSW DPWS (1993) *Capital Project Procurement Manual*, NSW Government, Sydney.

NSW DPWS (1996) *Report on the Project Performance Outcomes of Partnered and Non-partnered Projects*, NSW Department of Public Works and Services, February, Sydney.

NSW Government (1993) *Capital Works Investment: Risk Management Guidelines*, Policy Division, Department of Public Works and Services, NSW Government.

O'Rourke, J (1998) Union's plea: racism not all right on site, *Herald Sun*, 11 October: 19, Sydney.

Oakland, JS and Sohal, AS (1995) *Total Quality Management: Pacific Rim Edition*, Butterworth-Heinemann, Melbourne.

OHS (2001) *Occupational Health and Safety Regulation 2001*, WorkCover NSW, Sydney.

Organ, DW and Greene, CN (1974) Role ambiguity, locus of control, and work satisfaction, *Journal of Applied Psychology*, February: 101–2.

Oskamp, S (ed.) (1988) Television as a social issue, *Applied Social Psychology, Annual Editions*, 8, London.

Pascale, RT (1991) *Managing on the Edge*, Penguin Books, London.

Perry, JG and Hayes, RW (1985) Risk and its management in construction projects, *Proceedings Institution of Civil Engineers*, Part 1, 78: 499–521.

Pettigrew, TF and Meertens, RW (1995) Subtle and blatant prejudice in Western Europe, *European Journal of Social Psychology*, 25(1): 57–75.

Philips, RC (1988) Managing changes before they destroy your business, *Training and Development Journal*, 42(9): 66–71.

Phillis, BD (1987) Risk analysis in project financial appraisal, *Australian Project Manager*, 10(8): 28–30.

Phinney, JS (1996) When we talk about American ethnic groups, what do we mean? *American Psychologist*, 51(5): 918–27.

Pijnenburg, B and Van Duin, MJ (1990) The Zeebrugge ferry disaster, *Contemporary Crises*, 14(4): 321–49.

PMBOK (1996) *A Guide to the Project Management Body of Knowledge*, Project Management Institute, Upper Darby PA.

Pollert, A (1991) *Farewell to Flexibility?* Basil Blackwell, London.

Porter, CE (1981) Risk Allocation in Construction Contracts, MSc thesis, University of Manchester, UK.

Essentials of Construction Project Management

Posner, BZ (1988) What it takes to be a good manager? *Australian Project Manager*, March: 26–8.

Preece, C and Male, S (1997) Promotional literature for competitive advantage in the UK construction firms, *Construction Management and Economics,* 15(1): 59–69.

RCBCI (2002), *Workplace Health and Safety in the Building and Construction Industry*, Royal Commission into the Australian Building and Construction Industry, Commonwealth Government, Melbourne.

RDA (1975) *Racial Discrimination Act*, Commonwealth Government, Canberra.

Richardson, B (1996) Modern management's role in the demise of sustainable society, *Journal of Contingencies and Crisis Management*, 4(1): March: 20–31.

Robbins, S and Mukerji, D (1994) *Managing Organisations: New Challenges and Perspectives*, 2nd edn, Prentice Hall, Sydney.

Robbins, S, Bergman, R, Stagg, I and Coulter, M (2003) *Management*, 3rd edn, Prentice Hall, Sydney.

Robbins, SP, Waters-Marsh, T, Cacioppe, R and Millet, B (1994) *Organisational Behaviour: Concepts, Controversies and Applications: Australia and New Zealand*, Prentice Hall, Sydney.

Robbins, SR (1974) *Managing Organisational Conflict: A Non-Traditional Approach*, Prentice Hall, Englewood Cliffs NJ.

Robbins, SR (1978) Conflict management and conflict resolution are non synonymous terms, *California Management Review*, 21, Winter: 43–60.

Robinson, J (1987) Comparison of tendering procedures and contractual arrangements, *International Journal of Project Management*, 5(1): 19–24.

Roethlisberger, FJ and Dickson, WJ (1939) *Management and the Worker*, Harvard University Press, Cambridge MA.

Rogers, JP (1991) Crisis negotiation codes and crisis management, In George (ed.) *Avoiding War*, 79–92.

Rowlinson, S and Root, D (1996) *The Impact of Culture on Project Management*, British Council Report.

Rowlinson, S, Ho, TKK and Po-Hung, Y (1993) Leadership style of construction managers in Hong Kong, *Construction Management and Economics*, 11(6): 455–65.

Rowse, T (1993) *After Mabo: Interpreting Indigenous Traditions*, Melbourne University Press.

Runeson, G (1997) The role of theory in construction management research: comment, *Construction Management and Economics*, 15(3): 299–302.

Rwelamila, PD (1994) Group dynamics and the construction project manager, *ASCE Journal of Construction Engineering and Management*, 120(1): 3–10.

Sagan, SD (1991) Rules of engagement, In George (ed.) *Avoiding War*, 93–106.

Salminen, S (1995) Serious occupational accidents in the construction industry, *Construction Management and Economics*, 13(2): 299–306.

Sawacha, EO (1993) An Investigation into Safety Attitudes and Safety Performance in the Construction Industry, PhD thesis, Brunel University.

Sawczuk, B (1996) *Risk Avoidance for the Building Team*, E & FN Spon, London.

Sawin, G (1995) How stereotypes influence opinions about individuals, In CP Harvey and MJ Allard, *Understanding Diversity*, HarperCollins, New York, 27–39.

Schmucker, KJ (1984) *Fuzzy Sets, Natural Language, Computations, and Risk Analysis*, Computer Science Press, Annapolis MA.

Schumacher, EF (1993) *Small is Beautiful: A Study of Economics as if People Mattered*, Vintage Books, London.

Selznick, P (1957) *Leadership in Administration*, Row, Peterson, Evanson IL.

Seymour, D and Rooke, J (1995) The culture of the industry and the culture of research, *Construction Management and Economics*, 13(3): 511–23.

Seymour, DE, Hoare, DJ and Itau, L (1992) Project management leadership styles: problems of resolving the continuity-change dilemma, *Proceedings of 11th Internet World Congress on Project Management: Project Management Without Boundaries*, Florence, 16–19 June 2: 487–98.

Shaw, J (1995) *Cultural Diversity at Work: Utilising a Unique Australian Resource*, Business & Professional Publishing, Australia.

Shaw, ME (1954) Group structure and the behaviour of individuals in small groups, *Journal of*

Psychology, 38: 139–49.

Shaw, ME (1976) *Contemporary Topics in Social Psychology*, General Learning Press, Morristown NJ, 350–51.

Sheldrake, J (1996) *Management Theory: From Taylorism to Japanisation*, Thomson Business Press, London.

Shimmin, S, Leather, PJ and Wood, J (1981) *Attitudes and Behaviour about Safety on Construction Work*, Department of Behaviour in Organisations, University of Lancaster.

Shrivastava, A and Fryer, B (1991) Widening access: women in construction, In *Proceedings of 7th Annual ARCOM Conference:* 179–90.

Simon, HA (1948) *Administrative Behaviour*, Macmillan, New York.

Simon, HA (1960) *The New Science of Management Decision*, Harper & Row, New York.

Singh, G (2000) Why construction managers can't ignore ethnic minorities, *Construction Manager*, September: 11–12.

Skinner, BF (1953) *Science and Human Behaviour*, Macmillan, New York.

Skinner, BF (1971) *Contingencies of Reinforcement*, Appleton-Century-Crofts, Norwalk OH.

Skitmore, RM, Stradling, SG and Tuohy, AP (1989) Project management under uncertainty, *Construction Management and Economics*, 7(1): 103–13.

Spence, J (1980) Modern risk management concepts, *BSFA Conference Risk Management in Building*, Sydney: 21–9.

Spence, R and Mulligan, H (1995) Sustainable development and the construction industry, *Habitat International*, 19(3): 279–92.

Spooner, JE (1974) Probabilistic estimating, *Journal of the Construction Division*, 100(CO1): 65–77.

Sprott, WJH (1958) *Human Groups*, Penguin Books, London.

Stogdill, RM and Coons, AE (eds) (1951) Leader behaviour: its description and measurement, *Research Monograph No 88*, Ohio State University, Bureau of Business Research, Columbus.

Szilagyi, AD Jr and Wallace, MJ Jr (1987) *Organizational Behaviour and Performance*, 4th edn, Scott, Foresman & Co, Glenview IL.

Tannenbaum, R and Schmidt, WH (1958) How to choose a leadership pattern, *Harvard Business Review*, 36(2): 95–101.

Thompson, N (1997) *Anti-Discriminatory Practice*, Macmillan, London.

Thomson, JW (1980) Is selective tendering fair? *Chartered Builder*, 29: 25–7.

Tjosvold, DW and Tjosvold, MM (1991) *Leading the Team Organization*, Lexington Books, New York.

Toakley, AR and Aroni, S (1997) The challenge of sustainable development and the role of universities, *Proceedings of 5th International Symposium on the Role of Universities in Developing Areas*, Lulea, Sweden.

Trompenaar, F (1993) *Riding the Waves of Culture*, Economist Books, London.

Tsoukas, H (1996) *New Thinking in Organisational Behaviour*, Butterworth-Heinemann, Oxford.

Tucker, RL and Scarlett, BR (1986) *Evaluation of Design Effectiveness*, Publication No 8-1, and Source Document No 16, Construction Industry Institute, Austin TX.

Turkman, BW (1965) Development sequence in small groups, *Psychological Bulletin*, 63: 384–99.

Tyson, T (1998) *Working with Groups*, 2nd edn, Macmillan Education, Melbourne.

Uff, J (1995) Contract documents and the division of risk, In J Uff and AM Odams (eds) *Risk Management and Procurement in Construction*, Centre for Construction Law and Management, London, 49–69.

Uher, TE (1988) Bidding practice in Australian building, *Australian Project Manager*, 11(3): 35–38.

Uher, TE (1993) A general risk classification model for construction, *AIB Papers*, 5, 1993–94: 25–33.

Uher, TE (1994) *Partnering in Construction*, 2nd edn, School of Building, UNSW, Sydney.

Uher, TE (1996) Cost estimating practices in Australian construction, *Engineering, Construction and Architectural Management*, 3(1&2): 83–95.

Uher, TE (1999) Partnering performance in Australia, *Journal of Construction Procurement*, 5(2): 163–76.

Uher, TE (2003) *Programming and Scheduling Techniques*, UNSW Press, Sydney.

Uher, TE and Davenport, P (2002) *Fundamentals of Building Contract Management*, UNSW Press, Sydney.

Essentials of Construction Project Management

Uher, TE and Toakley, AR (1997) *Risk Management and the Conceptual Phase of the Project Development Cycle*, Risk Management Research Unit, Faculty of the Built Environment, UNSW, Sydney.

USACE (1990) *A Guide to Partnering for Construction Projects*, US Army Corps of Engineers Mobile District, January, Portland OR.

Vasta, E (1993) Multiculturalism and ethnic identity: the relationship between racism and resistance, *Australia and New Zealand Journal of Sociology*, 29(2): 209–25.

Vicere, AA and Fulmer, RM (1997) *Leadership by Design*, Harvard Business School Press, Boston MA.

Victor, DA (1992) *International Business Communications*, HarperCollins, New York.

Vleeming, RG (1979) Machiavellianism: a preliminary review, *Psychological Reports*, February: 295–310.

Vroom, V (1964) *Work and Motivation*, Wiley, New York.

Walker, A (1996) *Project Management in Construction*, 3rd edn, BSP Professional, Oxford.

Walker, C and Smith, AJ (eds) (1995) *Private Infrastructure: The Build Operate Transfer Approach*, Thomas Telford, London.

Walters, KJ (1995) Control hazard studies for process plants, In RE Melchers and MG Stewart (eds) *Proceedings of the Integrated Risk Assessment Conference*, Balkema, Rotterdam, 155–62.

Ward, SC and Chapman, CB (1991) Extending the use of risk analysis in project management, *International Journal of Project Management*, 9(2): 117–23.

Waring, A (1992) Developing a safety culture, *Safety and Health Practitioner*, 10(4): 42–4.

Weston, JF and Brigham, EF (1968) *Essentials of Managerial Finance*, Holt, Rinehart & Winston, New York.

Wetherell, M and Potter, J (1992) *Mapping the Language of Racism: Discourse and the Legitimation of Exploitation*, Harvester Wheatsheaf, London.

Whitfield, J (1994) *Conflicts in Construction*, Macmillan, London.

Whittington, C, Livingston, A and Lucas, DA (1992) *Research into Management, Organisational and Human Factors in the Construction Industry*, HSE Contract Research Report No 45/1992, HMSO, London.

Wildavsky, A (1988) Searching for safety. Transaction New Brunswick, cited In U Rosenthal and A Kouzmin (1993) Globalisation: an addenda for contingencies and crisis management, An editorial statement, *Journal of Contingencies and Crisis Management*, 1(1): 1–11.

Winch, G (1989) The construction firm and construction project: a transaction cost approach, *Construction Management and Economics*, 7: 331–45.

Wood, JT (1997) Gendered media: the influence of media on views of gender, In MH Davis (ed.) *Social Psychology Annual Editions, 97/98*: 162–71.

Woodward, J (1958) *Management and Technology*, HMSO, London.

Woodward, JF (1997) *Construction Project Management: Getting it Right First Time*, Thomas Telford, London.

Wright, P, Pringle, CD and Kroll, MJ (1992) *Strategic Management: Text and Cases*, 2nd edn, Allyn & Bacon, London.

Zeleny, M (1982) *Multiple Criteria Decision Making*, McGraw-Hill, New York.

INDEX

Essentials of Construction Project Management

Essentials of Construction Project Management